Advances in Animal Anatomy

Advances in Animal Anatomy

Editors

Matilde Lombardero Fernández
María del Mar Yllera Fernández

MDPI • Basel • Beijing • Wuhan • Barcelona • Belgrade • Manchester • Tokyo • Cluj • Tianjin

Editors
Matilde Lombardero Fernández
University of Santiago de Compostela—Campus of Lugo
Spain

María del Mar Yllera Fernández
University of Santiago de Compostela—Campus of Lugo
Spain

Editorial Office
MDPI
St. Alban-Anlage 66
4052 Basel, Switzerland

This is a reprint of articles from the Special Issue published online in the open access journal *Animals* (ISSN 2076-2615) (available at: https://www.mdpi.com/journal/animals/special_issues/anatomy).

For citation purposes, cite each article independently as indicated on the article page online and as indicated below:

LastName, A.A.; LastName, B.B.; LastName, C.C. Article Title. *Journal Name* **Year**, *Volume Number*, Page Range.

ISBN 978-3-0365-7328-1 (Hbk)
ISBN 978-3-0365-7329-8 (PDF)

Cover image courtesy of María del Mar Yllera Fernández.

Contents

About the Editors

Matilde Lombardero Fernández

Matilde Lombardero Fernández, DVM, PhD, is a Professor at the Faculty of Veterinary Sciences of Lugo, University of Santiago de Compostela, Spain. She teaches subjects related to the anatomy of domestic and exotic species, and veterinary citology and histology. Her previous academic positions were Associate Professor (1998–2005) and Assistant Professor (2006–2010). Her research activity started 30 years ago, and her scientific interests include studies on various topics from macroscopic to microscopic morphology, as well as new non-toxic preservation methods for anatomical specimens used in practical sessions of veterinary anatomy. She has undertaken various postdoctoral research stays at the Edinburgh University (Scotland, UK), Mayo Clinic (Rochester, Minnesota, USA), and at Toronto University (Toronto, Ontario, Canada).

María del Mar Yllera Fernández

María del Mar Yllera Fernández, DVM, PhD, studied Veterinary Medicine at Complutense University of Madrid (Spain) from 1979 to 1984. She received a PhD in Veterinary Medicine in 1988 from the University of Santiago de Compostela (Spain). Her research is focused on the study of the embryonic development of domestic animals and the anatomy of both domestic and exotic mammals, as well as new non-toxic preservation methods for anatomical specimens used in practical sessions of Veterinary Anatomy. At present, she is a Professor in the Department of Anatomy, Veterinary Medicine and Animal Production at the Faculty of Veterinary Sciences of Lugo, University of Santiago de Compostela (Spain). She teaches both the anatomy of domestic and exotic mammals, embryology and histology. Her previous academic positions include the following: pre- and postdoctoral fellow of the Spanish government (1985–1988), Assistant Professor (1988–1992) and Associate Professor (1992–1994). She has undertaken various postdoctoral research stays at the Sorbonne University (1989) and at the Institute National pour la Recherche Agronomique (INRA) in Jouy-en Josas (France) (1989–1990). She has also worked as the Veterinarian in Charge for the research animal facilities of the University of Santiago de Compostela and held the position of the President of the Bioethics Committee of the same institution.

Preface to "Advances in Animal Anatomy"

During the last few years, the subject of veterinary anatomy has been gradually reduced in the successive academic programs. However, the number of species that require veterinary attention is increasing. In an attempt to bridge this gap, this Special Issue entitled Advances in Animal Anatomy comprises a set of research articles and reviews focused on the importance of the applied anatomy of diverse anatomical structures in a wide range of species, from domestic animals to wild species, including new companion animals. Therefore, this Special Issue is mainly directed at veterinary students and professionals, researchers and technicians who want to update their knowledge in their daily practice in both veterinary clinics and conservation centers.

Matilde Lombardero Fernández and María del Mar Yllera Fernández
Editors

Editorial

Advances in Animal Anatomy

Matilde Lombardero * and María del Mar Yllera

Department of Anatomy, Animal Production and Clinical Veterinary Sciences, Unit of Anatomy, Faculty of Veterinary Sciences, Campus of Lugo, University of Santiago de Compostela, 27002 Lugo, Spain
* Correspondence: matilde.lombardero@usc.es

This Special Issue was the result of reviewing Leonardo da Vinci's anatomical drawings of the bear foot and the horse trunk (among others) [1]. Since then, we were challenged to propose an interesting topic under the title "Advances in Animal Anatomy". We were convinced that we had to opt for more than just descriptive anatomy; we chose applied anatomy since, at present, the time dedicated to teaching anatomy has been drastically reduced in the undergraduate programs of veterinary degrees in favor of subjects from the clinical field/area. We also decided to address both domestic and exotic animals, which are so prevalent today in veterinary practice, as well as those animals that must be kept in rehabilitation and conservation centers. Unfortunately, veterinarians are not very familiar with the anatomy of the latter species (which always tends to be considered similar to that of domestic animals), their physiology and more frequent pathologies, since these are not part of the core training in the veterinary degrees. This deficiency is probably not exclusive to the veterinary undergraduate programs. Researchers and specialized technicians who work in wildlife recovery centers and zoos and are involved in exotic medicine and the welfare of these species that are under their surveillance, also need specialized information regarding these species' anatomies and physiologies.

In order to shed light on the applied anatomy of domestic, exotic, and wild species, we introduce the fourteen manuscripts (eleven research articles and three reviews) that are compiled in this Special Issue (Figure 1), covering current research trends in applied anatomy by using a wide range of techniques. Computed-tomographic imaging was used in four articles, magnetic resonance imaging in three of them, radiography in two articles, and ultrasound techniques and endoscopy in addition to the traditional gross dissections, morphological, and morphometrical studies were also employed (Figure 2).

The horse is the species that most articles, including three research articles and one review, in this Special Issue study. Phalanges from the equine hand were studied by Gündemir et al. [2] by using X-ray imaging. They took radiogrametic measurements of the forelimb phalanges of Arabic and throughoutbred horses. Based on the left manus data, they stated that the proximal and middle phalanges could show sexual dimorphism. These two phalanges can also be used to differentiate breeds, especially, considering the depth of the caput of the proximal phalanx, which can reaching an accuracy level of breed classification of almost 90%. Moreover, all the radiometric measurements made to the thoracic limb phalanges of 75 horses are available as valuable reference data/material when evaluating the manus digital bones in horses.

The second research article about horses focuses on equine dentistry. The most accessible maxillary teeth of the horse, the incisors, were studied by Miró et al. [3] in 25 skulls from horses aged between 12–42 months. Combining visual inspection, radiographic study and computed-tomography imaging, they investigated the development of the deciduous incisors, the dental germs of permanent incisors, and the surrounding bone, just before and up to the beginning of teeth shedding. Measurements of three lengths and two proportions or relative lengths were also analyzed. The development of deciduous and permanent maxillary incisors, in addition to their alveoli, was described in detail at different ages. It

Citation: Lombardero, M.; Yllera, M.d.M. Advances in Animal Anatomy. *Animals* **2023**, *13*, 1110. https://doi.org/10.3390/ani13061110

Received: 17 March 2023
Accepted: 20 March 2023
Published: 21 March 2023

was concluded that the dental germ of the first permanent incisors appears at 12 months, although their crown mineralization starts a few months later.

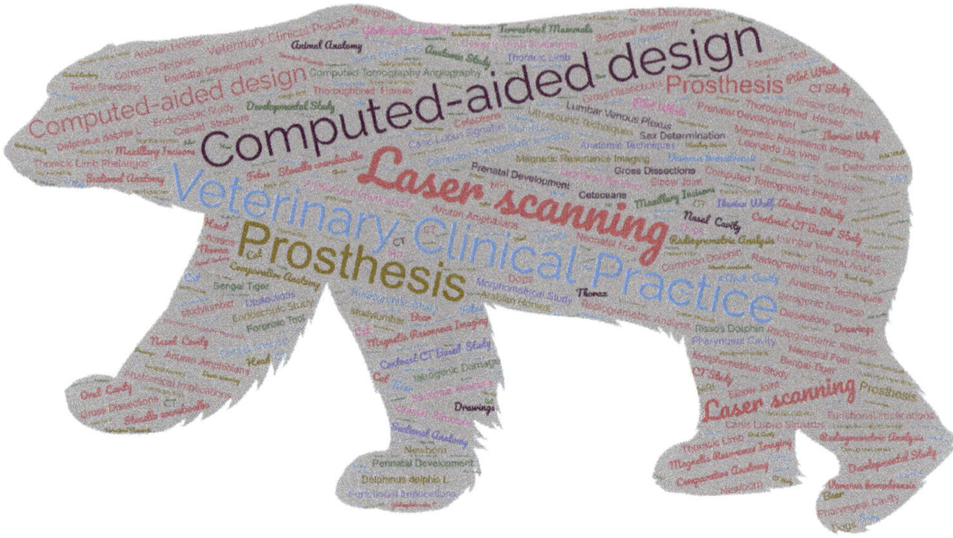

Figure 1. A tag cloud including the titles and keywords of the articles published in this Special Issue.

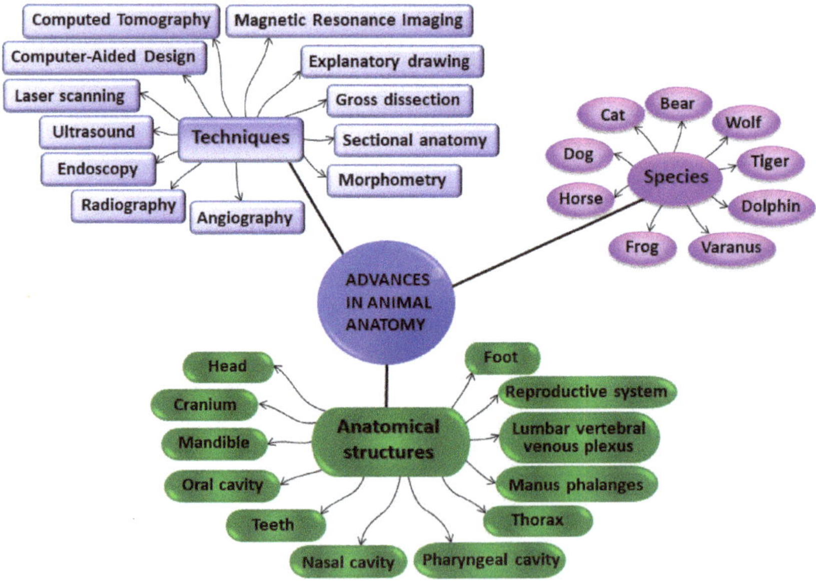

Figure 2. A mind map of this Special Issue summarizing the techniques used to study the different anatomic structures from the diverse species.

The third manuscript that focuses on horses addresses the study of the thorax in neonatal foals. Authored by Arencibia et al. [4], their study was based on computed-tomography angiography (CTA) complemented with the gross dissections and sectional anatomy of the same specimens used as anatomical references. With intravascular contrast,

CTA gives an overview of the thoracic morphology, providing information on the size and position of the heart, as well as the heart chambers, in addition to the information on the main thoracic blood vessels. This information could be used as reference data when comparative morphology is needed to reach a diagnosis in foals with thoracic disease. A dorsal view is preferable to visualize midline thoracic vascular structures (including pulmonary vessels and the brachiocephalic trunk), while lateral views are better to reveal the relationship between the heart chambers and the main blood vessels.

The other species of domestic mammals studied in three of the articles were the dog and the cat. The research article by Ariete et al. [5], which referred to the canine internal vertebral venous plexus (IVVP) of the lumbar segment, used contrast-enhanced computed-tomography (CT) to conduct a morphometrical study of the aforementioned vessels, the dural sac, and the vertebral canal. As part of the vertebral venous plexus, this IVVP is made of a paired thin-walled and valveless venous vessels with a rhomboidal pattern, placed on the floor of the vertebral canal, and included in the lumbar epidural space. It drains blood from the vertebral column, spinal cord and its meninges, including the paravertebral muscles. As a result, when bursting, spontaneous spinal epidural hematomas are produced and could generate spinal cord compression and progressive neurological disorders. IVVP is also supposed to be involved in the pathogenesis of many other pathologies. Consequently, the accurate morphometric study of the lumbar IVVP (and surrounding structures) in healthy and alive animals is necessary to obtain reference values and to, afterwards, assess their alterations in pathological states. The measurements of the cross sectional area of the vertebral canal, dural sac, epidural space, and the right and left IVVP of six dogs were obtained, in addition to the percentage occupied by the IVVP in relation to the vertebral canal and epidural space, and the proportion of the vertebral canal occupied by the dural sac. Regarding all the measurements, there was a turning point between L4 and L5, showing a clear change in the trends at that level, which is in concurrence with the emergence of the nervous roots of the lumbosacral plexus.

Regarding the domestic cat, two review articles assessed its mandible anatomy and its manipulation when a mandibular pathology had to be treated. The first of them, Part I by Lombardero et al. [6], analyses the mandible anatomy in detail in order to obtain a deep knowledge of its morphology and avoid iatrogenic damage, caused by the fact that they are usually treated as those of small dogs. The cat mandible has fewer dental pieces than the mandible of dogs, and most of its mandibular body is filled up with dental roots and the mandibular canal (with the neurovascular supply), leaving little bone surface for the safe placement of screws in a fracture resolution. It should be emphasized that the mandibular canal is not a medullary canal and intramedullary pins should be avoided. The innervation areas of the different branches of the inferior alveolar nerve (with afferences to both skin/mucosal and tooth/periodontal structures) were also considered to serve as a reference when a mental nerve blocking must be performed. In addition, other specific considerations include the angular process in the ventrocaudal part of the mandibular ramus when the mouth remains wide open (with mouth-gags) for a considerable time, which presses on the maxillary artery. This compression reduces its blood flow (mainly directed to the brain since a functional internal carotid artery is missing in cats) producing temporary or permanent neurological disorders due to a cerebral ischaemia. Consequently, in order to prevent this complication, the use of spring-loaded mouth gags to keep their mouth wide open should be avoided in cats. Complementary to this information, Part II, also by Lombardero et al. [7], described different mandibular fractures and temporomandibular joint dislocations and how they could be solved when possible non-invasive techniques should be considered first. Otherwise, it was recommended that simple jaw fractures should be used to repair caudal to rostral, preferably, using a ventral approach. Diverse surgical methods were discussed to maintain biomechanical functionality when repairing mandibular fractures. However, taking into account that the use of rigid fixation methods, such as osteosynthesis plates, are challenging due to the scarce availability of bone surface to fix the screws onto, a new prosthesis design was proposed by the authors to repair simple mandibular body fractures. This prosthesis proposal was a conceptual design to provide an acceptable rigid biomechanical stabilization while minimizing dental root and neurovascular damage, hence reducing patient suffering and speeding up their recovery. This prothesis would be custom-designed and would have to be manufactured in a biocompatible and resistant material, such as titanium. Its shape would look like two horizontal "Y" partially overlapped, and it would support three points of fixation with small screws, each one strategically placed to avoid damaging any tooth, periodontal structure or branch of the inferior alveolar nerve. The fourth fixation point was a flat hook-like flap embracing the body's ventral border, caudal to the mandibular fracture, thus keeping the integrity of the mandibular canal and its neurovascular supply, thus contributing to the patients' welfare. Custom-designed prostheses are utterly dependent on reliable technology, from diagnostic imaging (such as virtual surgical planning—VSP—or cone-beam computed tomography—CBCT) to computer-aided design/computer-aided manufacturing (CAD/CAM) technology. All together, they are the current trend and the trend of the near future.

The Bengal Tiger (*Panthera tigris tigris*) belongs to the same family as the domestic cat, *Felidae*, but it is exotic and much bigger. The elbow joints of the Bengal Tiger (*Panthera tigris tigris*) were studied by Encinoso et al. [8] using magnetic resonance imaging (MRI) in combination with traditional gross dissections. The elbow joint is complex and consists of a hinge joint between the humerus (whose distal end is a trochlea) and the proximal ends of the radius and ulna (with a reciprocal shape), as well as the pivot joint between the proximal ends of the antebrachial bones, all enclosed by a single capsule filled with synovial fluid. Unfortunately, not much information is currently available regarding the regional anatomy of the tiger elbow. Thus, gross dissections of the elbow joint, complemented by the brachial and antebrachial muscles and tendons, and their visualization/identification in MRI images would help us understand the normal tiger elbow anatomy, which can help us discern whether there is any elbow joint disorder in this species. The musculoskeletal system is usually studied by MRI, as it avoids ionizing radiation and it provides good image resolution and good contrast, even in soft tissues. In addition to bones, muscles, ligaments, and articular cavities filled with synovial fluid were observed with MRI, and all of these joint structures were confirmed by gross dissection. The study of the tiger elbow joint by means of MRI is essential for the proper training of experts involved in Bengal tiger conservation (such as veterinarians and researchers), as it will allow them to acquire a deep knowledge of the healthy elbow joint, and consequently, to be able to identify any alteration, reaching an accurate diagnosis, treating the patient successfully, minimizing suffering, and promoting animal welfare.

Additionally, the wild relative of the domestic dog, the Iberian wolf, was studied in this Special Issue in a study of their teeth, using a morphological and morphometric approach. This interesting dental analysis, led by Toledo et al. [9], is a comprehensive study of the dental type (incisors, canines, premolars, and molars from maxilla and mandible) morphology and morphometry, including any sexual dimorphism, and was carried out on 45 skulls of Iberian wolves (males and females) with permanent dentition. Up to 36 dental variables (including superficial and deep bite marks) were assessed and statistically analyzed to obtain a valuable morphometric data collection to establish their age and sex in a population control. In addition, the results of the analysis of their bite mark patterns (based on the tooth mark dimensions, distribution, and proportions) could be used as a database reference to differentiate between domestic dogs and Iberian wolves in forensic cases when an identification of a bite mark is needed in cases of livestock attack. This can aid in receiving economic compensation. From here on we will address the topic of mammals that live in the sea. This Special Issue includes three research articles that address the anatomical study of the whole head of dolphin fetuses and newborn dolphins, and the developmental and comparative study of some species of dolphins focused on different regions of the head (oral, pharyngeal and nasal cavities). Regarding García de los Ríos et al.'s [10] the developmental study on the head of striped dolphins during the fetal and perinatal periods, they used different techniques, from gross dissection and sectional anatomy to diagnostic imaging (CT and MRI), to study the anatomical and physiological inferences affecting the heads of marine and land mammals. The development of the head reveals the evolutionary changes of Cetaceans that helped them adapt to the marine environment. These changes include modifications in the feeding apparatus, the caudal relocation of the nasal opening, and other structures that were reduced or even removed. Currently, there are not many methods to estimate the age of dolphin specimens during the fetal period, thus they used CT and MRI in correlation with transversal sections to study the anatomical changes of the heads of the four specimens, which occurred in chronological order during gestation. The results were organized into nine subsections covering oral cavity, rostrum, melon, nasal cavity and paranasal sinuses, orbit and eyeball, central nervous system, ear, larynx, and cranial cavity. This article is richly illustrated with photographs, so they can be used as a head morphology atlas for these species of odontocetes, giving valuable information about the changes produced in the head during prenatal and postnatal development, which are discussed in detail under an anatomical and functional perspective. This general study serves as an introduction to the next two articles, which address different structures of the head in more detail. The second article by García de los Ríos et al. [11] studied the comparative anatomy of the nasal cavity in three species of dolphins, covering the external nose and nasal cavity in different stages of development. They used endoscopy, MRI, anatomic sections and dissections, and histology, as well as the CT to generate 3D reconstructions of the nasal cavity and the bones that delimit it. The external nose is protected by two nasal lips (rostral and caudal) becoming gradually more waterproof during the development to adulthood. Their nasal cavity, usually with vertical orientation, has a single nostril continuous with a vestibule (with two vestibular folds or phonic lips, two diverticula, and two incisive recesses) and two nasal cavities (right and left) at both sides of the nasal septum, conducting the air towards the choanae. Their nasal cavity has no nasal cornettes or conchae, and it is divided in two parts: respiratory and olfactory. However, no olfactory epithelium was identified. All findings were profusely illustrated with 30 figures, each of them composed of a panel of photographs. The latest article by García de los Ríos et al. conducted a developmental study of the oral and pharyngeal cavities [12] of five species of dolphins, also based on endoscopy, MRI, anatomic sections and dissections, and histology. They increased the number of species compared to their previous work on the nasal cavity [11] and studied specimens from a wide range of ages, from fetuses, through to newborns, juveniles, and adults. They described in detail an oral cavity divided in a vestibule and an oral cavity proper, with a roof and a floor, the tongue, and teeth. They reported the pharyngeal cavity divided in three parts and confirmed the existence of the pair of pharyngeal diverticula of the auditory tubes from the early stages of development, with two areas inside: the cavity lined by respiratory epithelium (in contact with air) and its walls with a pharyngeal vascular plexus that would help to eject the air into the adjoining nasopharynx, relevant in decompression. Taken together, these three articles [10–12] present valuable descriptive and visual information about the anatomy of the head structures studied in some species of dolphins during development, relevant in veterinary medicine and some fields of biology.

Moving from mammals to reptiles, this Special Issue contains a study of the head of the monitor lizard by Pérez et al. [13], which includes advanced imaging techniques, such as CT and 3D head reconstruction in order to explore the complex anatomy of the head and associated structures of these animals. They used sagittal and transverse CT images, volume-rendered reconstructed CT images, as well as maximum intensity projection (MIP) images to map the head of the Komodo dragon. This is valuable visual work, with all the main structures of the head identified, including the comparative images from CT bone and soft tissue windows at different transversal levels, the 3D volume-rendered reconstruction images from dorsal, ventral, and lateral views, complemented with the dorsal and ventral MIP images, display the head bone structure. Apart from the bones and other structures, including muscles, nervous structures (brain, cerebellum, and brain stem), eye (different structures, as well as sclera ossicles), glands (labial, nasal and Harderian glands), and air-filled structures (oral and nasal cavities, larynx, and trachea), the existence of cephalic osteoderms was also verified with the techniques used. This kind of morphological study based on CT imaging is very useful for people in charge of wildlife conservation centers, such veterinarians, researchers, and technicians, who have to deal with the day-to-day prevention and treatment of threatening diseases in this species in captivity.

Turning to amphibians, and to conclude the list of research articles, Ruiz-Fernández et al. [14], using non-invasive methods, such as benchtop MRI (BT-MRI) and high-resolution ultrasound (HR-US) techniques, managed to identify the biological sex of two species of Anuran in sexually mature specimens when sexual dimorphism is not apparent. BT-MRI is based on expensive equipment and the patients should be under anesthesia. The resulting images were more accurate, allowing for the identification of their sexual organs. Nevertheless, the equipment for HR-US is more affordable for veterinary clinics and zoo facilities, and has some advantages, such as that patients do not need anesthesia and that it is less time consuming. Using it, they identified the gonads of both species. The ovaries were clearly distinguished (as they appeared hyperechoic, with plenty of hypoechoic foci—follicles); however, the testes were not so evident due to their homogeneous echotexture. Therefore, amphibian sexing using non-invasive techniques is a significant advance in the conservation of these species and in reproduction programs for them.

To conclude the editorial of this Special Issue, we return to the initial article about Leonardo da Vinci's drawings representing the anatomy of the foot of the bear and the horse [1]. The bear foot was considered one of his early scientific drawings and, apart from being a very common animal in Italian mountains, he probably chose to dissect the bear foot because of its plantigrade gait. Later on, his masterpieces on human anatomy were based on his lifelong learning. He was an outstanding Renaissance scientist, as he expressed throughout his life a continuous interest in learning and exploring the world around him. He also did not hesitate to go a little further. This is the purpose of this Special Issue as well, moving forward, displaying the advances of applied anatomy. We do hope to have achieved this and we wish that we all maintain our interest in broadening our knowledge in applied animal anatomy, parallel to the advances in diagnostic technology.

Acknowledgments: We are very grateful to all the authors for submitting their research to this Special Issue. In the same way, we thank the reviewers and Academic Editors for their expertise and wise comments that helped improve all the manuscripts.

Conflicts of Interest: The author declares no conflict of interest.

References

1. Lombardero, M.; Yllera, M.d.M. Leonardo Da Vinci's Animal Anatomy: Bear and Horse Drawings Revisited. *Animals* **2019**, *9*, 435. [CrossRef] [PubMed]
2. Gündemir, O.; Szara, T.; Pazvant, G.; Erdikmen, D.O.; Duro, S.; Perez, W. Radiogrametric Analysis of the Thoracic Limb Phalanges in Arabian Horses and Thoroughbred Horses. *Animals* **2021**, *11*, 2205. [CrossRef] [PubMed]
3. Miró, F.; Manso, C.; Diz, A.; Novales, M. Maxillary Incisors of the Horse before and at the Beginning of the Teeth Shedding: Radiographic and CT Study. *Animals* **2020**, *10*, 1618. [CrossRef] [PubMed]
4. Arencibia, A.; Corbera, J.A.; Ramírez, G.; Díaz-Bertrana, M.L.; Pitti, L.; Morales, M.; Jaber, J.R. Anatomical Assessment of the Thorax in the Neonatal Foal Using Computed Tomography Angiography, Sectional Anatomy, and Gross Dissections. *Animals* **2020**, *10*, 1045. [CrossRef] [PubMed]
5. Ariete, V.; Barnert, N.; Gómez, M.; Mieres, M.; Pérez, B.; Gutierrez, J.C. Morphometrical Study of the Lumbar Segment of the Internal Vertebral Venous Plexus in Dogs: A Contrast CT-Based Study. *Animals* **2021**, *11*, 1502. [CrossRef] [PubMed]
6. Lombardero, M.; Alonso-Peñarando, D.; Yllera, M.M. The Cat Mandible (I): Anatomical Basis to Avoid Iatrogenic Damage in Veterinary Clinical Practice. *Animals* **2021**, *11*, 405. [CrossRef] [PubMed]
7. Lombardero, M.; López-Lombardero, M.; Alonso-Peñarando, D.; Yllera, M.d.M. The Cat Mandible (II): Manipulation of the Jaw, with a New Prosthesis Proposal, to Avoid Iatrogenic Complications. *Animals* **2021**, *11*, 683. [CrossRef] [PubMed]
8. Encinoso, M.; Orós, J.; Ramírez, G.; Jaber, J.R.; Artiles, A.; Arencibia, A. Anatomic Study of the Elbow Joint in a Bengal Tiger (Panthera Tigris Tigris) Using Magnetic Resonance Imaging and Gross Dissections. *Animals* **2019**, *9*, 1058. [CrossRef] [PubMed]

9. Toledo González, V.; Ortega Ojeda, F.; Fonseca, G.M.; García-Ruiz, C.; Navarro Cáceres, P.; Pérez-Lloret, P.; Marín García, M.d.P. A Morphological and Morphometric Dental Analysis as a Forensic Tool to Identify the Iberian Wolf (Canis Lupus Signatus). *Animals* **2020**, *10*, 975. [CrossRef] [PubMed]
10. García de los Ríos y Loshuertos, Á.; Arencibia Espinosa, A.; Soler Laguía, M.; Gil Cano, F.; Martínez Gomariz, F.; López Fernández, A.; Ramírez Zarzosa, G. A Study of the Head during Prenatal and Perinatal Development of Two Fetuses and One Newborn Striped Dolphin (Stenella Coeruleoalba, Meyen 1833) Using Dissections, Sectional Anatomy, CT, and MRI: Anatomical and Functional Implications in Cetaceans and Ter. *Animals* **2019**, *9*, 1139. [CrossRef] [PubMed]
11. García de los Ríos y Loshuertos, A.; Soler Laguía, M.; Arencibia Espinosa, A.; López Fernández, A.; Covelo Figueiredo, P.; Martínez Gomariz, F.; Sánchez Collado, C.; García Carrillo, N.; Ramírez Zarzosa, G. Comparative Anatomy of the Nasal Cavity in the Common Dolphin Delphinus Delphis L., Striped Dolphin Stenella Coeruleoalba M. and Pilot Whale Globicephala Melas T.: A Developmental Study. *Animals* **2021**, *11*, 441. [CrossRef] [PubMed]
12. García de los Ríos y Loshuertos, Á.; Soler Laguía, M.; Arencibia Espinosa, A.; Martínez Gomariz, F.; Sánchez Collado, C.; López Fernández, A.; Gil Cano, F.; Seva Alcaraz, J.; Ramírez Zarzosa, G. Endoscopic Study of the Oral and Pharyngeal Cavities in the Common Dolphin, Striped Dolphin, Risso's Dolphin, Harbour Porpoise and Pilot Whale: Reinforced with Other Diagnostic and Anatomic Techniques. *Animals* **2021**, *11*, 1507. [CrossRef] [PubMed]
13. Pérez, S.; Encinoso, M.; Corbera, J.A.; Morales, M.; Arencibia, A.; González-Rodríguez, E.; Déniz, S.; Melián, C.; Suárez-Bonnet, A.; Jaber, J.R. Cranial Structure of Varanus Komodoensis as Revealed by Computed-Tomographic Imaging. *Animals* **2021**, *11*, 1078. [CrossRef] [PubMed]
14. Ruiz-Fernández, M.J.; Jiménez, S.; Fernández-Valle, E.; García-Real, M.I.; Castejón, D.; Moreno, N.; Ardiaca, M.; Montesinos, A.; Ariza, S.; González-Soriano, J. Sex Determination in Two Species of Anuran Amphibians by Magnetic Resonance Imaging and Ultrasound Techniques. *Animals* **2020**, *10*, 2142. [CrossRef] [PubMed]

Review

Leonardo da Vinci's Animal Anatomy: Bear and Horse Drawings Revisited

Matilde Lombardero * and María del Mar Yllera

Unit of Veterinary Anatomy and Embryology, Department of Anatomy, Animal Production and Clinical Veterinary Sciences, Faculty of Veterinary Sciences, University of Santiago de Compostela—Campus of Lugo, 27002 Lugo, Spain

* Correspondence: matilde.lombardero@usc.es

Received: 10 April 2019; Accepted: 16 June 2019; Published: 10 July 2019

Simple Summary: Leonardo da Vinci was an outstanding artist of the Renaissance. He depicted numerous masterpieces and was also interested in human and animal anatomy. We focused on the anatomical drawings illustrating different parts of bear and horse bodies. Regarding Leonardo's "bear foot" series, the drawings have previously been described as depicting a bear's left pelvic limb; however, based on the anatomy of the *tarsus* and the *digit* (finger) arrangement, they show the right posterior limb. In addition, an unreported rough sketch of a dog/wolf *antebrachium* (forearm) has been identified and reported in detail in one of the drawings of the "bear's foot" series. After a detailed anatomical analysis, the drawing "The viscera of a horse" has more similarities to a canine anatomy than to a horse anatomy, suggesting that it shows a dog's trunk. Besides, the anatomies of the drawings depicting the horse pelvic limb and the human leg were analyzed from the unprecedented point of view of movement production.

Abstract: Leonardo da Vinci was one of the most influencing personalities of his time, the perfect representation of the ideal Renaissance man, an expert painter, engineer and anatomist. Regarding Leonardo's anatomical drawings, apart from human anatomy, he also depicted some animal species. This comparative study focused only on two species: Bears and horses. He produced some anatomical drawings to illustrate the dissection of "a bear's foot" (Royal Collection Trust), previously described as "the left leg and foot of a bear", but considering some anatomical details, we concluded that they depict the bear's right pelvic limb. This misconception was due to the assumption that the bear's *digit I* (1st toe) was the largest one, as in humans. We also analyzed a rough sketch (not previously reported), on the same page, and we concluded that it depicts the left *antebrachium* (forearm) and *manus* (hand) of a dog/wolf. Regarding Leonardo's drawing representing the horse anatomy "The viscera of a horse", the blood vessel arrangement and other anatomical structures are not consistent with the structure of the horse, but are more in accordance with the anatomy of a dog. In addition, other drawings comparing the anatomy of human leg muscles to that of horse pelvic limbs were also discussed in motion.

Keywords: bear pelvic limb; dog antebrachium; horse trunk; horse and human comparative anatomy

1. Introduction

Leonardo da Vinci was one of the most important renaissance personalities of his time, and the fifth centenary of his death will be commemorated in 2019. Being the illegitimate son of a notary, he did not continue the family saga and was educated privately. He had no formal education, thereby not conditioning his curiosity about the world around him. The erudite texts of his time were written in Latin and Greek, languages he did not master, and his access to the literature was therefore limited.

He was an artist and a scientist. As a painter, scientist, engineer and theorist, he produced thousands of drawings [1], personifying the 'Renaissance man' skilled and versed in arts and sciences [2].

His interest in anatomy was overwhelming, proven by the numerous sheets dedicated to his anatomical studies, with abundant notes and drawings, exemplifying Leonardo's principle that anatomic parts and organs should be represented in multiple views. Considering that dissections of human corpses outside Universities were not considered appropriate by the ecclesiastical authorities, he performed some dissections of animals. According to the Royal Collection Trust [3], at the outset of Leonardo's anatomical investigations, he was unable to procure much human material. Hence, many of his dissections were therefore of animals.

Practically his entire collection of anatomical drawings was compiled in the Windsor Codex, property of Her Majesty Queen Elizabeth II. These drawings of the human body were exhibited in an unprecedented exhibition in 2012 at the Queen's Gallery, Buckingham Palace (London, UK). Although previous access to the collection was highly restricted, nowadays, the Royal Collection Trust offers the possibility of free access to these drawings in high resolution on its website, which greatly enables the observation of these masterpieces and their details.

Several works have been published based on these anatomical drawings, the most exhaustive ones are those from the collection of three volumes from Clark [4,5], compiling all the inventory information, the book from O'Malley and Saunders [6] and its posterior editions in 1983 and 2003 [7,8], and the official book of the exhibition. Clayton and Philo [9] and another book published in 2013 [10], the two latter reviews, mainly referred to human anatomy, although they also include comments on some animal anatomy drawings. Apart from books, there are numerous scientific articles sharing the same subject: Leonardo da Vinci's anatomical drawings, mainly intended to some areas of expertise, such as those from Schultheiss et al. [11], Jose [12], Ganseman and Broos [13], Pasipoularides [2], Sterpetti [14], Bowen et al. [15] and West [16], among others.

It is well known that Leonardo dissected numerous animals [17]. As a result, many endeavors have been made to identify the animal of which the individual anatomical drawings have been made. In some cases, such identification is easy, while in others it is impossible [17]. Leonardo da Vinci's methods of acquiring knowledge were observation and experiment, and for him, the study of anatomy became a science, combining both the study of structure and function [12].

Reviewing the work of several authors on the description of Leonardo's animal anatomy drawings, and comparing them with the high-resolution images available on the website of the Royal Collection Trust [3], it can be noted that some of them were not properly described elsewhere, with some inaccuracies or misunderstandings that deserve to be discussed, probably due to the fact that the consulted authors were experts in human anatomy and, therefore, had no deep understanding of animal anatomy. Hence, it is important to point out that human anatomy could be considered similar, although with some differences, to animal anatomy. For major details, all of Leonardo da Vinci´s anatomical drawings can be accessed on the Royal Collection Trust website [3].

Regarding Leonardo's anatomical drawings, apart from human anatomy, he also depicted some animal species such as dogs, bears, pigs, horses, oxen and monkeys. The main aim of this comparative study on anatomy was focused only on two species: Bears and horses.

According to Forlani-Tempesti [1], da Vinci mentioned bears in his notes for his anatomical treatise: "I will discourse of the hands of each animal to show in what they vary; as in the bear which has the ligatures of the toes joined above the instep", and again: "Here is to be depicted the foot of the bear or ape or other animals to show how they vary from the foot of man or, say, the feet of certain birds" [1]. Bears are also the protagonists of other drawings of da Vinci: A bear walking (Robert Lehman Collection) and three other studies of a bear's (or a wolf's or dog's) paws (1490–1495) and head (Colville Collection) [1]. However, those images of paws cannot be from a bear, simply because bears have a *manus* (hand) with five *digits* (fingers) with their *distal phalanxes* (claws) in contact with the floor. In contrast, a dog and wolf *manus* only have four *digits* ending in *unguicula* (claws) in contact to the ground, plus the *digit I* (dewclaw) medially placed at the level of the metacarpal bones.

2. Flaws of the Anatomy of the Bear's *Pes* (Foot) and the Hidden *Antebrachium* (Forearm) and *Manus* from a Dog/Wolf

The set of drawings that made us realize that some inaccuracies were made in terms of their description was that of the bear's foot (Royal Collection Inventory Number—RCIN 912372-5). Regarding RCIN 912372 (Figure 1), the first reference to it was stated by Professor William Wright in 1919 [18], in a section entitled 'Leonardo as an Anatomist', published by the Burlington Magazine to commemorate the Quartercentenary of Leonardo da Vinci [18] as 'one of the finest of Leonardo's anatomical drawings, the hind foot of a plantigrade carnivorous animal—probably a bear, a view supported by the fact that in one of the manuscripts, a reference is made to a bear's foot'.

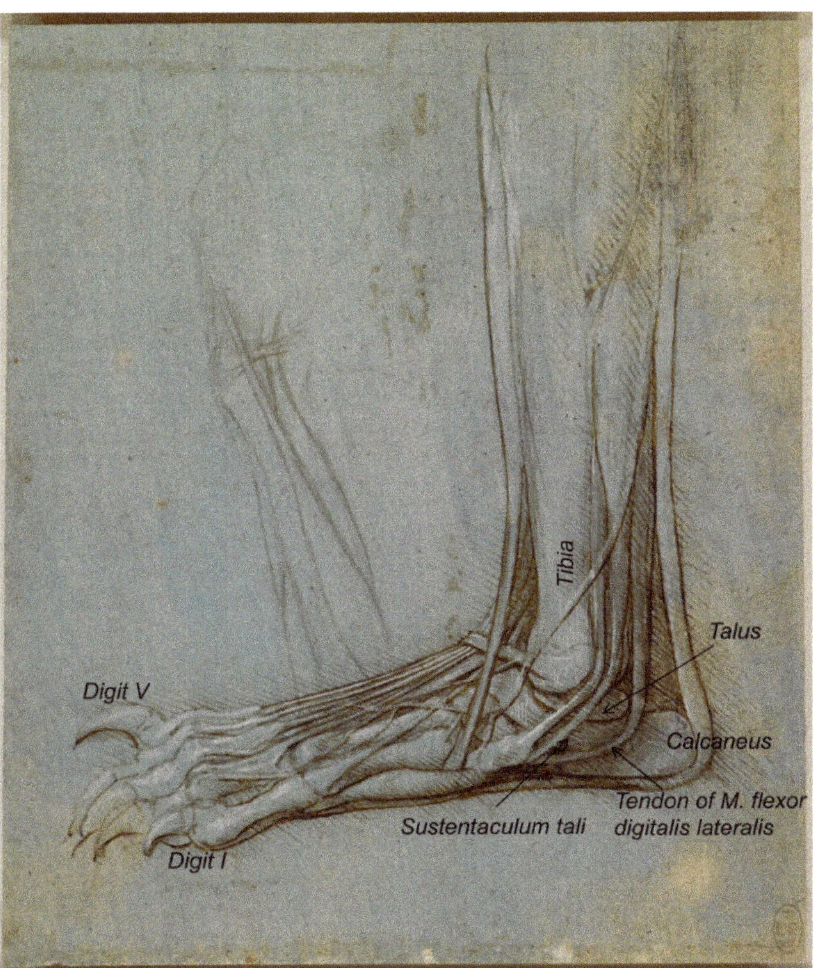

Figure 1. Bear's foot series—Number 1. Bear distal right pelvic limb/*pes*, medial aspect. A bear's foot c.1488–1490. Modified from www.rct.uk/collection/912372 (Royal Collection Trust [3]). This image is credited as Royal Collection Trust/© Her Majesty Queen Elizabeth II 2019.

The reference made to this figure (RCIN 912372) by Castiglioni [19] describes it as 'Studio di anatomia del piede umano, con artigli al posto delle unghie', a statement that clearly evidences its misinterpretation. Previously, it was catalogued as the foot of a monster, as reported later by Clark [4]. The description of Leonardo's bear's foot drawings (1488–1490), made by Clayton and Philo [9] (RCIN 912372; Figure 1), indicated that "Leonardo dissected the left hind leg of a bear ... The drawings show the bones, muscles and tendons of the lower leg and foot", in accordance to O'Malley and Saunders [6], whose Comparative Anatomy Chapter states that "the drawings represent a dissection of the left leg and foot of a bear as originally pointed out by the anatomist, William Wright. There can be no question that the identification is correct". It seems, indeed, to be a bear's foot. In accordance, the description of the same drawing at the Royal Collection Trust website [3] states "this drawing shows with some accuracy the bones, muscles and tendons of a bear's lower leg and foot, with the big toe, claw raised, away from the viewer". The bear, as a plantigrade animal, walks like humans, with the whole plantar face of the *pes* (the sole with the heel) touching the ground. However, in contrast to humans, the shortest *digit* (toe) is not the fifth one (the lateral one), but the medial one, that is the first *digit* [20,21]. Hence, to the best of our knowledge, we support that the bear's foot depicted by Leonardo corresponds to the right hind limb instead of the left one, as previously reported by O'Malley and Saunders [6] and Clayton and Philo [9] as well as at its description at the Royal Collection Trust website [3], maybe in resemblance to humans. In addition, the *calcaneus* bone of the *tarsus* is always in a lateral position to the *talus*, and the medial projection of the *calcaneus* bone to support the *talus* is quite visible, the so-called *sustentaculum tali*, with a groove to the tendon of the muscle *flexor digitalis lateralis* [22]. Figure 1 also shows, on the left-centre, a rough sketch of some muscles that continues beneath the represented bear's *pes*. If observed upside-down (Figure 2), it seems an *antebrachium* (forearm) and *manus* of another animal. Considering all the elements depicted (muscles' shape, the bone and the carpal orientation as well as the preliminary draft of the *manus*), we think it corresponds to a caudomedial/palmar view of the left *antebrachium* and the palmar view of a *manus* provided of short *digits* (at least *digits* shorter than those of humans, without an opposable *digit I* (thumb)). At the *carpus* level, the *retinaculum flexorum* is still patent to some extent, as a thick transverse fascial band from the medial carpal bones to the *os carpi accessorium*, forming the *canalis carpi* (canal between the proximal row of carpal bones and *retinaculum flexorum*) [22], although it has partially been removed to liberate the tendon of the muscle that seems to be the *flexor carpi radialis*, which appears to be attached at the base of the medial metacarpal bones. However, its deepest part still keeps the tendon of what seems to be the *M. flexor digitalis superficialis* at its place. This *antebrachium* and *manus* sketches are quite slim, suggesting being depicted from a dog or a wolf more than from a bear, with a more robust *antebrachium*. The proportions of the *antebrachium/manus* length are also more coincident with those of a dog/wolf than of a bear (with a relatively short *antebrachium*, but longer *manus*). The oblique line crossing the medial face of the radius might represent the *Vena cephalica*, which joins the *V. cephalica accessoria* (it runs dorsomedial to the *metacarpus* and *carpus*) to conform the *V. cephalica*, which continues cranially along the antebrachium. However, if that line represents the *V. cephalica*, it should have been a bit more distal (closer to the carpal joint). To the best of the authors' knowledge, this sketch was not previously described in any of the consulted literature, maybe because the ultimate detailed representation of the bear *pes* drawing catches the entire attention of the observer.

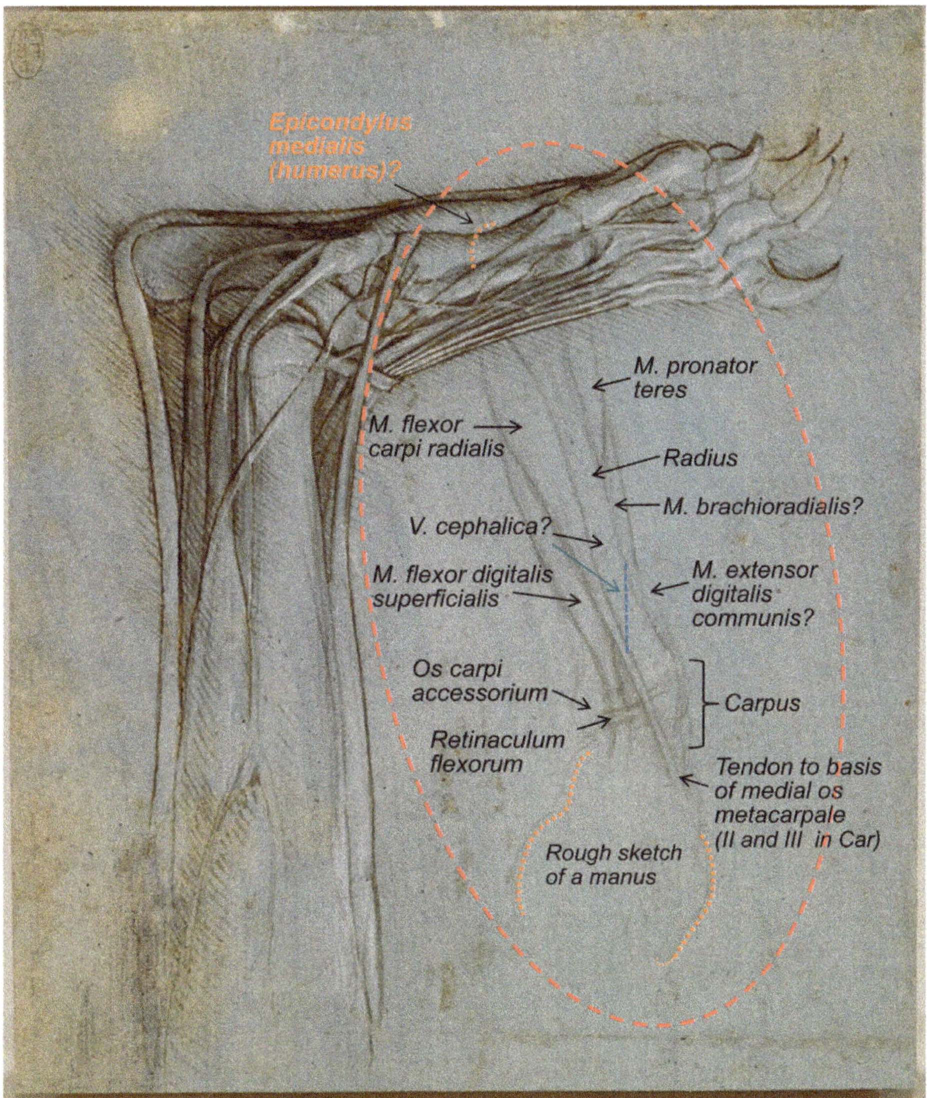

Figure 2. An *antebrachium* of a dog/wolf. Dog/wolf *antebrachium* and *manus*, caudomedial/palmar view. A bear's foot c.1488–1490. Modified from www.rct.uk/collection/912372 (Royal Collection Trust [3]). This image is credited as Royal Collection Trust/© Her Majesty Queen Elizabeth II 2019.

Another one of Leonardo da Vinci's bear *pes* drawings, RCIN 912373 recto (Figure 3), described at the Royal Collection Trust [3] as "a study of the dissection of the leg and foot of a bear, viewed in profile to the left", corresponds to the bear's right *pes* shown in a medial aspect (from the left). This sheet also includes an inset in the upper left, representing one *digit* viewed in profile, where the tendon of *M. flexor digitalis superficialis* fixes at two points in the second *phalanx*, between which a hole is provided, allowing the tendon of the *M. flexor digitalis profundus* to fix distally on the flexor surface of the third *phalanx* of the *pes digit*, a structure perfectly depicted by Leonardo da Vinci.

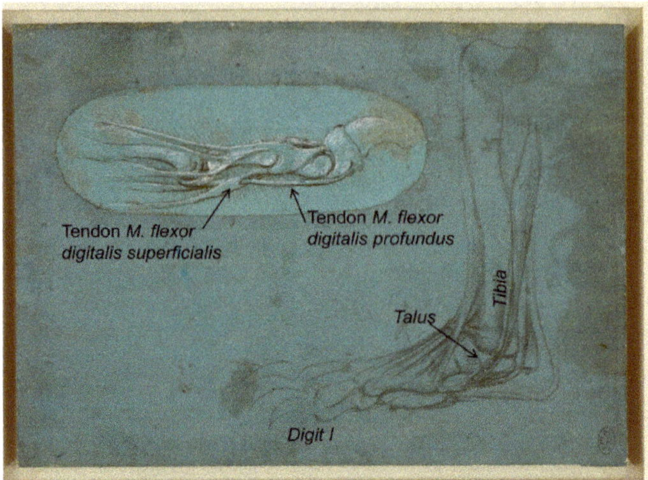

Figure 3. Bear's foot series—Number 2. Upper left: Bear *pes digit*. Down right: Bear distal right pelvic limb/*pes*, medial aspect. A bear's foot c.1488–1490. Modified from www.rct.uk/collection/912373 Recto (Royal Collection Trust [3]). This image is credited as Royal Collection Trust/© Her Majesty Queen Elizabeth II 2019.

The metalpoint drawing RCIN 912374 (1488–1490; Figure 4), "a study of the dissection of the leg and foot of a bear, viewed in profile to the right", shows the lateral side, with the perfectly defined *fibula* and three tendons associated to its distal end (*malleolus*), one in the lateral face belonging to the *M. peroneus longus* and two sliding caudal tendons of *M. peroneus brevis* and *M. extensor digitalis lateralis*. In addition, the lateral *digit* is one of the longest. Taking all this information into account, it could be concluded that the bear's *pes* represented in this drawing is again the right one.

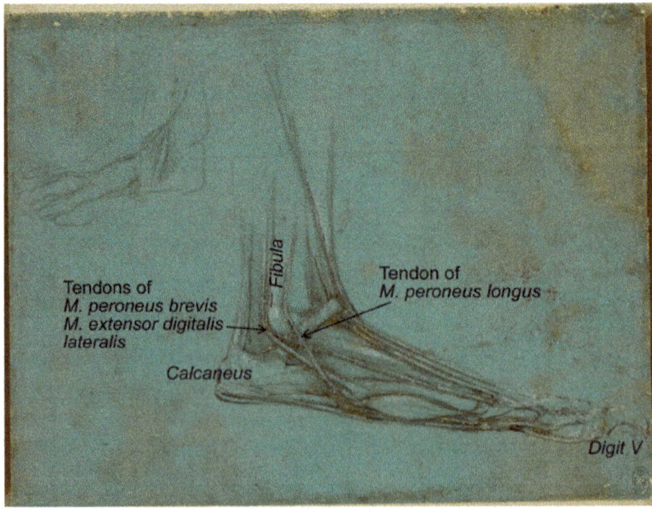

Figure 4. Bear's foot series—Number 3. Bear distal right pelvic limb/*pes*, lateral view. A bear's foot c.1488–1490. Modified from www.rct.uk/collection/912374 (Royal Collection Trust [3]). This image is credited as Royal Collection Trust/© Her Majesty Queen Elizabeth II 2019.

The last one of da Vinci's bear *pes* drawings of this series, catalogued as RCIN 912375: "A bear's foot" (Figure 5), shows an "outside view of the foot, partially from below" [9], where the tibia is visible and the *calcaneus* is lateral to it, corresponding also to the right bear's *pes*, an aspect not revealed in former descriptions.

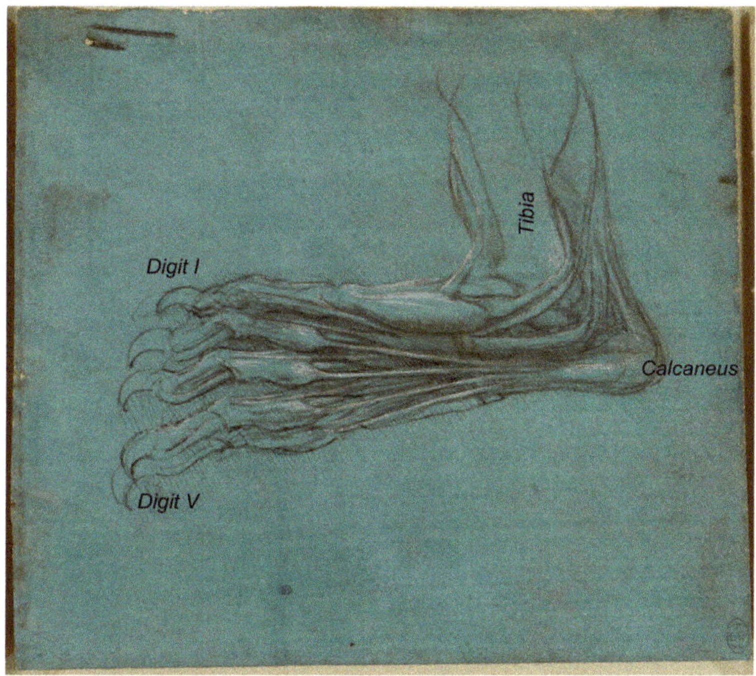

Figure 5. Bear's foot series—Number 4. Bear distal right pelvic limb/*pes*, plantaromedial oblique view. A bear's foot c.1488–1490. Modified from www.rct.uk/collection/912375 (Royal Collection Trust [3]). This image is credited as Royal Collection Trust/© Her Majesty Queen Elizabeth II 2019.

In summary, all bear's *pes* sheets depicted by Leonardo da Vinci correspond to the right pelvic limb. However, the preliminary draft of the *antebrachium* and *manus* of a dog/wolf illustrates the left one. It is clear that Leonardo da Vinci, on many occasions, used the sheets without an order. This fact leads to the interpretation made by Keele [17] relative to Leonardo's method of research: "he used pages of notes on particular subjects as we would use a filing system, returning to the same page at intervals of weeks, months, in some cases as long as twenty years later to record further drawings or verbal notes on the same subject".

3. Anatomy of the Horse Trunk that Turned into a Dog's Trunk

Later on, according to Clayton and Philo [10] "Writing in the mid-sixteenth century, the biographer Giorgio Vasari stated that Leonardo compiled a treatise on the anatomy of the horse. One drawing of the viscera of a large quadruped, probably a horse, does survive from this period, suggesting that Leonardo conducted full dissections to investigate the internal anatomy of the beast. But Vasari also stated that the treatise on the horse was lost when Milan was invaded by French forces in 1499. Ludovico Sforza was overthrown, and soon afterwards, Leonardo left the city and returned to Florence" [10]. This text refers to the drawing RCIN 919097-recto, entitled 'The viscera of a horse' (1490–1492; Figure 6), and described at the Royal Collection Trust [3] as: "an anterior view of the arteries, veins and the genito-urinary system of an animal, probably a horse," implying that Leonardo did not name this drawing. The drawing represents the ventral aspect of the trunk of an animal (supposedly, a horse) with the lungs and the *canalis alimentarius* (esophagus, stomach and intestines) removed. The large vessels depicted at the centre, all the figure down, represent the *aorta* (on the right of the figure) and the vena *cava caudalis* (on the left of the image). The way they ramify helps us to discard the idea that this drawing represents a human being. In humans, the *aorta* ends up dividing into two branches: *Aa. iliaca commune* (*dextra* and *sinistra*). In contrast, in animals, the end of the *aorta abdominalis* (at the level of the pelvic limbs) produces two branches (*external iliac arteries—dextra* and *sinistra*), well depicted, and subsequently continues and produces two more (*internal iliac arteries*), well represented in the drawing, ending as the *arteria sacral median*, not depicted. Regarding the veins, the *vena cava caudalis* is formed by the confluence of two *Vv. Iliaca commune—dextra* and *sinistra*, each one resulting from the junction of the *V. iliaca externa* and the *V. iliaca interna*, following a similar pattern both in humans and domestic mammals.

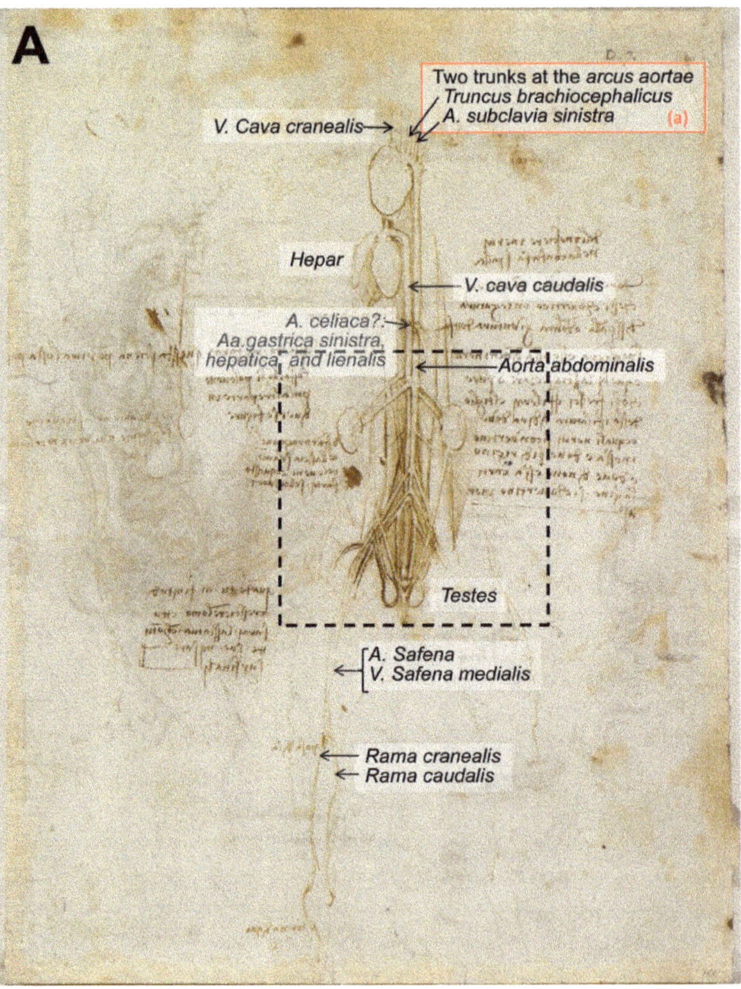

Figure 6. *Cont.*

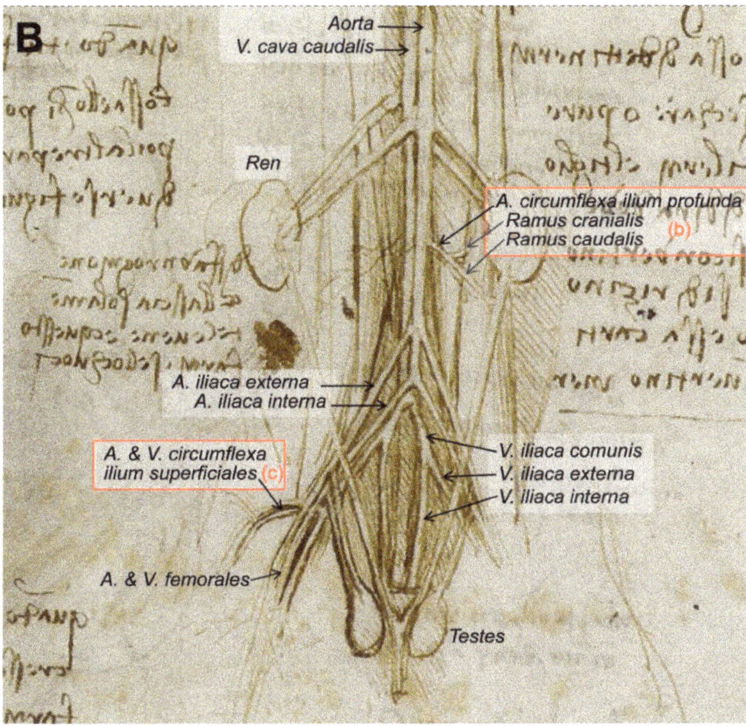

Figure 6. The viscera of a horse? (**A**) Whole drawing representing the ventral aspect of the trunk of an animal with the *canalis alimentarius* and lungs removed. (**B**) Inset at higher magnification depicting the lumbar and pelvic regions. The viscera of a horse c.1490–1492. Modified from www.rct.uk/collection/919097 recto (Royal Collection Trust [3]). This image is credited as Royal Collection Trust/© Her Majesty Queen Elizabeth II 2019.

Regarding the blood vessels, the drawing (Figure 6) provides three key points: (a) The first huge vessel (on the left of the image), reaching the heart, could be the *Vena cava cranealis*, and the other curved vessel going down is the *aorta* and its *arcus aortae*, with two big arteries leaving the aortic arch and some smaller ones (2–3) once the arch finishes and continues to the descendent aorta (*aorta descendens*). Large domestic mammals (horses—Eq, and ruminants—Ru) only have one artery deriving from the aortic arch, the *truncus brachiocephalicus*, which is then divided into a *truncus bicaroticus* (could be absent in carnivores—dogs and cats) and two *Aa. subclaviae*. In contrast, carnivores and pigs (Su) have two arteries leaving the *arcus aortae*: The *truncus brachiocephalicus* first and secondly the *A. subclavia sinistra*. (b) On the other hand, at the kidney level, there are two arteries perfectly outlined in the drawing stemming from the *aorta*: The *A. circumflexa ilium profunda* (*dextra* and *sinistra*), exclusive to carnivores [22] and dividing into the *rami craniales* and *caudales*. In contrast, the *Aa. circumflexa ilium profunda* derives from the *A. iliaca externa* in Su, Ru and Eq [22], similar to humans [23], not stemming directly from the aorta. (c) The arteria and vena *circumflexa ilium superficiales*, the first branches of the *A. femoralis* and *V. femoralis*, respectively, are exclusive to carnivores [22]. They leave their main vessels cranially oriented, at the medial and proximal part of the thigh. These vessels (a–c) in this drawing are the main clue to determine the species. Consequently, the horse representation/provenance of this drawing could be discarded. However, the horse is the unique domestic species in which the aorta does not end caudally as an *arteria sacral median*, which is not represented in the illustration.

In order to elucidate the identity of the depicted specimen, more elements were analyzed. The testicles in that position (*regio urogenitalis*-ventral pelvic region) could be mainly those of a dog. Male cats have the scrotum placed at the perineum, similar to boars, while the scrotum position of horses and bulls is more inguinal. The kidneys are not depicted as those from a horse (heart-shaped the *ren dexter* and more irregular the *ren sinister*). Those from pigs are more symmetrical and flattened, while those depicted are very similar to those from a dog. In contrast, cat kidneys have some vessels on the surface, with a radial arrangement toward the renal hilus, called capsular veins (*venae capsulares*), exclusive to cats [22], not represented in the drawing.

The hanging organ below the heart must be the liver. Accordingly, it has some very deep *incisurae interlobares* common for pigs and dogs/cats. At first glance, in between the liver and the renal arteries, it seems to be the lumbar part of the diaphragm (*crura*), but at higher magnification, it looks like an odd branch stemming from both the *aorta* and the *V. cava caudalis* (it is not clear in the drawing, maybe da Vinci had his doubts about this issue) that splits into three. There is a possibility that this vessel represents the *A. celiaca* with its three branches illustrated (*A. gastrica sinistra*, *A. hepatica* and *A. lienalis*). In that case, the organ below the heart cannot be the liver or was represented misplaced to give leadership to other, more relevant structures. Similarly, the penis has also been removed to expose the pelvic organs.

In conclusion, these details led us to confirm the hypothesis that this drawing does not represent the anatomy of a horse, as previously reported, since most of the anatomical elements are consistent with the open chest, abdomen and pelvis of a carnivore, probably a dog rather than a cat.

4. More on the Comparative Anatomy of Humans and Horses: The Case of the Horse and Human Anatomy of Their Pelvic Limb and Leg, Both Standing and Walking Forward

Continuing with horses, Leonardo da Vinci drew some sketches comparing the horse and human anatomy in terms of their pelvic limbs and legs, both standing and walking forward. The drawing entitled 'The leg muscles and bones of man and horse' (RCIN 912625; Figure 7) is described as "The muscles of a man's legs are here studied in several views, together with the bones of the pelvis and legs, with 'cords' indicating the lines of action of the muscles". At the lower centre is a diagram of the same structures in the horse, with the astute note that "to match the bone structure of a horse with that of a man you will have to draw the man on tip-toe" [3]. In 1919, and according to Wright [18], these drawings "serve to illustrate Leonardo's methods, referred to previously, of analysing a region into its elements and of making use of comparative anatomy". However, Wright described those illustrations as "the left hindlimb of an animal, probably a dog, and the left lower limb of a man, both drawn in the natural standing posture and both showing in the upper parts strips of corresponding muscles" [18]. Although the mentioned species of dog does not properly match with the skeleton morphology represented in this sheet, mainly due to the coxal bone (*os coxae*) shape, the presence of the third trochanter and the long metatarsal bone. Referring to these drawings, Keele [17] reported "Leonardo's interest in movement extended from those of man to animals", and relative to these figures, Leonardo "compares the bones of a man's leg with those of a horse when standing, saying that to compare the two, the man must be shown standing on tiptoe. When they walk forward, this becomes even more evident". Later on, Clayton and Philo [9] stated that "Leonardo's study of the horse here was to some degree compromised by his knowledge of human anatomy: the pelvis is too upright and not long enough, and the femur is too long and thin". We are in accordance with those authors: The pelvis is represented shorter than it should be, and its natural position should be a bit more horizontal. In addition, the femur should be more robust (Figure 7).

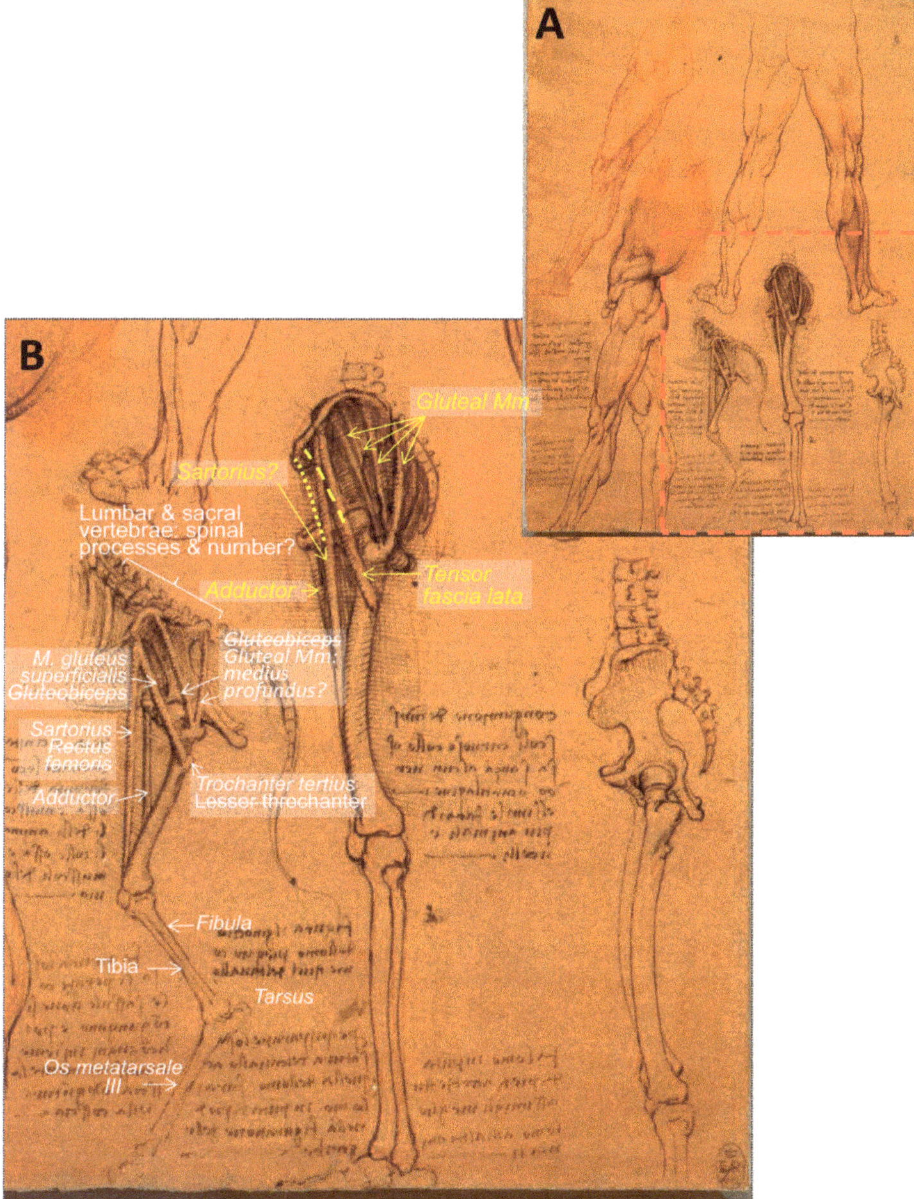

Figure 7. Series of comparative anatomy of man's leg and horse's pelvic limb—Number 1. (**A**) Whole drawing. (**B**) Inset at higher magnification showing a horse left pelvic limb (on the left) and a human left leg (centre and right), lateral aspect. The leg muscles and bones of man and horse c.1506–1508. Modified from www.rct.uk/collection/912625 (Royal Collection Trust [3]). This image is credited as Royal Collection Trust/© Her Majesty Queen Elizabeth II 2019.

Besides, the lumbar vertebrae, which should be five or six, have large horizontal transverse processes that are not illustrated, neither are the high spinous processes present at the lumbar vertebrae and at the *os sacrum* shown, composed of five fused sacral vertebrae. In addition, the number of the lumbar and sacral vertebrae is not accurately illustrated: Three to four at the lumbar region and also at the sacrum. Moreover, the tibia and fibula are represented as bones of the horse's leg (crus), but quite inaccurately, showing the fibula as long as the tibia, although horses only have a head and rudimentary body of the fibula (*caput* and *corpus fibulae*) [22] in a lateral position of the tibia, which barely reaches the half tibia and never articulates with the tarsus. O'Malley and Saunders [6] stated that "These three figures are for comparative purposes ... The supposedly corresponding muscles of the horse are shown ... but very inexactly. Presumably the cords represent the adductor, sartorius, tensor fasciae latae, gluteus superficialis or gluteus medius and gluteo-biceps of the horse". Clayton and Philo [9] also stated that "A few muscles are represented by threads: rectus femoris from the anterior iliac spine to the patella; tensor fasciae latae from roughly the same point towards the lesser trochanter of the femur (in the horse this trochanter is much more prominent than in the human); and the gluteal muscles, represented by a number of threads (two in the horse, four in the human) running from the iliac crest towards the greater trochanter". It is indeed the caudal part of a horse skeleton in profile because of the *os coxae* morphology, and its femur has a third trochanter (*Trochanter tertius*) specific for horses [22] (Figure 7). Hence, Clayton and Philo's description [9] is quite inaccurate, because the lesser trochanter is placed medially with respect to the major trochanter (below the head and neck of the femur, so it cannot be seen from this lateral view), and the detail they described as lesser trochanter is, in fact, the third trochanter (*Trochanter tertius*), in which only one muscle attaches: The *M. gluteus supeficialis* [22]. On the other hand, in horses, the *M. gluteus supeficialis* should extend from the *fascia glutea* and the *os sacrum* to the *trochanter tertius*, it has a cranial head originating from the *M. tensor fasciae latae* [22]. As the unique muscle that attaches to the third trochanter is the *M. gluteus superficialis*, it is obvious that Leonardo has represented only one of its first attachments (at the *tuber coxae*, where the *M. tensor fasciae latae* comes from). Consequently, in Eq, there could be some confusion in terms of the extent of the *M. tensor fasciae latae* as it starts together with one of the heads of the *M. gluteus superficialis* (fixing at the third trochanter) to the *fasciae latae*, but in quadrupeds, never to a bone fixation as reported by Clayton and Philo [9], because it tenses the *fasciae latae*, which is one of the *M. biceps femoris* fixations.

According to Schaller [22], the *M. gluteobiceps* described by O'Malley and Saunders [6] is inexact because only Su and Ru present it (resulting from the fusion of the *M. gluteus superficialis* and the cranial portion of the *M. biceps femoris*); consequently, it does not appear in horses.

In addition, in domestic mammals, the *M. rectus femoris* (as part of the *M. quadriceps femoris*) arises from cranial to acetabulum areas [22], but not from the ventral area of the *tuber coxae* (*os ilium*), as Leonardo depicted and as Clayton and Philo stated [9]. However, it could be the *M. sartorius* as reported by O'Malley and Saunders [6], whose attachments, from the *fascia iliaca* (in Ungulates) to the medial side of the proximal portion of the tibia and *fascia cruris* [22], match Leonardo's drawing. In humans, the *M. sartorius* runs from the *spina iliaca anterior superior* to medial to the *tuberositas tibiae* [23]. Hence, the longest thread depicted by Leonardo does not perfectly match the description. Regarding the vertical inner muscle (thread), it seems to start at the *symphysis pelvina* and to end medially and distally in the femur. This location is compatible with (a) the *M. gracilis* (flat superficial adductor of the thigh) from the *symphysis pelvina* (by *tendo symphysialis*) to the *fascia cruris* on the medial surface of the proximal portion of the crus [22], but it is placed caudal to the femur or, (b) the possibility that the *M. adductor* 'undivided in ungulates, from the *tendo symphysialis*, *ramus caudalis ossis pubis* and *ramus ossis ischia* to the *fascies aspera*, in Eq also *condylus medialis femoris*' [22]. In accordance to O'Malley and Saunders [6], it seems the latter (*M. adductor*) is the muscle that better fits to Leonardo's representation. In humans, there are the *M. adductor longus* and *magnus* from near the *symphysis* and the *tuber ischiadicum* and *ramus ischii*, respectively, to the *labium mediale* of *linea aspera* both, and the *magnus* additionally to the *condylus medialis* of the femur [23]. As in a profile view from lateral, it is not possible to see the distal attachments of the represented threads; it is therefore difficult to infer which ones of the adductor muscles are depicted.

Clayton and Philo [9] also stated that "the drawing at lower right has been called an 'anatomical fantasy', blending the bones of a horse with those of a man. It is much more likely that Leonardo intended it to be purely human, but incorporated errors (in particular, the extended ischium below the coccyx; cf. no. 64b, in which the error is corrected), derived from his superior knowledge at that date, of equine anatomy-the study of human bones and muscle threads to the left display the same errors". Should it say: ... because of his superior knowledge of human anatomy. O'Malley and Saunders [6] referred to this figure as 'the elongation of the innominate bone and the length of the coccyx suggest the pelvis of an animal rather than of a man. In addition, the observer will note a trochanter tertius below the greater trochanter. From these appearances, this figure seems to have been derived by the expansion of animal bones to the approximate proportions of the human'. Regarding RCIN 919012 recto: The skeleton c.1510–1511 (figure not included), drawn 2 to 5 years later than RCIN 912625 recto (Figure 7), Clayton and Philo [9] stated that "Leonardo has here corrected the length of the ischium", referring to the lower right figure. According to Wright [18], these drawings "serve to illustrate Leonardo's methods, referred to previously, of analyzing a region into its elements and of making use of comparative anatomy".

There are other illustrations depicting the comparative anatomy between the man and the horse, belonging to the 'Manuscrit K' from the Bibliothèque de l'Institut de France [24]. One of them, folio 102 recto (Figure 8), compares the right horse hind limb (in a lateral view) with the right human leg in a frontal view. The drawing at the left centre (Figure 8B) is without doubt the laterocaudal view of the right pelvic limb of the horse, and again he has drawn a fibula that extends much too far distally along the caudal aspect of the tibia for the horse (this fact adds to the evidence that the limb depicted in Figure 7 is indeed an equine limb).

Nevertheless, the most impressive drawing in 'Manuscrit K' about this issue corresponds to Folio 109 verso (Figure 9), in which both sketches are reported to compare a man's leg with a horse´s hind limb when walking forward, for which the man should be on tiptoes [17]. The illustration shows part of the trunk and the left leg of a man seen in left profile and compared with the left posterior limb of a horse, but not keeping a relative size/proportion.

Here, however, it is not clear that the represented man is walking forward, because the flexed knee during walking is concurrent with tiptoe contact to the ground, but not the trunk inclination/angle as depicted. This position is similar to that when a man or a horse is taking impulse from below (flexing their legs or hind limbs) to be able to jump. These illustrations exemplify what O'Malley [7] stated about Leonardo's interests: "they are directed towards the structure of the body in relationship to its workings". Maybe Leonardo tried to show a parallelism between the relative positions of bones and muscles of the horse, captured and transferred to a human skeleton in order to describe and understand the combination of the flexion of the joints together with the muscle action to accumulate the energy needed on the impulse of jumping, as it is described in the text of the following page of the manuscript K [24] (Folio 110 (30) recto: 'saut de l'homme'; compare Figure 7 on standing position vs. Figure 9 on flexion; Figure 8 seems to be a previous sketch to Figure 9).

The main difference between the drawing at the Royal Collection Trust [3] (Figure 7) and those from 'Manuscrit K' [24] (Figures 8 and 9) is that the muscle is represented as an inverted 'V', illustrated as different threads in the Royal Collection Trust. The inverted 'V' muscle has its apex at the *tuber coxae*, and the shorter part fixes laterally at the third trochanter. The longest part goes to the *condylus medialis* of the tibia in one of them (Figure 9), but it is not clear in the other folio (Figure 8) because its fixation should have been out of the sheet (lower sketch) or is not drawn with enough detail (upper sketch), indicating a rough study previous to Figure 9. In addition, just analyzing the depicted muscles and threads of these illustrations, it is clear that in Manuscript K [24] (Figures 8 and 9), the muscles are represented as a wide thread with a sort of 'belly'. However, in the drawing RCIN 912625 (Figure 7), slimmer threads are depicted; hence, it could be deduced that the sketches from Manuscript K were made previously to the drawing of the Royal Collection Trust, just as an evolving progress in his studies, a trend in all artists´ lives, as a result of a lifelong development in their conceptualization and abstraction ability/aptitude. However, comparing the contents of both illustrations (Figures 7 and 9), it is not likely that Leonardo started to study the flexed hind limb to approach its standing position later. These facts should be taken into account when dating back to Leonardo's work, as those sheets are assigned more or less to the same periods of c.1506–1508 (Figure 7) and c.1503–1508 (Figures 8 and 9).

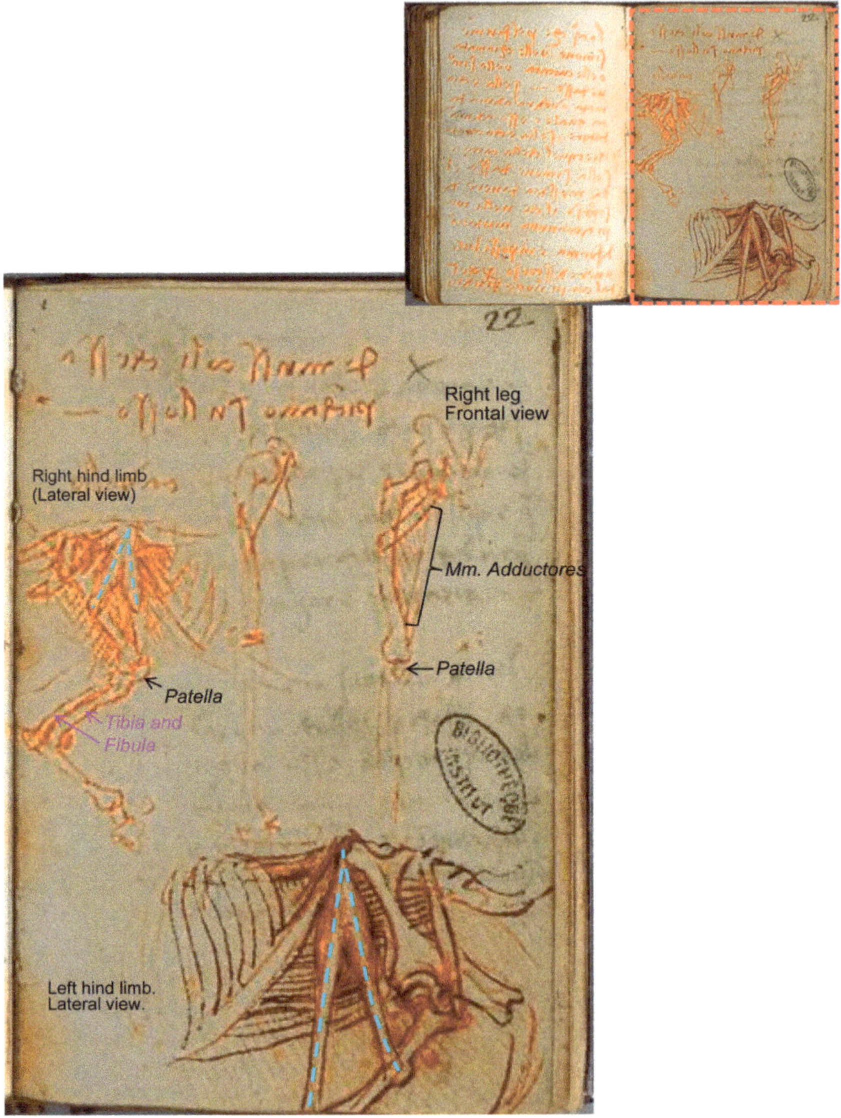

Figure 8. Series of the comparative anatomy of a man's leg and horse's pelvic limb—Number 2. (**A**) Two pages. (**B**) Inset: Higher magnification of Folio 102(22) recto: 'Anatomie, cheval'. Modified from Ms 2181-folio 102(22)-manuscrit K (Bibliothèque de l'Institut de France [24]) (1503–1508). This image presentation is authorized in limited time period by Réunion des Musées Nationaux-Grand Palais (R.M.N.-Grand Palais), and the Bibliothèque de L'Institut de France. Photo © RMN-Grand Palais (Institut de France)/René-Gabriel Ojéda.

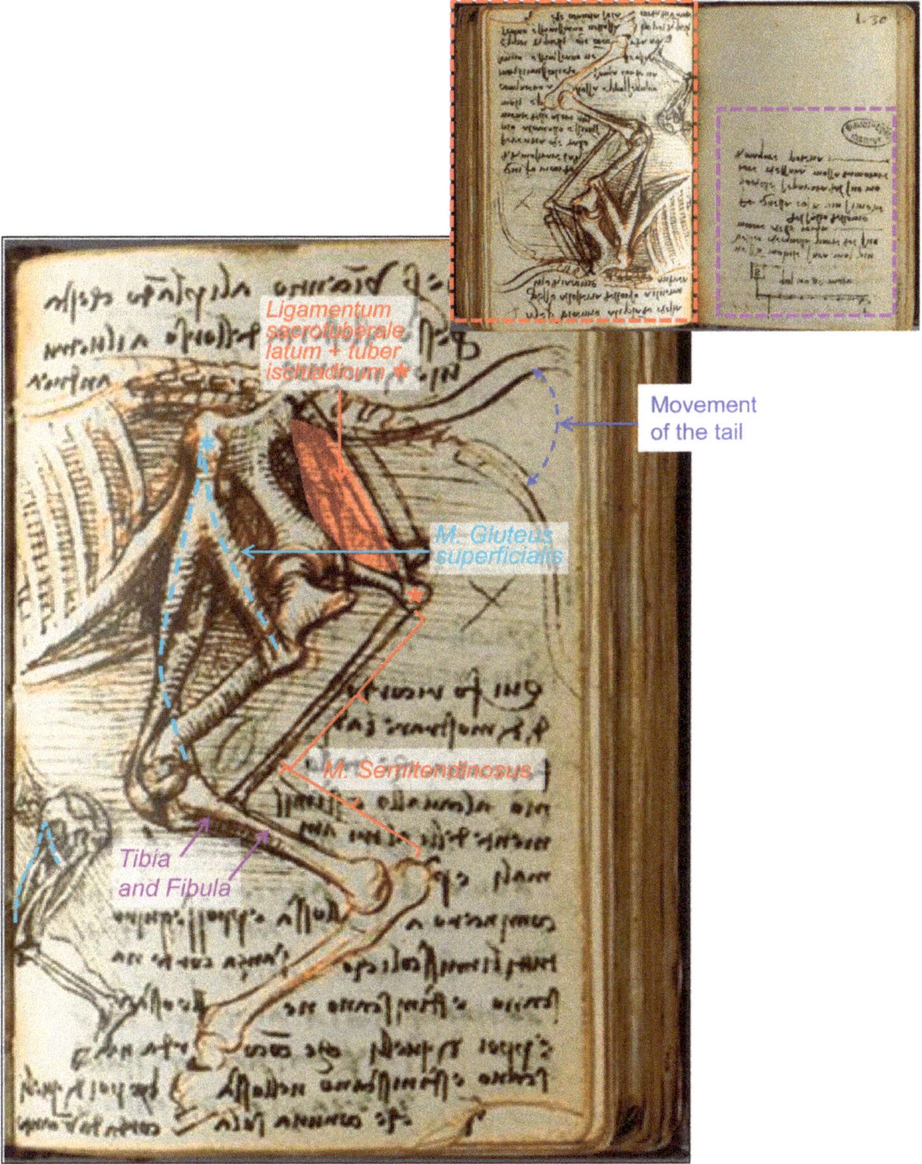

Figure 9. Series of the comparative anatomy of a man's leg and horse's pelvic limb—Number 3. (**A**) Two pages: Folio 109 verso (red) and folio 110 recto (violet inset with da Vinci notes about 'Leviers, movements, saut de l'homme'). (**B**) Folio 109 verso: 'Anatomie comparée de l'homme et des animaux (chevaux)'. Red inset upside down at higher magnification displaying the comparative between the man left leg (bottom, left) and the horse left pelvic limb, lateral aspect. Modified from Ms 2181-folio 109(29-30) and 110 (30)-manuscrit K (Bibliothèque de l'Institut de France [24]) (1503–1508). This image presentation is authorized in limited time period by Réunion des Musées Nationaux-Grand Palais (R.M.N.-Grand Palais), and the Bibliothèque de L'Institut de France. Photo © RMN-Grand Palais (Institut de France)/René-Gabriel Ojéda.

5. Conclusions

Leonardo da Vinci was an outstanding artist and scientist, looking for the reasons and mechanisms of all the aspects he had studied during his entire life. He tried to find the common clues shared by human and animal anatomy through comparative anatomy. He produced an impressive collection of anatomical drawings, which were, in general, quite accurate. However, some later studies have misunderstood some of them. Hence, a deep anatomical insight into the bear and horse anatomical drawing collections revealed some inaccuracies, which, for some anatomical elements, we identified and amended. Such is the case of the depicted 'bear's foot' series, described as the left hind limb, but based on their *digits* and *tarsus*, it is the right one. Regarding the horse anatomy, one of Leonardo's drawings illustrating the internal and dorsal parts of the horse trunk, if considered the different blood vessels depicted and other viscera, seems to be from a carnivore (probably a dog) rather than that of a horse, as previously reported. Other drawings illustrating the comparative anatomy of the horse hind limb and the human leg were also reconsidered in a new approach, assuming that Leonardo da Vinci used the comparative anatomy in order to understand the process to produce movement, especially when jumping.

Author Contributions: M.L. and M.d.M.Y. participated equally in the conception/design, interpretation, drafting the manuscript, critical review of the manuscript and approval of its final version.

Funding: The authors declared they received no financial support for their research and/or authorship of this article.

Acknowledgments: The authors are thankful to Royal Collection Trust website, Biblioteca Digitale from Museo Galileo-Firenze-Italia and Bibliothèque de l'Institut de France website for the possibility of free access to their Leonardo da Vinci´s digital collection. In addition, we thank the Royal Collection Trust for granting their images for free.

Conflicts of Interest: The authors declared that they had no conflict of interests with respect to their authorship or the publication of this article.

References

1. Forlani-Tempesti, A. *The Robert Lehman Collection V: Italian Fifteenth- to Seventeenth-Century Drawings;* The Metropolitan Museum of Art and Princeton University Press: New York, NY, USA; Princeton, NJ, USA, 1991; pp. 236–240.
2. Pasipoularides, A. Historical continuity in the methodology of modern medical science: Leonardo leads the way. *Int. J. Cardiol.* **2014**, *171*, 103–115. [CrossRef] [PubMed]
3. Royal Collection Trust Website. Available online: www.royalcollection.org.uk/collection (accessed on 18 May 2019).
4. Clark, K. *The Drawings of Leonardo da Vinci in the Collection of Her Majesty the Queen at Windsor Castle*, 2nd ed.; Blunt, A.F., Ed.; Phaidon Press Ltd.: London, UK, 1968; Volume 1.
5. Clark, K. *The Drawings of Leonardo da Vinci in the Collection of Her Majesty the Queen at Windsor Castle*, 2nd ed.; Blunt, A.F., Ed.; Phaidon Press Ltd.: London, UK, 1969; Volume 3.
6. O'Malley, C.D.; Saunders, J.B.D.C.M. *Leonardo da Vinci on the Human Body: The Anatomical, Physiological, and Embryological Drawings of Leonardo da Vinci;* Henry Schuman: New York, NY, USA, 1952.
7. O'Malley, C.D.; Saunders, J.B.D.C.M. *Leonardo da Vinci on the Human Body: The Anatomical, Physiological, and Embryological Drawings of Leonardo da Vinci;* Crown Publications: Victoria, BC, Canada, 1983.
8. O'Malley, C.D.; Saunders, J.B.D.C.M. *Leonardo da Vinci on the Human Body: The Anatomical, Physiological, and Embryological Drawings of Leonardo da Vinci;* Gramercy: New York, NY, USA, 2003.
9. Clayton, M.; Philo, R. *Leonardo da Vinci Anatomist;* Royal Collection Publications: London, UK, 2012.
10. Clayton, M.; Philo, R. *Leonardo da Vinci. The Mechanics of Man;* Royal Collection Trust: London, UK, 2013.
11. Schultheiss, D.; Laurenza, D.; Götte, B.; Jonas, U. The Weimar anatomical sheet of Leonardo da Vinci (1452–1519): An illustration of the genitourinary tract. *BJU Int.* **1999**, *84*, 595–600. [CrossRef] [PubMed]
12. Jose, A.M. Anatomy and Leonardo da Vinci. *Yale J. Biol. Med.* **2001**, *74*, 185–195. [PubMed]
13. Ganseman, Y.; Broos, P. Leonardo da Vinci and Andreas Vesalius; the shoulder girdle and the spine, a comparison. *Acta Chir. Belg.* **2008**, *108*, 477–483. [CrossRef] [PubMed]

14. Sterpetti, A.V. Anatomy and physiology by Leonardo: The hidden revolution? *Surgery* **2016**, *159*, 675–687. [CrossRef] [PubMed]
15. Bowen, G.; Gonzales, J.; Iwanaga, J.; Fisahn, C.; Loukas, M.; Oskouian, R.J.; Tubbs, R.S.; da Vinci, L. Leonardo da Vinci (1452–1519) and his depictions of the human spine. *Childs Nerv. Syst.* **2017**, *33*, 2067–2070. [CrossRef] [PubMed]
16. West, J.B. Leonardo da Vinci: Engineer, bioengineer, anatomist, and artist. *Am. J. Physiol. Lung Cell. Mol. Physiol.* **2017**, *312*, L392–L397. [CrossRef] [PubMed]
17. Keele, K.D. Leonardo da Vinci's 'Anatomia Naturale' the inaugural John Fulton Lecture. *Yale J. Biol. Med.* **1979**, *52*, 369–409. [PubMed]
18. Ochenkowski, H.; Wright, W. The quatercentenary of Leonardo da Vinci. *Burlingt. Mag. Connoiss.* **1919**, *34*, 186–203.
19. Castiglioni, A. Leonardo da Vinci anatomo e fisiologo. In *Il Volto di Ippocrate: Istorie di Medici e Medicine D'altri Tempi*; Società Editrice Unitas: Milano, Italy, 1925; pp. 172–209.
20. Sims, M.E. *Comparison of Black Bear Paws to Human Hands and Feet*; Identification Guides for Wildlife Law Enforcement No. USFWS; National Fish and Wildlife Forensics Laboratory: Ashland, OR, USA, 2007.
21. Dogăroiul, C.; Dermengiu, D.; Viorel, V. Forensic comparison between bear hind paw and human feet. Case report and illustrated anatomical and radiological guide. *Rom. J. Leg. Med.* **2012**, *20*, 131–134. [CrossRef]
22. Schaller, O. *Illustrated Veterinary Anatomical Nomenclature*, 2nd ed.; Enke Verlag: Stuttgart, Germany, 2007.
23. Feneis, H. *Nomenclatura Anatómica Ilustrada*; Salvat Editores: Barcelona, Spain, 1988.
24. Bibliothèque de l'Institut de France. Available online: http://www.bibliotheque-institutdefrance.fr/content/les-carnets-de-leonard-de-vinci (accessed on 18 May 2019).

MDPI

Article

Radiogrametric Analysis of the Thoracic Limb Phalanges in Arabian Horses and Thoroughbred Horses

Ozan Gündemir [1], Tomasz Szara [2,*], Gülsün Pazvant [1], Dilek Olğun Erdikmen [3], Sokol Duro [4] and William Perez [5]

1 Department of Anatomy, Faculty of Veterinary Medicine, Istanbul University-Cerrahpasa, Istanbul 34500, Turkey; ozan.gundemir@istanbul.edu.tr (O.G.); gulsun@iuc.edu.tr (G.P.)
2 Department of Morphological Sciences, Institute of Veterinary Medicine, Warsaw University of Life Sciences—SGGW, 02-776 Warsaw, Poland
3 Department of Surgery, Faculty of Veterinary Medicine, Istanbul University-Cerrahpasa, Istanbul 34500, Turkey; dilekolg@istanbul.edu.tr
4 Faculty of Veterinary Medicine, Agricultural University of Tirana, 1029 Tirana, Albania; durosokol@ubt.edu.al
5 Unidad de Anatomía, Facultad de Veterinaria, Universidad de la República, Montevideo 11600, Uruguay; vetanat@gmail.com
* Correspondence: tomasz_szara@sggw.edu.pl; Tel.: +48-2259-362-02

Simple Summary: The aim of the research was to determine the radiogrametric features differentiating the phalanges of the thoracic limbs of Arabian horses and Thoroughbred horses including sexual dimorphism. Nine traits and three indexes were analyzed. Radiological measurements of phalanges showed that sexual dimorphism is not clearly marked and differences between breeds manifest themself mainly in the proximal phalanx measurements. None of the parameters tested in Thoroughbred horses differed significantly between males and females. The discriminant analysis enabled the correct classification of 89.33% of the proximal phalanx samples to the exact breed. This percentage was 77.33% in the case of the middle phalanx and 54.67% for the distal phalanx, respectively. The data obtained from this study can be used as a reference material for the radiogrametric evaluation of the skeleton of the manus in horses.

Abstract: In this study, it was aimed to determine the statistical differences between Arabian horses and Thoroughbred horses based on X-ray images of forelimb digital bones. Latero-medial X-ray images of digital bones of thoracic limbs were taken of 25 Arabian horses and 50 Thoroughbred healthy horses. The difference between males and females within the breed was statistically analyzed as well. Nine measurements and three indexes taken from phalanges of thoracic limbs were used. Thoroughbred horses did not differ significantly between sexes, as indicated by the ANOVA. For the Arabian horses, the length of the middle of the proximal phalanx ($p < 0.05$), the length of the middle of the middle phalanx ($p < 0.001$), and the length of the dorsal surface of the distal phalanx ($p < 0.05$) measurement points were found to be differentiated between sexes. In the analysis made between Thoroughbred horses and Arabian horses with no respect to sex, the critical measurement was the depth of the caput of the proximal phalanx. The discriminant analysis enabled the correct classification of 89.33% of the proximal phalanx samples to the exact breed. The correct classification rate was 77.33% in the case of middle phalanx and 54.67% in the case of distal phalanx. Measurement results of the distal phalanx were found to be insignificant between both breeds and sexes. The radiological measurements of digital bones showed that sexual dimorphism was not too expressed and that decisive differences were found between the breeds.

Keywords: horse; phalanx; radiogrametric study; veterinary anatomy

Citation: Gündemir, O.; Szara, T.; Pazvant, G.; Erdikmen, D.O.; Duro, S.; Perez, W. Radiogrametric Analysis of the Thoracic Limb Phalanges in Arabian Horses and Thoroughbred Horses. *Animals* **2021**, *11*, 2205. https://doi.org/10.3390/ani11082205

Academic Editor: Chris W. Rogers

Received: 27 May 2021
Accepted: 23 July 2021
Published: 26 July 2021

Publisher's Note: MDPI stays neutral with regard to jurisdictional claims in published maps and institutional affiliations.

1. Introduction

Radiographic imaging techniques have become more readily available, and are now widely used in veterinary environments. Beside clinical use, these image data are used in veterinary anatomy as educational tools, and for differentiating breeds [1–4]. Quantitative radiography allows not only subjectively assessment of the condition of the digital organ in particular individuals, but to learn about the changes that the phalanges undergo with age [5]. Radiological examination is crucial in revealing the source of orthopedic problems in horses [6,7]. In particular, a significant number of cases of lameness in thoracic limbs may be due to the non-optimal alignment of phalanges [8]. Various studies have been carried out to examine the position of phalanges relative to each other and the angular positions between them radiogrametrically [9–11]. In addition, X-ray images have been used to evaluate the general joint conformation of healthy animals, not just in suspicion of lameness [12]. Radiographic techniques can provide information regarding relationships between the hoof capsule conformation and the geometry of the distal phalanx [13,14]. Cohen et al. [15] attempted to determine prospectively the association between the abnormal radiographic findings and performance indicators in English Thoroughbred horses.

As stated above, the positions and conformations of phalanges are very important for clinical diagnosis. In this regard, it is thought that the reference information obtained from healthy animals will be valuable for future studies on this subject [16]. In addition, while collecting radiometric reference data on digital bones of the thoracic limbs of healthy animals, we tried to answer the following study questions:

Is there any difference in radiometric measurements of digital bones of the thoracic limbs of Arabian horses and Thoroughbred horses?

Can gender determination be made in radiometric measurements of digital bones of the thoracic limbs?

2. Materials and Methods

The study was conducted with 25 Arabian horses (7 females, 18 males) and 50 (15 females, 35 males) Thoroughbred horses aged between 2 and 8 years, brought to the Jockey Club of Turkey for control purposes. We used the unit of age as years. Horses that had no musculoskeletal or orthopedic problems before and actively participated in races and training were used in the study. Latero-medial radiological images of the digital bones of their thoracic limbs were acquired with Gierth X-ray (TR9030). The difference between the right–left manus was ignored, and only the images of the digital bones of the left thoracic limbs were used for analyses.

Three measurements of each of the proximal phalanx, middle phalanx, and distal phalanx were taken (Figure 1). Measurements were recorded using the Radiant DICOM Viewer (version 2020.2.2) software:

Proximal phalanx:

1. Depth of the basis (DBP), greatest depth.
2. Depth of the caput (DCP), greatest depth.
3. Length of the middle (LMP).
4. Index 1 (DBP × DCP/LMP).

Middle phalanx:

1. Depth of the basis (DBM), greatest depth.
2. Depth of the caput (DCM), greatest depth.
3. Length of the middle (LMM).
4. Index 1 (DBM × DCM/LMM).

Distal phalanx:

1. Length of the dorsal surface (LD), length of facies parietalis of the distal phalanx.
2. Greatest solear length (GSL), distance between the crena marginis solearis and the processus palmaris.
3. Length of the caudo-dorsal surface (FA), distance between the processus extensorius of the distal phalanx and the processus palmaris.
4. Index 3 (LD × GSL/FA).

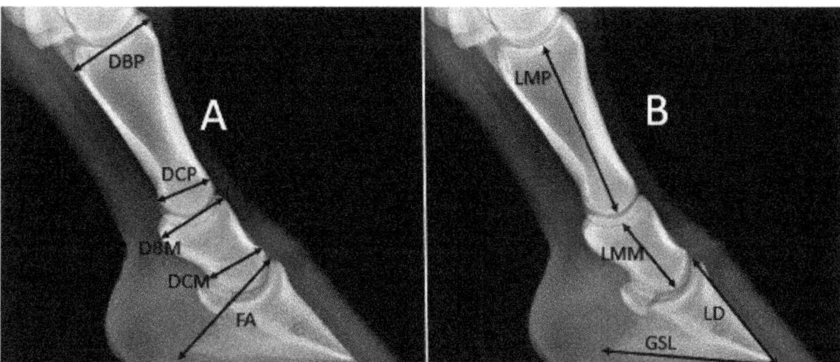

Figure 1. Measurement (4-year-old male Arabian horse). (**A**): Depth of the basis of the proximal phalanx (DBP), depth of the caput of the proximal phalanx (DCP), depth of the basis of middle phalanx (DBM), depth of the caput of the middle phalanx (DCM), length of the caudo-dorsal surface of the distal phalanx (FA). (**B**): Length of the middle of the proximal phalanx (LMP), length of the middle of the middle phalanx (LMM), length of the dorsal surface of the distal phalanx (LD), greatest solear length of the distal phalanx (GSL).

SPSS (version 22) was used for statistical calculations. Differences between Arabian and Thoroughbred horses in terms of the morphometric features were evaluated separately for proximal phalanx, middle phalanx, and distal phalanx. Male and female samples of the two breeds were evaluated separately. In addition, measurement differences between the two horse breeds were analyzed regardless of gender. ANOVA was used for these analyses. Homogeneity of variances was examined, and p values were obtained. Discriminant function analysis was applied to reveal the differences between Arabian horses and Thoroughbred horses. Eigenvalue and Wilks' lambda values were examined. A confusion table was used to evaluate the validity of DFA results. The correctly classified values and a discriminant distribution table were obtained with the Past (4.01) statistics program. Correlations between variables of the phalanges and between phalange traits and horse age were examined as well.

3. Results

The mean values of particular traits and standard deviation determined separately for the males and females of Arabian horses and Thoroughbred horses are presented in Table 1. The measurement points selected for Thoroughbred horses had no significant effect on the difference between sexes, as indicated by ANOVA results. The distribution of the samples between both breeds and sexes was homogeneous. Trait values in males were found to be higher for the proximal phalanx and middle phalanx. In Arabian horses, LMP ($p < 0.05$), LMM ($p < 0.001$), and LD ($p < 0.05$) were found to be significantly different between sexes. Values of all traits and indexes were found to be higher in the males of Arabian horses. According to the index results used, index 3 was a determinant between sexes only in Arabian horses ($p < 0.05$).

Table 2 shows the mean values, standard deviations, and statistical differences in phalangeal traits between Arabian and Thoroughbred horses regardless of sex. All traits and index values of the proximal phalanx and middle phalanx can be used in breed distinction. The difference in phalangeal traits was significant. The proximal phalanx and middle phalanx values were higher in Thoroughbred horses. The distal phalanx values were not statistically significantly different between the breeds. The GSL and FA values were found to be higher in Arabian horses, unlike other length measurements.

Table 1. Mean values of measurements, standard deviations, and *p* values determined between sexes of Arabian horses and Thoroughbred horses (ANOVA).

		Thoroughbred Horses				Arabian Horses			
Measurement	Sex	*n*	Mean (cm)	SD	*p* Value	*n*	Mean (cm)	SD	*p* Value
DBP	Female	15	4.72	0.25	0.84	7	4.30	0.14	0.06
	Male	35	4.74	0.30		18	4.46	0.20	
DCP	Female	15	2.99	0.13	0.05	7	2.73	0.10	0.07
	Male	35	3.09	0.19		18	2.83	0.13	
LMP	Female	15	9.82	0.38	0.44	7	8.99	0.39	0.03
	Male	35	9.93	0.47		18	9.39	0.37	
Index 1	Female	15	1.44	0.10	0.32	7	1.31	0.06	0.37
	Male	35	1.48	0.15		18	1.35	0.11	
DBM	Female	15	4.02	0.19	0.23	7	3.69	0.25	0.11
	Male	35	4.11	0.24		18	3.85	0.19	
DCM	Female	15	3.02	0.22	0.53	7	2.70	0.10	0.06
	Male	35	3.06	0.21		18	2.86	0.20	
LMM	Female	15	4.28	0.18	0.71	7	3.67	0.32	0.00
	Male	35	4.31	0.29		18	4.00	0.19	
Index 2	Female	15	2.85	0.31	0.36	7	2.73	0.17	0.75
	Male	35	2.93	0.26		18	2.76	0.28	
LD	Female	15	6.81	0.37	0.77	7	6.40	0.41	0.03
	Male	35	6.84	0.44		18	6.92	0.52	
GSL	Female	15	8.63	0.59	0.96	7	8.45	0.53	0.06
	Male	35	8.62	0.62		18	8.92	0.52	
FA	Female	15	6.28	0.36	0.33	7	6.19	0.48	0.38
	Male	35	6.16	0.43		18	6.37	0.43	
Index 3	Female	15	9.36	0.81	0.38	7	8.75	0.51	0.02
	Male	35	9.61	0.93		18	9.72	0.98	

Table 2. Mean values of measurements, standard deviations, and *p* values determined for Arabian horses and Thoroughbred horses (ANOVA).

Measurement	Breed	*n*	Mean (cm)	SD	*p* Value
DBP	Arabian	25	4.42	0.19	0.00
	Thoroughbred	50	4.73	0.28	
DCP	Arabian	25	2.80	0.13	0.00
	Thoroughbred	50	3.06	0.18	
LMP	Arabian	25	9.28	0.41	0.00
	Thoroughbred	50	9.89	0.45	
Index 1	Arabian	25	1.34	0.10	0.00
	Thoroughbred	50	1.47	0.14	
DBM	Arabian	25	3.81	0.22	0.00
	Thoroughbred	50	4.09	0.23	
DCM	Arabian	25	2.82	0.19	0.00
	Thoroughbred	50	3.05	0.21	
LMM	Arabian	25	3.90	0.27	0.00
	Thoroughbred	50	4.30	0.26	
Index 2	Arabian	25	2.75	0.25	0.03
	Thoroughbred	50	2.90	0.27	
LD	Arabian	25	6.78	0.54	0.63
	Thoroughbred	50	6.83	0.42	
GSL	Arabian	25	8.79	0.55	0.26
	Thoroughbred	50	8.63	0.60	
FA	Arabian	25	6.32	0.44	0.24
	Thoroughbred	50	6.19	0.41	
Index 3	Arabian	25	9.45	0.97	0.71
	Thoroughbred	50	9.53	0.90	

Results of the discriminant function analysis between Arabian and Thoroughbred horses are given in Table 3. Separate tests were performed for proximal phalanx, middle phalanx, and distal phalanx. Separate formulas were obtained for each bone. The highest canonical correlation value in the breed discrimination was obtained for the proximal phalanx. Multivariate discriminant function analysis score equations were as follow:

Proximal phalanx: (DBP × −16.584) + (DCP × −28.755) + (LMP × 6.686) + (Index 1 × 50.261) + 25.973.

Middle phalanx: (DBM × 0.931) + (DCM × 2.555) + (LMM × 2.187) + (Index 2 × −1.026) − 17.503.

Distal phalanx: (LD × −1.969) + (GSL × 1.706) + (FA × 0.454) + (Index 3 × 0.012) − 4.331.

Table 3. Stepwise discriminant function analysis for Arabian horses and thoroughbred horses.

	V	UC	Constant	SM	WL	E	GC	CC
Proximal phalanx	DBP DCP LMP Index 1	−16.584 −28.755 6.686 50.261	25.973	−0.614 −0.801 −0.705 −0.520	0.524	−0.909	A: 1.330 T: −0.665	0.690
Middle phalanx	DBM DCP LMM Index 2	0.931 2.555 2.187 −1.026	−17.503	0.780 0.721 0.951 0.353	0.638	0.567	A: −1.051 T: 0.525	0.602
Distal phalanx	LD GSL FA Index 3	−1.969 1.706 0.454 0.012	−4.331	−0.267 0.632 0.666 −0.208	0.958	0.044	A: 294 T: −1.147	0.206

V: variables, UC: unstandardized coefficient, SM: structure matrix, WL: Wilks' lambda, E: eigenvalue, GC: group centroids, CC: canonical correlation. A: Arabian horses T: thoroughbred horses.

For the proximal phalanx, the most distinctive feature in the discriminant function analysis was DCP (structure matrix: −0.801). In turn, LMM (structure matrix: 0.951) was the most distinguishing feature for the middle phalanx and FA (structure matrix: 0.666) for phalanx distalis. The eigenvalue value for the proximal phalanx was −0.909. It allowed correct classification of 89.33% of the samples used in the study (Table 4). The distinguishing features were the most tangible in the proximal phalanx; which was indicated by the highest percentage of the correctly classified bones. In the case of middle phalanx and distal phalanx, the correct classifications reached 77.33% and 54.67%, respectively. Traits of the distal phalanx were insignificant in the breed distinction. Wilks' lambda value was very high (0.958).

Table 4. Confusion matrix. Percentage of initial classifications that were correct, shown by breed.

		Arabian	Thoroughbred	
Proximal phalanx	Arabian	23	2	89.33%
	Thoroughbred	6	44	
Middle phalanx	Arabian	20	5	77.33%
	Thoroughbred	12	38	
Distal phalanx	Arabian	15	10	54.67%
	Thoroughbred	24	26	

Percentages within rows sum to 100%.

The distribution frequency of Arabian horses and Thoroughbred horses is shown in Figure 2. This figure demonstrates that the division into breeds based on the proximal phalanx data was more distinct. It can be seen that overlaps were more frequent in the distal phalanx. Almost none of the characteristics of the distal phalanx in Arabian horses differed from those of Thoroughbred horses.

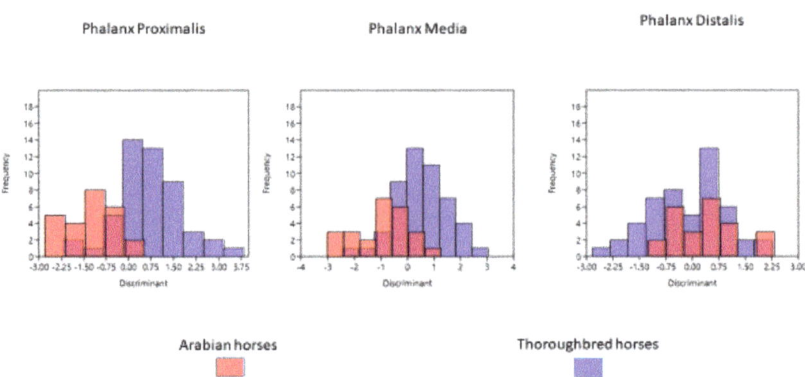

Figure 2. Distribution frequency of Arabian horses and Thoroughbred horses after discriminant analysis.

Correlation values are presented in Table 5. There was usually a significant positive correlation between the individual measurement values. The length measurements of all three phalanges correlated strongly with the other parameters. The strongest correlation was found between the DCP and DBM.

Table 5. Coefficients of correlation between measurements of digital bones and between these measurements and horse age.

	DBP	DCP	LMP	DBM	DCP	LMM	LD	GSL	FA
Age	−0.021	−0.020	0.076	−0.064	−0.121	−0.030	0.204	0.381 **	0.227
DBP		0.609 **	0.612 **	0.671 **	0.456 **	0.698 **	0.425 **	0.142	0.055
DCP			0.508 **	0.692 **	0.686 **	0.667 **	0.339 **	0.020	−0.043
LMP				0.491 **	0.355 **	0.691 **	0.280 *	0.234 *	0.080
DBM					0.794 **	0.749 **	0.352 **	0.104	0.104
DCP						0.615 **	0.272 *	0.069	0.085
LMM							0.423 **	0.128	0.039
LD								0.559 **	0.299 **
GSL									0.760 **

** Correlation is significant at 0.01 level. * Correlation is significant at 0.05 level.

4. Discussion

The phalanges of thoracic limbs of 75 horses were examined in the study. The LMP, LMM, and LD values were statistically different between males and females in Arabian horses, while no significant difference was found between males and females of Thoroughbred horses. Between the breeds, DCP was found to be more determinant for the proximal phalanx. In turn, LMM was the most distinguishing morphometric feature for the middle phalanx. It was observed that in Thoroughbred horses, most length measurement results were higher than those of Arabian horses (DBP, DCP, LMP, DBM, DCM, LMMLD, and FA). The reason for this is that Thoroughbred horses are generally taller than Arabian horses and this might be related to the length of individual sections of the limb skeleton [17]. The traits of the distal phalanx provide limited possibilities for breed differentiation.

Only radiogrametric measurements of digital bones of the left thoracic limbs of Arabian horse and Thoroughbred horse were used in this study. The difference between the right and left bones was ignored; only bone measurements were used to examine whether there was a sex and breed distinction. Previous studies have proved the lateralization of the skeleton of the manus. Alrtib et al. [15] compared right and left proximal phalanx measurements in their study using 10 Thoroughbred horses, five Standardbred horses and eight ponies. It was said that the medial length of proximal phalanx on the right side was larger than the left side. This difference was statistical. According to Kummer et al. [11], the left LMM is greater than the right and this difference is statistically significant. He reported that the left LMM before trimming was 4.6 ± 0.24 cm. In addition, he stated that the LMM correlates positively with the height at the withers. In contrast, Linford et al. [12] did not observe significant differences between the left and right phalanges in any radiographic determination. In our study, LMM was 3.9 ± 0.27 cm in Arabian horses and 4.3 ± 0.26 cm in Thoroughbred horses.

Measurements and 3D-modeling using imaging systems has become an alternative to traditional anatomical morphometry. In the study of Dos Reis et al. [2], it was stated that the 3D models successfully reflect the anatomical characteristics of bones and that measurements taken using these models approximate the actual bone measurements. In their study using 3D printing, the length of the proximal phalanx was 8.42 cm and the length of the middle phalanx was 4.48 cm. No information was given about the breed of horses used in this research, but the length of phalanx media was close to the respective value found in this study for Thoroughbred horses.

The mean value of the length of the proximal phalanx was 9.28 cm and the length of the middle phalanx media was 3.9 cm in Arabian horses, whereas in the Thoroughbred horses the respective values reached 9.89 cm and 4.3 cm. We examined whether there was a difference between the two breeds in radiogrametric measurements taken of phalangeal bones. Results of the proximal phalanx were found to be more distinctive than those of the other digital bones in breed differentiation. Dzierzęcka and Komosa [18] also revealed a statistically significant ($p \leq 0.01$) difference in the greatest length of the proximal phalanx between warmblood horses and coldblood horses. They stated that the length of the proximal phalanx plays an important role in a certain morphological classification of animals.

In this study performed with Arabian horses and Thoroughbred horses, a correlation between the age of animal and osteometric features of phalanges was also evaluated. Analyses conducted demonstrated a positive correlation between the greatest solear length of the distal phalanx and age. Some studies can be found in literature addressing the effects of increasing age in horses on digital radiological measurements. Mullard et al. [19] proved that the ratio of the hoof distal phalanx distance to the length of the palmar aspect of the distal phalanx in horses decreases with increasing age, and that this correlation is statistically significant. In this study, the correlation of age with GSL was positive and only this value was statistically significant (R = 0.381). Several traits were negatively correlated with age, but these values were statistically insignificant. Anderson et al. [20] proved that there was no significant increase of length measurements in horses from age 2 to 3 years, suggesting that the growth rate either slowed or reached a plateau between ages 2 and 3 years. The lack of relationship of most analyzed parameters with age in our study is presumably due to the fact that most of the growth and development in the distal limbs of the horses ended at the age before the youngest of the measured horses. According to Sadek et al. [21], correlations between body measurements in Arabian horses vary a lot for both genders. Cruz et al. [22] examined the magnetic resonance images taken on these bones and stated that the exercise caused a positive change in the width parameters of the distal phalanx. In turn, Turek et al. [23] reported that different factors, including age, feeding, exercise and breed, might determine mechanical properties of the proximal phalanx in horses.

Osteometric measurements of phalanges were taken on equine bone remains obtained from excavations. Forsten [24] measured bones of horses in the Steinheim region. The length of the proximal phalanx found in this study was 9.13 ± 0.08 cm, and the length of the middle phalanx was 5.1 ± 0.14 cm. In another study, horses' remains were examined in Üzüür Gyalan Tomb, and measurements of proximal phalanx and middle phalanx were taken of these bone remains. The greatest length of the proximal phalanx from this excavation was 8.47 cm and the greatest length of the phalanx media was 5.00 cm [25]. In the present study conducted with Arabian horses and Thoroughbred horses, reference measurement values were obtained radiogrametrically from contemporary samples. These reference data may suggest that in terms of taxonomy, they can be helpful in interpreting osteoarchaeological studies. These data are particularly comparable with bone remnant measurements obtained from excavations, and can provide information on animals of ancient periods. Although, when using these data, it should be kept in mind that there may be differences between measurements taken of bones and measurements taken via radiological imaging, as shown in the reference study [26].

The authors are aware that the lack of significance of differences between breeds and genders is partly due to the low abundance of research material. Another limitation is the lack of zoometric data, including the height at the withers, concerning the tested horses.

5. Conclusions

In this study, the statistical significance of radiogrametric measurements of the skeleton of the manus was examined in terms of sex and breed. Measurements of male individuals were generally greater. However, it can be concluded that the proximal phalanx and the middle phalanx manifest sexual dimorphism. The traits of the distal phalanx used in the study did not discriminate sex or breed. These data can serve as a reference material for radiogrametric studies in animals. The present study allowed obtaining reference radiogrametric measurements of the digits of the manus of Arabian horses and Thoroughbred horses.

Author Contributions: Conceptualization, O.G. and G.P.; methodology, D.O.E., O.G. and G.P.; software, O.G., S.D. and T.S.; investigation, D.O.E. and O.G.; validation, T.S., W.P.; writing—original draft preparation, S.D., T.S., W.P. and O.G.; writing—review and editing, T.S. and W.P. All authors have read and agreed to the published version of the manuscript.

Funding: This research received no external funding.

Institutional Review Board Statement: The study was approved by the Local Ethics Committee of the Faculty of Veterinary Medicine, İstanbul University-Cerrahpaşa (approval number: (2019/43)).

Informed Consent Statement: Not applicable.

Data Availability Statement: Not applicable.

Acknowledgments: We would like to thank Turkey's Jockey Club and Hülya Hartoka for their support in conducting the study.

Conflicts of Interest: The authors declare no conflict of interest.

References

1. Demircioglu, I.; Yilmaz, B.; Gündemir, O.; Dayan, M.O. A three-dimensional pelvimetric assessment on pelvic cavity of gazelle (*Gazella subgutturosa*) by computed tomography. *Anat. Histol. Embryol.* **2021**, *50*, 43–49. [CrossRef] [PubMed]
2. Dos Reis, D.D.A.L.; Gouveia, B.L.R.; Júnior, J.C.R.; de Assis Neto, A.C. Comparative assessment of anatomical details of thoracic limb bones of a horse to that of models produced via scanning and 3D printing. *3D Print. Med.* **2019**, *5*, 1–10.
3. Ketelsen, D.; Schrödl, F.; Knickenberg, I.; Heckemann, R.A.; Hothorn, T.; Neuhuber, W.L.; Grunewald, M. Modes of information delivery in radiologic anatomy education: Impact on student performance. *Acad. Radiol.* **2007**, *14*, 93–99. [CrossRef]
4. Oheida, A.H.; Alrtib, A.M.; Shalgum, A.A.; Shemla, M.E.; Marzok, M.A.; Davies, H. Radiographic Comparison of Carpal Morphometry in Thoroughbred and Standardbred Race horses. *AJVS* **2019**, *61*, 74–82. [CrossRef]
5. Burd, M.A.; Craig, J.J.; Craig, M.F. The palmar metric: A novel radiographic assessment of the equine distal phalanx. *Open Vet. J.* **2014**, *4*, 78–81.
6. Holroyd, K.; Dixon, J.J.; Mair, T.; Bolas, N.; Bolt, D.M.; David, F.; Weller, R. Variation in foot conformation in lame horses with different foot lesions. *Vet. J.* **2013**, *195*, 361–365. [CrossRef]
7. Wilson, A.; Agass, R.; Vaux, S.; Sherlock, E.; Day, P.; Pfau, T.; Weller, R. Foot placement of the equine forelimb: Relationship between foot conformation, foot placement and movement asymmetry. *Equine Vet. J.* **2016**, *48*, 90–96. [CrossRef]
8. Page, B.T.; Hagen, T.L. Breakover of the hoof and its effect on stuctures and forces within the foot. *JEVS* **2002**, *22*, 258–264.
9. Faramarzi, B.; McMicking, H.; Halland, S.; Kaneps, A.; Dobson, H. Incidence of palmar process fractures of the distal phalanx and association with front hoof conformation in foals. *Equine Vet. J.* **2015**, *47*, 675–679. [CrossRef]
10. Pauwels, F.E.; Rogers, C.W.; Wharton, H.; Flemming, H.; Wightman, P.F.; Green, R.W. Radiographic measurements of hoof balance are significantly influenced by a horse's stance. *Vet. Radiol. Ultrasound* **2017**, *58*, 10–17. [CrossRef] [PubMed]
11. Kummer, M.; Geyer, H.; Imboden, I.; Auer, J.; Lischer, C. The effect of hoof trimming on radiographic measurements of the front feet of normal Warmblood horses. *Vet. J.* **2006**, *172*, 58–66. [CrossRef]

12. Linford, R.L.; O'Brien, T.R.; Trout, D.R. Qualitative and morphometric radiographic findings in the distal phalanx and digital soft tissues of sound thoroughbred racehorses. *Am. J. Vet. Res.* **1993**, *54*, 38–51. [PubMed]
13. Dyson, S.J.; Tranquille, C.A.; Collins, S.N.; Parkin, T.D.H.; Murray, R.C. An investigation of the relationships between angles and shapes of the hoof capsule and the distal phalanx. *Equine Vet. J.* **2011**, *43*, 295–301. [CrossRef] [PubMed]
14. Kalka, K.; Pollard, D.; Dyson, S.J. An investigation of the shape of the hoof capsule in hindlimbs, its relationship with the orientation of the distal phalanx and comparison with forelimb hoof capsule conformation. *Equine Vet. Educ.* **2020**, *33*. [CrossRef]
15. Alrtib, A.M.; Philip, C.J.; Abdunnabi, A.H.; Davies, H.M.S. Morphometrical study of bony elements of the forelimb fetlock joints in horses. *Anat. Histol. Embryol.* **2013**, *42*, 9–20. [CrossRef] [PubMed]
16. Cohen, N.D.; Carter, G.K.; Watkins, J.P.; O'Conor, M.S. Association of racing performance with specific abnormal radiographic findings in Thoroughbred yearlings sold in Texas. *J. Equine Vet. Sci.* **2006**, *26*, 462–474. [CrossRef]
17. Brooks, S.A.; Makvandi-Nejad, S.; Chu, E.; Allen, J.J.; Streeter, C.; Gu, E.; McCleery, B.; Murphy, B.A.; Bellone, R.; Sutter, N.B. Morphological variation in the horse: Defining complex traits of body size and shape. *Anim Genet.* **2010**, *41*, 159–165. [CrossRef] [PubMed]
18. Dzierzęcka, M.; Komosa, M. Variability of the proximal phalanx in warmblood and coldblood horses-morphological and structural analyses. *Belg. J. Zool.* **2013**, *143*, 119–130.
19. Mullard, J.; Ireland, J.; Dyson, S. Radiographic assessment of the ratio of the hoof wall distal phalanx distance to palmar length of the distal phalanx in 415 front feet of 279 horses. *Equine Vet. Educ.* **2020**, *32*, 2–10. [CrossRef]
20. Anderson, T.M.; McIlwraith, C.W. Longitudinal development of equine conformation from weanling to age 3 years in the Thoroughbred. *Equine Vet. J.* **2004**, *36*, 563–570. [CrossRef] [PubMed]
21. Sadek, M.H.; Al-Abound, A.Z.; Ashmawy, A.A. Factor analysis of body meaurements in Arabian horses. *J. Anim. Breed. Genet.* **2006**, *123*, 369–377. [CrossRef] [PubMed]
22. Cruz, C.D.; Thomason, J.J.; Faramarzi, B.; Bignell, W.W.; Sears, W.; Dobson, H.; Konyer, N.B. Changes in shape of the Standardbred distal phalanx and hoof capsule in response to exercise. *Comp. Exerc. Physiol.* **2006**, *3*, 199–208. [CrossRef]
23. Turek, B.; Wajler, C.; Klos, Z.; Szara, T. Mechanical properties of the cortical bone of the proximal phalanx in horses. *Med. Weter.* **2013**, *69*, 120–123.
24. Forsten, A. The horses (genus Equus) from the Middle Pleistocene of Steinheim, Germany. *Deinsea* **1999**, *6*, 147–154.
25. Onar, V.; Küçük, S.; Galbadrakh, E.; Taşağıl, A.; Erdikmen, D.O. Horse sacrifice in the Üzüür Gyalan Tomb: An Altai Mountain Kurgan. *Art-Sanat Dergisi* **2019**, *11*, 275–298.
26. Carew, R.M.; Viner, M.D.; Conlogue, G.; Márquez-Grant, N.; Beckett, S. Accuracy of computed radiography in osteometry: A comparison of digital imaging techniques and the effect of magnification. *J. Forensic Radiol. Imaging* **2019**, *19*, 100348. [CrossRef]

Article

Maxillary Incisors of the Horse before and at the Beginning of the Teeth Shedding: Radiographic and CT Study

Francisco Miró [1,*], Carla Manso [2], Andrés Diz [1] and Manuel Novales [3]

1 Department Comparative Anatomy and Pathology, University of Córdoba, Ctra. de Madrid, 14071 Córdoba Ctra, Spain; an1dipla@uco.es
2 Veterinaries Specialist in Equine dentistry, Pedro Laín Entralgo 8, Boadilla del Monte, 28660 Madrid, Spain; cpmanso@hotmail.com
3 Diagnostic Imaging Service, Veterinary Teaching Hospital, University of Córdoba, Ctra. de Madrid, 14071 Córdoba, Spain; pv1nodum@uco.es
* Correspondence: an1mirof@uco.es; Tel.: +34957218143

Received: 24 August 2020; Accepted: 31 August 2020; Published: 10 September 2020

Simple Summary: Although much is known about equine dentistry, there is a period of the horse's life, prior to teeth shedding, in which there is lack of knowledge related to the development of deciduous incisors and dental germs of permanent incisors. To gain insight into the radiographic appearance of maxillary deciduous incisors and dental germs of maxillary permanent incisors during this period, a radiographic and computed tomography study of 25 horse skulls was made. Data regarding morphology and development were obtained. The results of the present study indicate that radiographic intraoral images are suitable to identify the grade of development of the dental germs of permanent incisors in horses before dental change. A detailed description of the radiographic appearance of deciduous incisors and dental germs of permanent incisors will help clinicians to expand their knowledge for diagnostic and treatment purposes.

Abstract: To gain insight into the radiographic appearance of maxillary deciduous incisors and dental germs of maxillary permanent incisors in the period prior to teeth shedding, radiographs and computed tomography (CT) of 25 horse skulls, with an estimated age of between 12 and 42 months, were studied. Data regarding morphology and development were obtained. Dental germs of first maxillary permanent incisors were identified radiographically as rounded radiolucent areas at the level of the apical parts of the first deciduous incisors, in skulls with an estimated age of twelve months. The first sign of crown mineralization of these dental germs appeared in skulls supposedly a few months older. Before teeth shedding, the unerupted, mineralized crowns of the first permanent incisor could be identified radiographically relatively caudal to the corresponding first deciduous incisors. The results of the present study indicate that radiographic intraoral images are suitable to identify the grade of development of the dental germs of maxillary permanent incisors. A detailed description of the radiographic appearance of deciduous incisors and dental germs of permanent incisors will help clinicians to expand their knowledge for diagnostic or treatment purposes.

Keywords: computed tomography; development; incisor; radiograph

1. Introduction

In horses, the grade of development of deciduous and permanent teeth, including incisors, has always been of great interest for veterinary clinicians to determine their age. Besides, knowledge of dental and periodontal regions is needed when dental and periodontal disorders, such as malocclusions, dental fractures, persistent deciduous teeth, supernumerary teeth, traumas, etc., are present in young animals [1,2]. Although visual examination of the mouth and radiography have always been the most-used methods by veterinary clinicians [3], recent diagnostic imaging procedures, such as computed tomography (CT), provide complementary and more precise information on dental examination in horses [4–9].

Some authors have reviewed and summarized the current knowledge of equine dental and periodontal anatomy [10–13] and the three-dimensional appearance of teeth, as well as their individual composition of dental hard structures [14]. Throughout the life of a horse, specific changes occur to the appearance of its teeth, and dental examination therefore provides the most convenient mean for age determination and diagnostic purposes. In this sense, there is information relating morphological characteristics of deciduous and permanent incisors observed by visual examination with the age of the horse [11,15–17]. Equine teeth, pertaining to the high crowned hypsodont teeth, are subjected to continuous dental wear [18], and the process of change from deciduous to permanent dentition involves a complex mechanism of development of the dental germs. Time of radiographic appearance and grade of development of dental germs are well-known for deciduous and permanent cheek teeth [19,20], but very few data are available for incisor teeth [21,22]. The latest studies focus on the disorders involving the incisive bone, maxillary incisors, and periodontal structures and show very few radiographic data of dental germs of the permanent incisors. These few data and the clinicians' experience indicate that there is a period of the horse's life in which the deciduous incisors and the dental germs of permanent incisors undergo different stages that could be identified radiographically. The referred period would extend from the complete eruption of deciduous incisors, some months before twelve months of age, to the eruption of the first permanent incisor, at the age of thirty months [11,15,23]. High quality intraoral dorsoventral radiographs are an excellent tool to study the maxillary incisors [1,23]. Moreover, modern imaging techniques such as computed tomography complement and overcome the limitations of two-dimensional radiographic images. To accurately diagnose and formulate a treatment plan, evaluation of dental disorders requires a complete maxillofacial–oral examination and supplemental imaging means [21]. Some authors have compared and validated the accuracy of CT and radiographic imaging in detecting cheek teeth disorders in horses [24]. As stated above, in horses, there is a lack of radiographic data on deciduous incisors and dental germs of permanent incisors in the period before and at the beginning of the teeth shedding. Hence, a detailed description of the radiographic appearance of the referred structures in the referred period will help clinicians to expand their knowledge for diagnostic or treatment purposes.

We hypothesized that, in young horses before shedding, the different grade of development of deciduous incisors and dental germs of permanent incisors can be identified radiographically. The purpose of the present study was to gain further insight into the radiographic appearance of deciduous incisors and dental germs of permanent incisors of horses in the period prior to and at the beginning of teeth shedding.

2. Materials and Methods

Twenty–five horse skulls were studied in the present study. Neither ponies nor draft horses were included in the study. Equine skulls, obtained from the Anatomy Department of the Veterinary School of Cordoba University, Spain, kept their complete dentition and surrounding bone without noticeable abnormalities.

2.1. Visual Inspection

Based on published aging guidelines for horses [11,15,23], the whole skulls were visually examined by one of the authors, a specialist on equine dentistry (C.M., board-certified specialist, European Veterinary Dental College). Eruption of deciduous and permanent maxillary and mandibular incisors, morphological characteristics of their erupted crowns, and the presence of some of the permanent cheek teeth were the main criteria for determining the estimated age. Incisors were named according to the anatomical nomenclature, such as first, second, and third [25]. Correspondence with the modified Triadan system, worldwide used in veterinary dentistry [26], was made, appearing between en-dash after the corresponding anatomical name. Skulls with all deciduous incisors erupted and their occlusal surfaces in the same plane, assessed as between 12 and 30 months, were included in group 1. Skulls maintaining their second and third deciduous incisors—02 and 03; Quadrants 5,6,7, and 8—and with first permanent incisors—01; Quadrants 1,2,3, and 4—as well as second and maybe third permanent premolars—06 and 07; Quadrants 1,2,3, and 4—erupted, assessed as between 30 and 42 months of age, were included in group 2.

2.2. Radiographic Study

After visual inspection the maxillary incisor arcades were radiographed with an X-ray machine (Odel model C306-20®, Monza, Italy). Radiographs were processed by computerized radiology (Fuji Computed Radiography, Capsule XL, CR-IR 356® Tokyo, Japan). Radiographs were obtained according to the standard procedures of intraoral dorsoventral technique, being the X-ray beam directed 90° to the plane that bisects the angle between the 1st maxillary incisors and the imaging plate [1,3,4,22,27]. Radiographs, being the incisors crown down and the horse's left side presented to the viewer's right, were then assessed [3]. Maxillary deciduous incisors were identified and the radiological characteristics of their erupted and nonerupted portions and surrounding bone were assessed. Dental germs of the maxillary permanent incisors were identified, and their radiological characteristics and surrounding bone were analyzed. In skulls in which maxillary permanent incisors were identified, their radiological characteristics of the erupted and non-erupted portions and surrounding incisive bone were assessed. A first analysis of the intraoral radiographs of group 1 skulls allowed us to identify the dental germs of the 1st maxillary permanent incisors—101 and 201—according to different grades of development. Some of the skulls—classified from now on as *group 1a*—revealed dental germs like round quite radiolucent areas and apparently no other notorious characteristics. Several skulls—organized from now on within *group 1b*—showed circumscribed radiopaque images within the respective radiolucent zones of the dental germs. In radiographic studies of some other skulls—from now on as *group 1c*—unerupted 1st permanent incisors—101 and 201—were identified as being short, conic, wide, and with hardly any observable radiopacity. A group of skulls—from now on *group 1d*—showed larger unerupted 1st permanent incisors—101 and 201—with greater radiopacity. Table 1 shows the distribution of the skulls in the above referred groups and some of the most remarkable radiographic characteristics of 1st permanent maxillary incisors—101 and 201—used to classify them.

Table 1. In 25 skulls of Spanish horses, most remarkable radiographic characteristics of 1st permanent maxillary incisors—101, 201—and number of skulls studied. Groups 1a to 1d, estimated age of between 12 and less than 30 months. Group 2, estimated age between 30 and 42 months.

Groups	Radiographic Appearance of 1st Maxillary Permanent Incisors—101, 201	Number of Skulls
1 1a	Dental germs, round radiolucent areas in the incisive bone No other notorious characteristics	5
1b	Dental germs, round radiolucent areas in the incisive bone Circumscribed radiopaque images inside	4
1c	Dental germs, unerupted crowns within the incisive bone, hardly any observable radiopacity	8
1d	Dental germs, larger unerupted crowns within the incisive bone, greater radiopacity	4
2	Erupted, quite large, and radiopaque	4

2.3. Computed Tomography Imaging

Subsequently, skulls were scanned with a helical CT scanner (CT Hi Speed CT/e Dual, General Electric Yokogawa Medical Systems LTD®, Hino, Japan) to obtain CT images from the incisors to the canines. Skulls were placed with the mandible on the CT scanning table. After image acquisition, CT slices from different planes of the incisors and adjacent structures were chosen, exported in DICOM format, and analyzed in Horos © software (open source for Apple, 64 bit, version 3.3.6). CT characteristics of deciduous incisors, dental germs of permanent incisors, permanent incisors (when they were present), and surrounding bone were analyzed.

2.4. Length Measurements and Proportions

Some length measurements and proportions were analyzed as a complementary study. Due to the small number of skulls per group, no statistical analysis was performed. Based on certain anatomical points (Figure 1), the following distances in mm were measured in sagittal scans at midlevel of the 1st deciduous incisors—501 and 601—by using Horos © software (open source for Apple 64 bit, version 3.3.6):

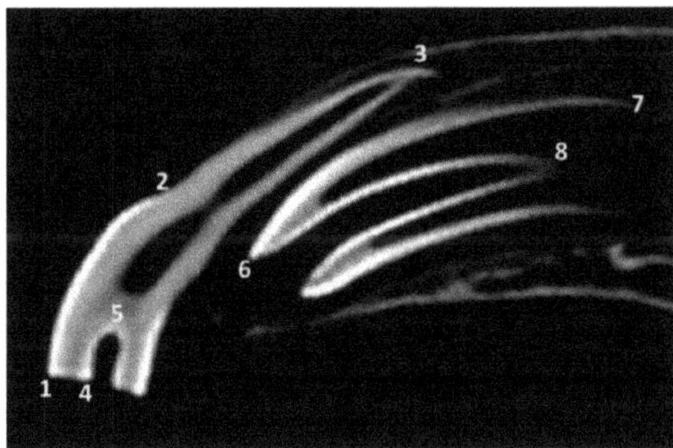

Figure 1. Sagittal computed tomography (CT) scan at midlevel of the 1st left maxillary deciduous incisor—601—of a skull of group 1d (supposedly the oldest among those with estimated age of between 12 and less than 30 months) showing the reference points for length measurements in the left maxillary 1st deciduous incisor—601— and in the unerupted 1st left permanent incisor—201. **1** and **6**, most occlusal points of the labial side of the teeth; **2** and **7**, most apical points of the labial enamel cover of the teeth; **3**, most apical point of the labial side of the 1st left maxillary deciduous incisor—60—; **4** and **6**, most occlusal points of the infundibular enamel, labial side; **5** and **8**, most apical points of the infundibular enamel, labial side.

Length of the tooth (LTOOTH): distance between the most occlusal point and the most apical point of the labial side of the tooth.

Length of the crown (LCR): distance between the most occlusal point of the labial side and the most apical point of the labial enamel cover (i.e., dental crown) of the tooth.

Length of the infundibulum (LINF): distance between the most occlusal point and the most apical point of the infundibular enamel, both on labial side.

Based on the above measurements, the following percentage measurements were also calculated:

Relative length of the crown (%CR): percent of length of the crown with respect to the total length of the same tooth.

Relative length of the infundibulum (%INF): percent of length of the infundibulum with respect to the length of the same tooth.

When the unerupted 1st first maxillary permanent incisors—101 and 201—were identified, the same measurements were taken. Until they erupt, the 1st permanent incisors—101 and 201—were considered as a whole tooth or crown. Once these teeth had erupted, the same measurements were taken on sagittal scans at midlevel of the tooth. Using the measurements for each group, the mean and standard deviations were calculated based upon the corresponding values of the right tooth and left tooth.

3. Results

3.1. Visual Inspection

Examination of the teeth of the skulls of group 1, classified as having an estimated dental age of between 12 and less than 30 months, revealed that all the deciduous incisors were clearly erupted and with wear on the occlusal surface of the 3rd incisors—03, Quadrants 5, 6, 7, and 8. They were of small size, with the oval occlusal surfaces in a mesiodistal orientation, and possessed shallow infundibula. Upon visual inspection, the most notable result available from the skull incisors of group 2, classified as having an estimated dental age of between 30 and less than 42 months, was the presence of the 1st permanent incisors—01, upon erupted Quadrants 1,2,3, and 4. In this group of skulls all other incisors were deciduous. In addition to the above, the skulls of group 2 had erupted the second permanent premolars or the second and third permanent premolars. In some of the skulls of group 2, there were the remains of the 1st deciduous incisors—01, Quadrants 5–7 and 8—in contact with the labial surface of the corresponding permanent tooth. Small oval openings, termed as gubernacular canals of the permanent incisors, were identified in the palatal region of the alveolar bone. They were present close to the 1st maxillary deciduous incisors—501 and 601—in groups 1a, 1b, and 1c (Figure 2A). In one of the skulls of group 1c the gubernacular canals were also present close to the 2nd deciduous incisors—502, 602. In group 1d the gubernacular canals were present close to the 1st and 2nd deciduous incisors—501, 502, 601, and 602—(Figure 2B). In group 2, there were gubernacular canals close to the 2nd and 3rd deciduous incisors—502, 503, 602, and 603—and in one of the skulls there was also a gubernacular canal close to the erupted 1st left permanent incisors—201.

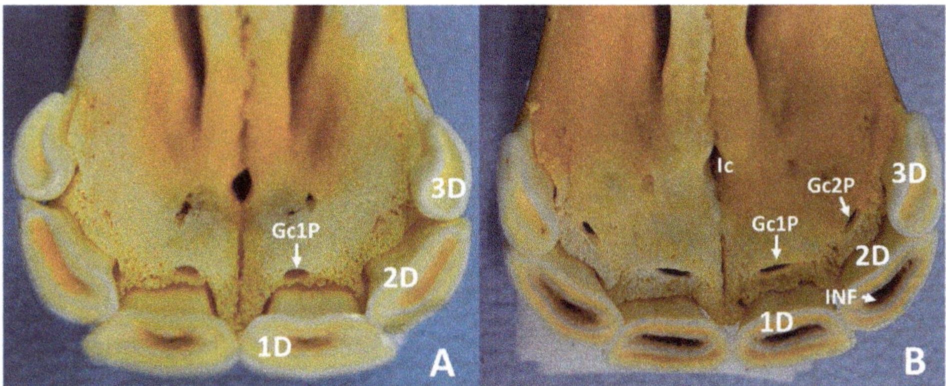

Figure 2. Ventral view of maxillary dental arcade, incisive part, of skull of group 1a (**A**) and of group 1d (**B**). **1D**, right 1st deciduous incisor—501; **2D**, right 2nd deciduous incisor—502; **3D**, right 3rd deciduous incisor—503; **Gc1P**, gubernacular canal of right 1st permanent incisor—101; **Gc2P**, gubernacular canal of right 2nd permanent incisor—102; **Ic**, incisive canal; **INF**, infundibulum of the right 2nd deciduous incisor—502.

3.2. Radiographic and CT Studies

In radiographic studies of skulls of *group 1a* (Figure 3A), the 1st deciduous incisors—501 and 601—were identified with an elongate shape. These teeth stick out approximately one third from the incisive bones. This part was widened in a mesiodistal direction. Approximately two thirds of these teeth were embedded in the incisive bone. The embedded parts of these teeth comprised a central elongated and radiolucent area with two collateral radiopaque fringes. They may be identified, respectively, as the pulp cavity (the cavity of the tooth that contains the pulp) and the surrounding hard component of the dental root. The apical portion of these teeth end at the level of the incisive canal. Lateral to the image of the 1st deciduous incisors—501 and 601—overlapping images of the second and third deciduous incisors—502, 503, 602, and 603. In the apical parts of the 1st deciduous incisors—501 and 601—rounded radiolucent zones identify the dental germs of the 1st permanent incisors—101 and 102. These zones almost medially reached the interincisive suture, and their caudal limits were just at the level of the incisive canal. Sagittal CT images at midplane of 1st deciduous incisors—501 and 601—(Figure 3B) showed these teeth convexly curved on the labial face and concavely curved on the lingual face. They tapered evenly from their occlusal part to their apex. Analysis of the structural density allowed us to identify in them the anatomical crowns (parts with peripheral enamel) and dental roots (parts without superficial enamel). The enamel cover (i.e., crown) extended more apically on the labial side than in the palatal side. Most parts of the crowns had erupted, and the all dental roots were enclosed within the alveolus. Scans showed shallow infundibula filled at their bottom with hyperdense tissue. Transversal and sagittal images (Figure 3B,C) showed round hypodense areas touching the palatal side of the root of the corresponding 1st deciduous incisors—501 and 601. These areas corresponded with the dental germs of the 1st permanent incisors—101 and 201. The dorsal tip of the dental germs reached the end of the apical height of the root of the 1st deciduous incisors—501 and 601—and it is limited caudally and ventrally by the more rostral portion of the hard palate.

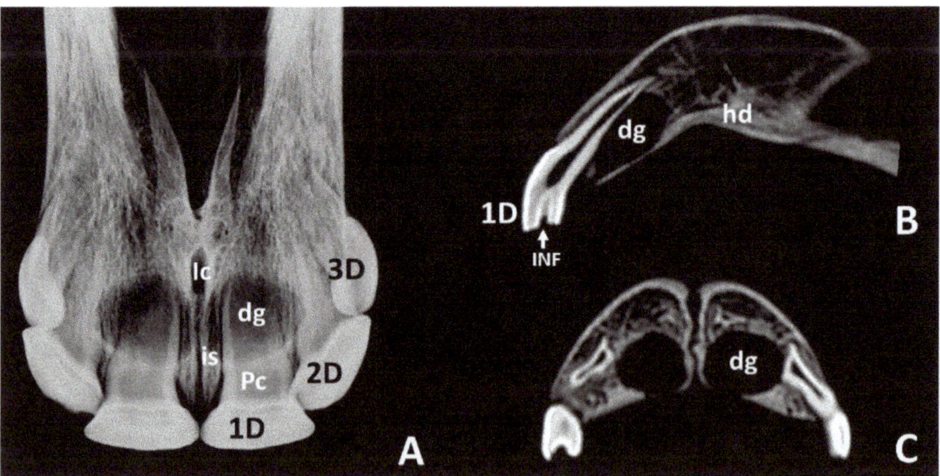

Figure 3. Intraoral radiographic image (**A**) and sagittal and transverse scans (**B**,**C**) of maxillary incisor arcade of a skull representative of group 1a. **1D**, left 1st deciduous incisor—601; **2D**, left 2nd deciduous incisor—602; **3D**, left 3rd deciduous incisor—603; **Pc**, pulp cavity; **dg**, dental germ of left 1st permanent incisor—201; **hd**, hard palate; **Ic**, incisive canal; **INF**, infundibulum of the left 1st deciduous incisor—601; **is**, interincisive suture.

Intraoral radiographs of skulls of *group 1b* (Figure 4A) did not show notable differences in images of the 1st deciduous incisors with that of the same teeth in group in 1a. However, it was notable that the radiolucent zones of dental germs of the 1st permanent incisors—101 and 201—were in this group smaller, more elongated, and caudally went beyond the incisive canal. The most remarkable aspect of the radiographic study in this group was the presence of circumscribed radiopaque images within the respective radiolucent zones of the dental germs (white arrow, Figure 4A). CT scans of skulls in this group (Figure 4B,C) showed, as in the group 1a, most of the crown of the 1st deciduous incisors—501 and 601—already erupted, and their dental roots enclosed within the alveolus. As it was in the previous group, the enamel cover extended more apically in the labial side than in the palatal side. There were characteristic hypodense and oval areas of dental germs in permanent incisors—101 and 201—touching the palatal sides of the roots of the corresponding deciduous teeth. Within the dental germs circumscribed hyperdense images could be clearly distinguished (white arrows, Figure 4B,C). Sagittal scans showed that the dorsal limit of the dental germs did not dorsally reach the more apical point of the root of the corresponding deciduous tooth.

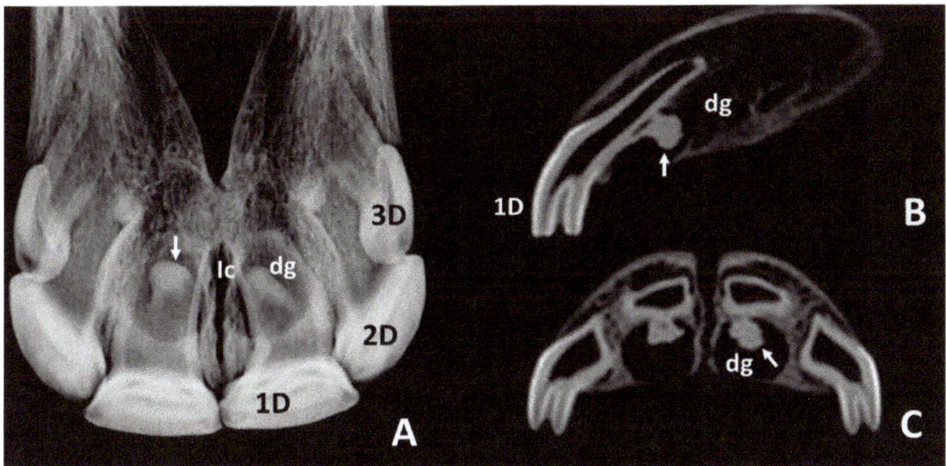

Figure 4. Intraoral radiographic images (**A**) and sagittal and transverse scans (**B**,**C**) of the maxillary incisor arcade of a skull representative of group 1b. **1D**, left 1st deciduous incisor—601; **2D**, left 2nd deciduous incisor—602; **3D**, left 3rd deciduous incisor—603; **dg**, dental germ of left 1st permanent incisor—201; **Ic**, incisive canal; **white arrows**, circumscribed radiopaque (**A**) and hyperdense (**B**,**C**) images within the dental germs of left 1st permanent incisor—201.

In intraoral radiographs of skulls of *group 1c* (Figure 5A), 1st deciduous incisors—501 and 601—showed a trapezoidal occlusal surface area. Wide radiolucent zones of dental germs of 1st permanent incisors—101 and 201—were visible in the body of every incisive bone. Their caudal limits exceeded the level of the incisive canal. Fine radiopaque lines enclosed the radiolucent areas of the germs in some of the skulls. They were considered the lamina dura of the alveolus. Wide and conic images of the unerupted crowns of 1st permanent incisors—101 and 201—could be identified within the radiolucent areas of the germs with hardly any observable radiopacity. Sagittal and transverse CT scans of skulls of this group (Figure 5B,C) showed the crown of the 1st deciduous incisors—501 and 601—to be erupted with the dental root within the alveolus. Images of the dental germs of the 1st permanent incisors—101 and 201—were touching the palatal sides of the roots of their corresponding deciduous teeth. In sagittal scans the dental germs of 1st permanent incisors—101 and 201—showed oval hypodense areas containing parallel hyperdense structures, which correspond to the unerupted crowns during early mineralization. The surrounding limits of the germs clearly exceeded dorsocaudally the apical point of the root of the corresponding deciduous tooth. It is noticeable that the ventral limits of the germs were disrupted caudally by a few millimeters to the corresponding deciduous incisors (Figure 5, Gc). This structure was identified as the gubernacular canal of the 1st permanent incisors—101 and 201. As stated above, within the hypodense area of the germs, the short crowns of the 1st permanent incisor—101 and 201—were identified. In them, the enamel did not reach the apical point of the infundibulum.

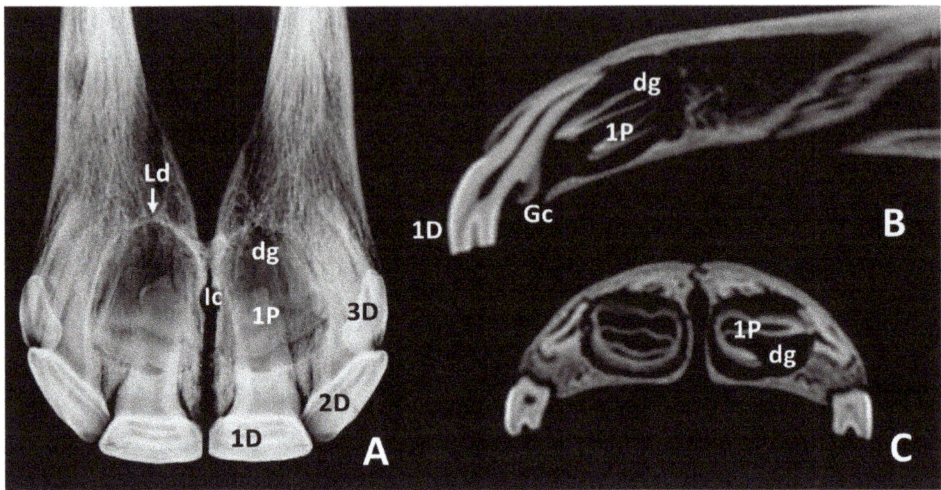

Figure 5. Intraoral radiographic images (**A**) and sagittal and transverse scans (**B**,**C**) of maxillary incisor arcade of one skull representative of group 1c. **1D**, left 1st deciduous incisor—601; **2D**, left 2nd deciduous incisor—602; **3D**, left 3rd deciduous incisor—603; **1P**, left unerupted 1st permanent incisor—201; **dg**, dental germ of left 1st permanent incisor—201; **Gc**, gubernacular canal of left 1st permanent incisor—201; **Ic**, incisive canal; **Ld**, lamina dura of the alveolus of the right 1st permanent incisor—101.

In intraoral radiographs of skulls of *group 1d* (Figure 6A), those images of the 1st deciduous incisors—501 and 601—had notably shorter lengths than those in all previously reported groups. The area corresponding to the occlusal surface was trapezoid. Relatively caudal to the 1st deciduous incisors—501 and 601—the crowns of unerupted 1st permanent incisors—101 and 201—were identified clearly (Figure 6, 1P). They were larger and of greater radiopacity than in the radiographs of skulls of group 1c. They tapered apically, where they were surrounded by radiolucent areas, smaller than in previous groups. Most parts of these radiolucent areas were caudal to the incisive canal. As it was also reported in some of the skulls of group 1c, a thin radiopaque line, the lamina dura, enclosed the radiolucent area of the dental germs in two of the three skulls of this group. The lamina dura was not identified in the radiograph shown in Figure 6. It was noticeable that lateral to the crowns of the unerupted 1st permanent incisors—101 and 201—two radiolucent images can be easily identified. They corresponded with the dental germs of the 2nd permanent incisors—102 and 202—(Figure 6, dg2). Sagittal CT scans at midplane of 1st deciduous incisors—501 and 601—(Figure 6A) showed the roots of these teeth hardly enclosed by osseous tissue. Hence, a great part of the roots were not included within the alveolus. As shown in previous groups, the enamel cover of the 1st deciduous incisors—501 and 601—i.e., crowns, extended more apically on the labial side than in the palatal side. Caudal to the root of the 1st deciduous incisors—501 and 601—the dental germs of the 1st permanent incisors—101 and 201—were shown, with large unerupted crowns occupying most of the space (Figure 6, 1P and dg). The length of the unerupted crowns of 1st permanent incisors—101 and 201—were apparently larger than in any of the previously reported groups. The infundibulum (Figure 6B, INF) of the unerupted incisors were quite large and completely formed by calcified tissue, from its occlusal part to the apical point. In sagittal scans at midlevel of the 2nd deciduous incisors—502 and 602—(not shown in the Figure 6) and in transverse scans (Figure 6C) two small hypodense areas were identified (Figure 6C, dg2) laterally to the dental germs of the 1st permanent incisors—101 and 201—and caudally to the roots of the 2nd deciduous incisors—502 and 602. They corresponded with the dental germs of the 2nd permanent incisors—102 and 202.

interaction with the surrounding mesenchyma leads to the formation the tooth germs [10]. After that, a complicated process of growth and remodeling of the tooth germs, and production of the tooth substances leads to the fully developed teeth [10,13], deep into the roots of the temporary set of equivalent teeth [28]. After completing the development of the permanent incisors, the deciduous incisors are displaced labially by the erupting permanent incisors [10], which usually erupt on the lingual aspect of the corresponding deciduous teeth [18]. The latest process involves continuous adjustment and remodeling of the alveoli and surrounding bone [28]. Intraoral radiographs of skulls obtained in the present study were suitable to identify the dental germs of the 1st permanent incisors—101 and 201—and 2nd the permanent incisors—102 and 202—when they were present. CT scans confirmed the position and morphology of these dental germs. Some authors have shown in equines that tooth germs of some permanent molar teeth that do not have predecessor deciduous teeth (4th cheek teeth) appear radiographically at 6 to 7 weeks of age and early crown mineralization of them is detected at 7 weeks of age [19]. The same authors have showed radiographic evidence of tooth germs in premolar teeth that have predecessor deciduous teeth (1st to 3rd cheek teeth) at 10 to 12 months of age and early signs of mineralization at 11 to 13 months of age. The dental germs of the 1st permanent incisors—101 and 201—in the present study were identified radiographically in the skulls of the horses of groups 1a (supposedly the youngest of the study), and the first sign of crown mineralization appeared in group 1b, supposedly being a few months older than the previous ones. In skulls of the horses of groups 1c and 1d (supposedly older than horses of groups 1a and 1b, but also with an estimated age of between 12 and less than 30 months), the crowns of the 1st permanent incisors—101 and 201—were mineralized. CT scans of these last skulls (groups 1c and 1d) showed the crowns of the 1st permanent incisors—101 and 201—are clearly hyperdense due to their grade of calcification. Intraoral radiographs of a yearling (between one and two years old) showed the dental germs of the 1st permanent incisors—101 and 201—to have a similar grade of development and mineralization as the skulls of group 1c of the present study [22]. All the above may indicate that the dental germs of the 1st permanent incisors—101 and 201—may appear radiographically shortly before 12 months of age, and that the first sign of mineralization would appear some months later.

It has been described that at the periphery of the dental germ, cell differentiation, and calcification organize the individual constituents of the periodontium, i.e., the cementum, the periodontal ligament, and the alveolar bone [13]. Periodontal research has demonstrated that the periodontium, in addition to fixing the tooth within the alveolus, it also possesses a shock-absorbing and reparatory function and also contains multiple elements, such as mesenchymal cells, capable of transforming into any of the periodontal tissues when repair is needed [9]. Alveolar bone is very flexible and constantly remodels to accommodate the changing shape and size of the dental structures it contains [10,18]. The intraosseous stage of tooth eruption needs resorption of bone in the direction of eruption and formation of bone on the opposite side [10]. These activities depend upon the adjacent parts of the true dental germ [30]. Sagittal CT scans at midplane of the 1st deciduous incisors—501 and 601—of the skulls of group 1d (Figure 6A) showed a substantial part of the root was not included within the corresponding alveoli. On the contrary, CT scans of the 1st permanent incisors—101 and 201—of group 2 (Figure 7A) showed these teeth enclosed by the corresponding alveoli on the dorsal and palatal sides. The above may demonstrate in equines that bone resorption and bone formation are also polarized around erupting incisor teeth. The inner wall of the alveolar bone, a thin layer of compact bone, has been shown to appear radiographically as a delicate radiopaque line called the lamina dura [10,11,13,18]. It has been stated that the lamina dura is not detectable on computed tomography [11]. We disagree with this statement, because in the present study, the lamina dura of the alveoli of the 1st permanent incisors—101 and 201—was detected radiographically in some of the skulls of groups 1c and 1d and in all the skulls of group 2. Confirmation of the presence of this osseous part of the alveoli was confirmed by the hyperdensity shown in the structure in sagittal CT scans.

By visual inspection of skulls of the present study, small oval openings were detected in the palatal region of the alveolar bone, close to the deciduous incisors. They were identified as the openings of the so-called gubernacular canals. In humans, the gubernacular canals have been defined as small channels that connect the bony crypt of the permanent unerupted teeth to the oral surface of the alveoli [30,31]. In the skulls of horses in the present study, they seemed to appear gradually from medial to lateral from group 1a to group 2, related to the grade of development and calcification of the corresponding dental germs found in radiographs. In one of the skulls of group 2, there was also a gubernacular canal close to the corresponding already erupted 1st permanent incisors—101 and 201. The gubernacular canal contains the gubernacular cord, a structure composed of conjunctive tissue, which links the tooth follicle (part of the dental germ) to the overlying gingiva [31]. Both the gubernacular canal and the cord were supposed to help guiding and directing the course of tooth eruption [31]. However, the existence and functions of the gubernacular canal and the cord in humans are still controversial and questioned [31]. To the authors' knowledge there are no references in the literature describing the presence and the function of the gubernacular canal in horses. Therefore, studies on this issue are needed to deepen the knowledge of the role of the gubernacular canal and the cord in the eruption of the incisors in equines. Anyway, this structure is present in the change of dentition of equines, and it may appear on radiographs (and CT scans) of the incisor region, as was the case for group 2 (Figure 7, Gc2). Differential diagnosis using a possible pathological image must be employed in the corresponding cases.

4.3. Deciduous and Permanent Incisors: Lengths and Proportions

To the authors' knowledge, there is no available data concerning tooth length of equine deciduous incisors. Hence, the length measurements and proportions obtained in the present study may complement the morphological data studied by radiographs and CT scans. However, due to the lack of statistical analysis, these results should be considered with caution. It has been studied that formation of permanent equine teeth is not finished once the teeth erupt [13]. Abrasion and mastication wear down the functional crown at a rate of 2 to 3 mm per year, but the reserve crown erupts continuously so that approximately 2 cm of exposed crown is maintained [11]. In the 1st deciduous incisors—501 and 601—of the present study, the relative lengths of the crown and the infundibulum decreased from groups 1a and 1b to groups 1c and 1d (i.e., the relative length of the root increased in the same manner), coincident with an increase of the total length of the tooth. These results may suggest the leading role of root elongation to outbalance the grade of occlusal wearing in deciduous incisors, as it had already been suggested for permanent incisors from two to four years post eruption [14]. The infundibulum of the equine incisors is an enamel infolding inside the occlusal surface [15] that, in horses, has a depth of 10–30 mm [10]. In the present study, the length of the infundibulum ranged between 4.5 ± 1.4 mm of group 1d and 35.9 ± 3.2 mm of group 2. Our results agree with what has been stated by some authors who affirmed that deciduous incisors contain shallower infundibula than their permanent successors [10]. Once formed, the length of the infundibulum cannot be shortened, except by being worn from the occlusal side [14]. Hence, it seems logical that, in the present study, the minimum length of the infundibulum of the deciduous incisors was found to exist in the supposedly older animals (group 1d). It has been affirmed that during eruption, hypsodont teeth have no true roots, i.e., with an apical enamel-free area [10]. We suppose that the above statement referred to permanent incisors, since in the present study the length of the root in the erupted 1st permanent incisors—101 and 201—(group 2) represented less than 13 % of the tooth length (relative length of the crown 87.4 ± 5.6 mm). However, the length of the root in the 1st deciduous incisors—501 and 601—of the supposedly youngest animals (group 1a) represented more than 50% of the length of the tooth (relative length of the crown 48.5 ± 4.1 mm).

From this study, veterinary clinicians might improve their knowledge of the radiographic appearance of deciduous incisors and dental germs in the permanent incisors of horses in the period prior to teeth shedding. These radiographic data may be considered when injuries or disorders affect the maxillary incisor arcade. The main limitations of the present study were the lack of knowledge of the precise age of the horses to which the skulls used to belong and the small number of skulls in each group to perform a statistical analysis of the dental measurements and proportions. Further radiographic and CT investigations on a representative number of horses of known ages would provide a more comprehensive understanding of the development of dental germs of permanent incisors.

5. Conclusions

Radiographic intraoral images are suitable to identify the grade of development of the dental germs of the maxillary permanent incisors. In horses conforming to the general dental aging system, the dental germs of 1st permanent incisors—101 and 201—(identified radiographically as rounded radiolucent areas located at the level of the apical parts of the 1st deciduous incisors—501 and 601—appear at the age of twelve months. The first sign of crown mineralization of these dental germs appears a few months later. Before the teeth shedding, the unerupted mineralized crowns of the 1st permanent incisor can be identified radiographically as being relatively caudal to the corresponding 1st deciduous incisors.

Author Contributions: Conceptualization, F.M. and M.N.; methodology, F.M., C.M., A.D. and M.N.; formal analysis, F.M., C.M. and M.N.; writing—original draft preparation, F.M. and M.N.; writing—review and editing, F.M. and M.N.; supervision, F.M., C.M., A.D. and M.N. All authors have read and agreed to the published version of the manuscript.

Funding: This research received no external funding.

Conflicts of Interest: The authors declare no conflict of interest.

References

1. Barrett, M.F.; Easley, J.T. Acquisition and interpretation of radiographs of the equine skull. *Equine Vet. Educ.* **2013**, *25*, 643–652. [CrossRef]
2. Rawlinson, J.T.; Earley, E. Advances in the Treatment of Diseased Equine Incisor and Canine Teeth. *Vet. Clin. Equine* **2013**, *29*, 411–440. [CrossRef] [PubMed]
3. Baratt, R.M. Advances in Equine Dental Radiology. *Vet. Clin. N. Am. Equine Pract.* **2013**, *29*, 367–395. [CrossRef] [PubMed]
4. Barakzai, S.Z. Dental imaging. In *Equine Dentistry*, 3rd ed.; Easley, J., Dixon, P.M., Schumacher, J., Eds.; Saunders Elsevier: London, UK, 2011; pp. 199–221.
5. Manso-Díaz, G.; García-López, J.M.; Maranda, L.; Taeymans, O. The role of head computed tomography in equine practice. *Equine Vet. Educ.* **2015**, *27*, 136–145. [CrossRef]
6. Selberg, K.; Easley, J.T. Advances Imaging in Equine Dental Disease. *Vet. Clin. Equine* **2013**, *29*, 397–409. [CrossRef] [PubMed]
7. Simhofer, H.; Boehler, A. Computed tomography. In *Equine Dentistry*, 3rd ed.; Easley, J., Dixon, P.M., Schumacher, J., Eds.; Saunders Elsevier: London, UK, 2011; pp. 221–230.
8. Saunders, J.; Windley, Z. Equine sinonasal and dental. In *Veterinary Computed Tomography*, 1st ed.; Schwarz, T., Saunders, J., Eds.; Wiley-Blackwell: West Sussex, UK, 2011; pp. 427–442.
9. Pearce, C.J. Recent developments in equine dentistry. *N. Z. Vet. J.* **2020**, *68*, 178–186. [CrossRef]
10. Dixon, P.M.; du Toit, N. Dental anatomy. In *Equine Dentistry*, 3rd ed.; Easley, J., Dixon, P.M., Schumacher, J., Eds.; Saunders Elsevier: London, UK, 2011; pp. 51–76.
11. Lowder, M.Q.; Mueller, P.O. Dental Embryology, Anatomy, Development, and aging. *Vet. Clin. N. Am. Equine Pract.* **1998**, *14*, 227–245. [CrossRef]
12. Singh, B. The Head and Ventral Neck of the Horse. In *Dyce, Sack and Wensing's Textbook of Veterinary Anatomy*, 5th ed.; Singh, B., Ed.; Elsevier: St. Louis, MO, USA, 2018; pp. 492–521.
13. Staszyk, C.; Suske, A.; Pöschke, A. Equine dental and periodontal anatomy: A tutorial review. *Equine Vet. Educ.* **2015**, *27*, 474–481. [CrossRef]
14. Schrock, P.; Lüpke, M.; Seifert, H.; Staszyk, C. Three-dimensional anatomy of equine incisors: Tooth length, enamel cover and age-related changes. *BMC Vet. Res.* **2013**, *9*, 249. [CrossRef]
15. Muylle, S. Aging. In *Equine Dentistry*, 3rd ed.; Easley, J., Dixon, P.M., Schumacher, J., Eds.; Saunders Elsevier: London, UK, 2011; pp. 85–96.

16. Muylle, S.; Simoens, P.; Lauwers, H. Age-related morphometry of equine incisors. *J. Vet. Med.* **1999**, *46*, 633–643. [CrossRef]
17. Tremaine, H.; Casey, M. A modern approach to equine dentistry 1. Oral examination. *Practice* **2012**, *34*, 2–10. [CrossRef]
18. Dixon, P.M. The gross, histological, and ultrastructural anatomy of equine teeth and their relationship to disease. In Proceedings of the Focus on Dentistry, American Association of Equine Practitioners, Orlando, FL, USA, 4–8 December 2002.
19. Misk, N.A.; Seilem, S.M. Radiographic studies on the development of cheek teeth in donkeys. *Equine Pract.* **1997**, *19*, 27–38.
20. Domingo, M.S.; Cantero, E.; García-Real, I.; Chamorro Sancho, M.J.; Martín Perea, D.M.; Alberdi, M.T.; Morales, J. First Radiological Study of a Complete Dental Ontogeny Sequence of an Extinct Equid: Implications for Equidae Life History and Taphonomy. *Sci. Rep.* **2018**, *8*, 8507. [CrossRef] [PubMed]
21. Earley, E.; Rawlinson, J.T. A New Understanding of Oral and Dental Disorders of the Equine Incisor and Canine Teeth. *Vet. Clin. Equine* **2013**, *29*, 273–300. [CrossRef]
22. Baratt, R.M. Equine Dental Radiography. In Proceedings of the Focus on Dentistry, American Association of Equine Practitioners, Albuquerque, NM, USA, 18–20 September 2011.
23. Tremaine, H. A modern approach to equine dentistry 3. Imaging. *Practice* **2012**, *34*, 114–127. [CrossRef]
24. Liuti, T.; Smith, S.; Dixon, P.M. A comparison of computed tomographic, radiographic, gross and histological, dental, and alveolar findings in 30 abnormal cheek teeth from equine cadavers. *Front. Vet. Sci.* **2018**, *4*, 236. [CrossRef]
25. World Association of Veterinary Anatomists (W.A.V.A.). *Nomina Anatomica Veterinaria*, 6th ed.; Editorial Committee: Columbia, SC, USA, 2017.
26. Floyd, M.R. The modified Triadan system: Nomenclature for veterinary dentistry. *J. Vet. Dent.* **1991**, *8*, 18–19. [CrossRef]
27. Limone, L.E.; Barrat, R.M. Dental Radiography of the horse. *J. Vet. Dent.* **2018**, *35*, 37–41. [CrossRef]
28. Singh, B. The Digestive Apparatus. In *Dyce, Sack and Wensing's Textbook of Veterinary Anatomy*, 5th ed.; Singh, B., Ed.; Elsevier: St. Louis, MO, USA, 2018; pp. 91–138.
29. Slolunias, N.; Danowitz, M.; Buttar, I.; Copee, Z. Hypsodont Crowns as Additional Roots: A new Explanation for Hypsodonty. *Front. Ecol. Evol.* **2019**, *7*, 135. [CrossRef]
30. Araújo, D.C. Gubernacular cord and canal. Does these anatomical structures play a role in dental eruption? *RSBO* **2013**, *10*, 167–171.
31. Marks, S.; Schroeder, H.E. Tooth Eruption: Theories and Facts. *Anat. Rec.* **1996**, *245*, 374–393. [CrossRef]

Article

Anatomical Assessment of the Thorax in the Neonatal Foal Using Computed Tomography Angiography, Sectional Anatomy, and Gross Dissections

Alberto Arencibia [1,*], Juan Alberto Corbera [2], Gregorio Ramírez [3], María Luisa Díaz-Bertrana [4], Lidia Pitti [4], Manuel Morales [2] and José Raduan Jaber [1]

1 Departamento de Morfología, Facultad de Veterinaria, Universidad de Las Palmas de Gran Canaria, Trasmontaña, Arucas, 35413 Las Palmas, Spain; joseraduan.jaber@ulpgc.es

2 Instituto Universitario de Investigaciones Biomédicas y Sanitarias (IUIBS), Facultad de Veterinaria, Universidad de Las Palmas de Gran Canaria, Trasmontaña, Arucas, 35413 Las Palmas, Spain; juan.corbera@ulpgc.es (J.A.C.); manuel.morales@ulpgc.es (M.M.)

3 Departamento de Anatomía y Anatomía Patológica, Universidad de Murcia, 30100 Murcia, Spain; grzar@um.es

4 Hospital Clínico Veterinario, Facultad de Veterinaria, Universidad de Las Palmas de Gran Canaria, Trasmontaña, Arucas, 35413 Las Palmas, Spain; luigi.bertrana@ulpgc.es (M.L.D.-B.); lpitti@fpct.ulpgc.es (L.P.)

* Correspondence: alberto.arencibia@ulpgc.es

Received: 27 May 2020; Accepted: 16 June 2020; Published: 17 June 2020

Simple Summary: This research aimed to describe the normal appearance of the thorax in neonatal foals by computed tomography angiography (CTA). The newborn foals were imaged using a 16-slice helical CT scanner after the administration of an iodinated contrast medium. CTA images and three-dimensional cardiac volume-rendered reconstructed images were obtained to enhance cardiovascular structures. In addition, thoracic anatomical sections and gross dissections were used as anatomical references. Clinically relevant anatomical structures were identified on the CTA images, anatomical sections, and gross dissections. These findings could serve as a reference to the CTA image assessment of the thorax of neonatal foals.

Abstract: The purpose of this study was to correlate the anatomic features of the normal thorax of neonatal foals identified by CTA, with anatomical sections and gross dissections. Contrast-enhanced transverse CTA images were obtained in three neonatal foals using a helical CT scanner. All sections were imaged with a bone, mediastinal, and lung windows setting. Moreover, cardiac volume-rendered reconstructed images were obtained. After CT imaging, the cadaver foals were sectioned and dissected to facilitate the interpretation of the intrathoracic cardiovascular structures to the corresponding CTA images. Anatomic details of the thorax of neonatal foals were identified according to the characteristics of CT density of the different organic tissues and compared with the corresponding anatomical sections and gross dissections. The information obtained provided a valid anatomic pattern of the thorax of foals, and useful information for CTA studies of this region.

Keywords: CT angiography; sections; dissections; thorax; anatomy; neonatal foal

1. Introduction

CTA is a minimally invasive imaging technique used to assess the organs of the respiratory and cardiovascular system. In humans, medicine has become the imaging modality of choice for diagnosis of abnormalities, injuries, and thoracic disease [1,2]. CTA displays the anatomical detail of specific tissue densities and blood vessels more precisely compared with radiography, ultrasonography, and magnetic resonance imaging. It is due to its high spatial resolution, shortened examination time, and improved visibility of vascular structures and pulmonary parenchyma [3,4]. CTA can be used to obtain high-quality two-dimensional images and three-dimensional cardiac reformatted images that delineate the morphologic features of the cardiovascular system without the superimposition of other structures [5,6].

In equine medicine, radiology has been the main imaging modality used to image the thorax of foals, but the superimposition of adjacent anatomical structures makes its interpretation difficult [7]. Nonetheless, an advanced imaging technique such as the ultrasound [8,9] has given to the equine practitioners the opportunity to obtain an accurate diagnosis. In addition, helical CTA has also been used in horses, but information is limited to reports on the head, spinal cord, musculoskeletal system, and lungs [10–19]. To the best of our knowledge, a detailed anatomical study using CTA with an iodinated contrast medium for imaging the thorax of neonatal foals has not been reported before.

An accurate anatomical interpretation of helical CTA of the thorax could be useful to aid in the diagnosis of different diseases described in neonatal foals such as thoracic trauma, pulmonary disorders, and congenital heart anomalies [20–25]. Therefore, the objectives of this study were: (1) To describe the normal cross-sectional anatomy of the thorax of neonatal foals using helical CTA images with the use of a vascular contrast medium, anatomic sections, and gross dissections, and (2) to obtain three-dimensional cardiac volume-rendered reconstructed images to assist in the understanding of the anatomy of the heart and the main intrathoracic vessels.

2. Materials and Methods

2.1. Animals

Three newborn crossbreed foals of 2, 5, and 6 days with weights ranging between 50–55 kg were selected from equine patients attending to the Veterinary Hospital of Las Palmas de Gran Canaria University between January to December 2019. The animals had neurological signs that included head tilt, seizures, circling, and ataxia. No other physical examination abnormalities were detected. After clinical evaluation, a combination of butorphanol 10 mg/mL at a dose of 0.4 mL (Torbugesic®; Zoetis S.L.U., Madrid, Spain), and dexmedetomidine 10 mg/mL at a dose of 0.3 mL (Dexdormitor®; Lab. Dr. Esteve SAU, Barcelona, Spain) injected IM were employed as a preanesthetic medication. Anesthesia using sevuflorane 98 (0.5 to 2%) (Sevoflo.; Abbot Laboratories SA, Madrid, Spain) was maintained during the procedure. After the head scan, the study of the thorax was performed. The euthanasia was done due to the diagnosis of CNS congenital abnormalities. The Animal Ethical Committee of Veterinary Medicine of Las Palmas de Gran Canaria University authorized the research protocol (MV–2016/04). The owners of these foals were informed of the study and signed a consent for participation in the study.

2.2. CTA Technique

Contrast-enhanced sequential transverse CT slices were performed using a 16-slice helical CT scanner (Toshiba Astelion, Toshiba Medical System, Madrid, Spain). The animals were positioned symmetrically in dorsal recumbency on the CT couch and a standard clinical protocol (120 kVp, 80 mA, 512 × 512 acquisition matrix, 283 × 283 field of view, a spiral pitch factor of 0.94, and a gantry rotation of 1.5 s) was used to acquire transverse CT thorax images, with 3-mm slice thickness. In addition, all foals received a bolus of iomeprol 300 mg/mL at a dose of 2 mL/kg (Iomeron.; Rovi S.A., Madrid, Spain) via the jugular vein. The transverse original data was stored and transferred to the CT workstation. To better evaluate the CT appearance of the thoracic structures, three CT windows were applied by adjusting window widths (WW) and window levels (WL): A bone window setting (WW = 1500; WL = 300), a mediastinal window setting (WW = 248; WL = 123), and a lung window setting (WW = 1400; WL = −500). The original data were used to generate cardiac volume-rendered reconstructed images from the right and left surfaces, and the base of the heart after manual editing of the transverse CT images to remove bone structures and other soft tissues using a standard dicom 3D format (OsiriX MD, Geneva, Switzerland).

2.3. Anatomic Evaluation

The interpretation of the CTA images was based on anatomical sections and gross dissections to facilitate the identification of the thoracic structures. At the end of the scanning procedure, the euthanized foals were frozen until solid. Later, two frozen cadavers were sectioned using an electric band saw to obtain sequential transverse anatomical sections. The other cadaver was used to perform the gross anatomical dissections within 24 h of death to minimize post-mortem changes. Thoracic structures studied in the CT images were correlated with those identified in the corresponding anatomical sections and gross dissections, evaluated according to the characteristics of CT density of different tissues and labelled to conform to anatomical texts [26,27].

3. Results

3.1. Transverse Computed Tomography Angiography Images

The results of the thoracic CTA images are presented in seven sequential transverse CTA images of the thorax at different levels that best correlated with the macroscopic sections (Figures 1–7). Each figure consists of four images: (a) Bone window, (b) mediastinal window, (c) lung window, and (d) anatomical section. Transverse CT images are presented in a cranial to caudal progression from the level of the brachiocephalic trunk (Figure 1) to the level of the left ventricle and apex of the heart (Figure 7).

The CTA images obtained with the use of the bone window setting (Figures 1A–7A), provided a good differentiation between the bones and the soft tissues of the thoracic cavity. Thus, the thoracic vertebrae (including the vertebral body and arch, and the corresponding articular, transverse, and spinous processes), the ribs (with its head tubercle, body, and costal cartilage), and the sternum were well visualized (Figures 1A–7A). In addition, the vertebral cortical and bone marrow fat were also delineated. However, the costovertebral, costochondral, and sternocostal joints and those muscles associated with the thorax such as epaxial (semispinalis, longissimus, and iliocostalis) and thoracic wall (external and internal intercostal, and pectoral) muscles appeared with an intermediate CT density (Figures 1A–7A). Other anatomical structures such as the thoracic duct (Figures 1A–7A), the right and left vagus nerves (Figures 1A–4A), and the dorsal and ventral vagal trunks (Figure 5A) were also identified.

The CT mediastinal window (Figures 1B–7B) showed good visualization of the bony thoracic wall structures. Moreover, this CT window also provided an excellent visualization of the heart with its chambers (atrium and ventricles) and the main arteries and veins, which appeared with a high attenuation due to the intravenous contrast medium (Figures 1B–7B). Thus, important associated vessels such as the cranial (Figures 1B and 2B), and the caudal vena cava (Figures 5B–7B) were seen leading into the right atrium. Additionally, the course of the right azygos vein (Figure 3B–7B), the pulmonary trunk (Figure 3B), and the left and right pulmonary arteries (Figure 4B) were observed. Other intrathoracic vessels, including the aortic root (Figures 2B–4B), ascending (Figure 2B) and descending aorta (Figures 2B–7B), and the internal thoracic arteries and veins (Figures 1B–6B) were also identified. By contrast, with the use of CT bone window images (Figures 1A–7A), these structures appeared with an intermediate CT density. Other structures such as the myocardial walls showed an intermediate CT density and were best visualized in the CT bone and mediastinal windows settings (Figures 1A–7A and 1B–7B).

Concerning the lungs, the CT bone (Figures 1A–7A) and mediastinal (Figures 1B–7B) window settings showed the bronchi and the vascular formations of the lungs, which were only clearly defined at the level of the hilus because of the deeper lumen and the use of intravenous contrast medium. In contrast, the CT lung window (Figures 1C–7C) allowed a better definition of the lobes and a better tomographical definition of the trachea, tracheal bifurcation, main bronchi, and lobar bronchi due to these structures that presented higher attenuation than the lungs. Moreover, it was possible to visualize the triad that comprises the lobar pulmonary vein, the lobar arterial branch, the lobar bronchus, and the pleural cavity.

3.2. Anatomical Sections

On transverse anatomical sections (Figures 1D–7D), additional morphologic and topographic information about the thoracic structures could be identified when compared with CTA images. All bones, cartilaginous structures, and associated muscles were identified. The respiratory tract structures, including the trachea (Figures 1D–3D) and its bifurcation (Figure 4D), the principal and lobar bronchi, and pulmonary parenchyma (Figures 5D–7D), were also well observed. Other intrathoracic structures such as the heart with its chambers and associated large vessels were likewise visible in Figures 1D–7D. Other anatomical structures such as the thoracic duct (Figures 1D–7D), the right and left vagus nerves (Figures 1D–7D) were identified.

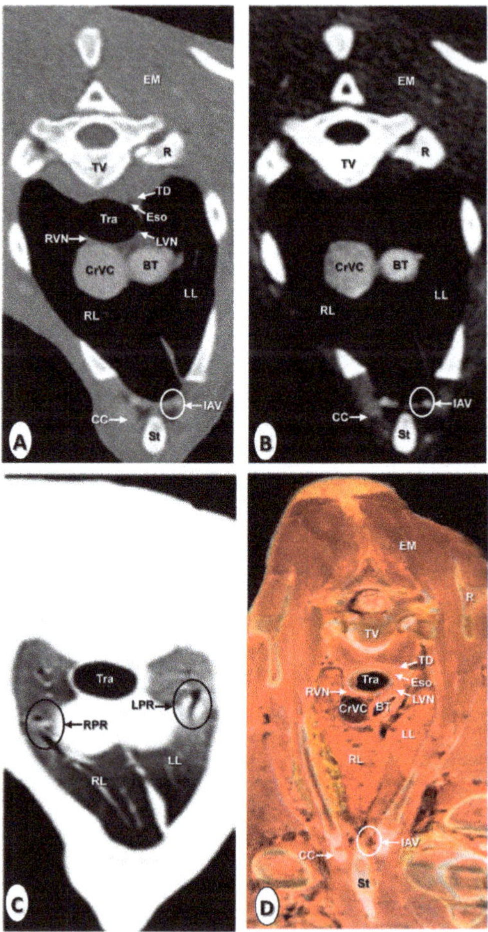

Figure 1. Transverse images at the level of the brachiocephalic trunk. (**A**) Computed tomography (CT) bone window; (**B**) CT mediastinal window; (**C**) CT lung window; and (**D**) anatomical section. These images are displayed so that the right side of the foal is to the viewer's left and the dorsal view is at the top. Brachiocephalic trunk (BT); cranial vena cava (CrVC); epaxial muscles (EM); esophagus (Eso); internal thoracic artery and vein (IAV); left lung (LL); left pulmonary root (LPR); left vagus nerve (LVN); rib: Costal bone (R); rib: Costal cartilage (CC); right lung (RL); right pulmonary root (RPR); right vagus nerve (RVN); scapula (Sca); sternum (St); thoracic duct (TD); thoracic vertebra (TV); and trachea (Tra).

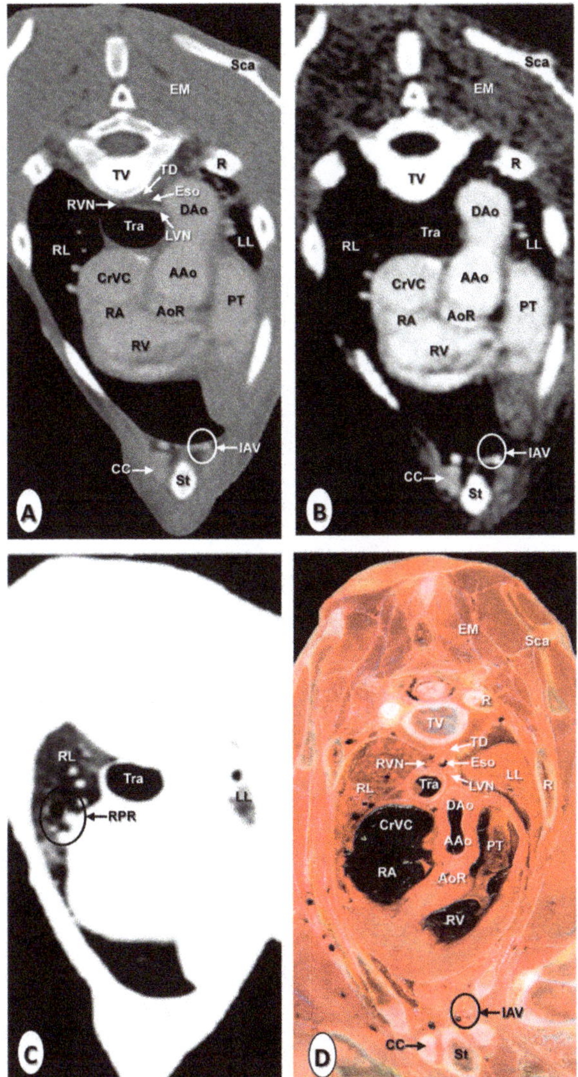

Figure 2. Transverse images at the level of the ascending aorta. (**A**) CT bone window; (**B**) CT mediastinal window; (**C**) CT lung window; and (**D**) anatomical section. These images are displayed so that the right side of the foal is to the viewer's left and the dorsal view is at the top. Aortic root (AoR); ascending aorta (AAo); cranial vena cava (CrVC); descending aorta (Dao); epaxial muscles (EM); esophagus (Eso); internal thoracic artery and vein (IAV); left lung (LL); left vagus nerve (LVN); pulmonary trunk (PT); rib: Costal bone (R); rib: Costal cartilage (CC); right atrium (RA); right lung (RL); right pulmonary root (RPR); right vagus nerve (RVN); right ventricle (RV); scapula (Sca); sternum (St); thoracic duct (TD); thoracic vertebra (TV); and trachea (Tra).

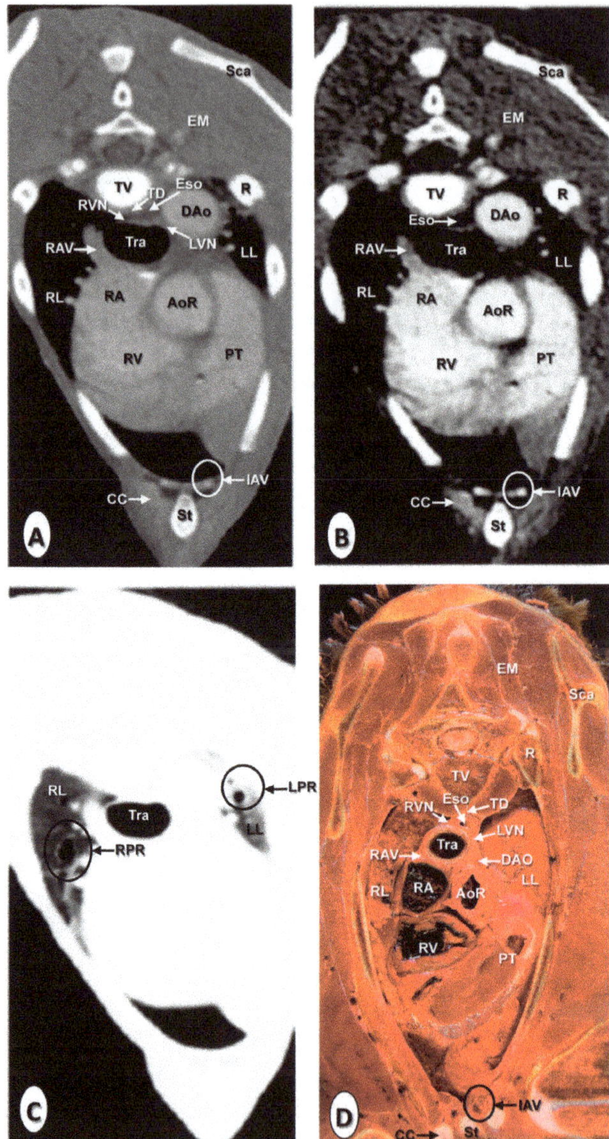

Figure 3. Transverse images at the level of the pulmonary trunk. (**A**) CT bone window; (**B**) CT mediastinal window; (**C**) CT lung window; and (**D**) anatomical section. These images are displayed so that the right side of the foal is to the viewer's left and the dorsal view is at the top. Aortic root (AoR); descending aorta (Dao); epaxial muscles (EM); esophagus (Eso); internal thoracic artery and vein (IAV); left lung (LL); left pulmonary root (LPR); left vagus nerve (LVN); pulmonary trunk (PT); rib: Costal bone (R); rib: Costal cartilage (CC); right atrium (RA); right azygos vein (RAV); right lung (RL); right pulmonary root (RPR); right vagus nerve (RVN); right ventricle (RV); scapula (Sca); sternum (St); thoracic duct (TD); thoracic vertebra (TV); and trachea (Tra).

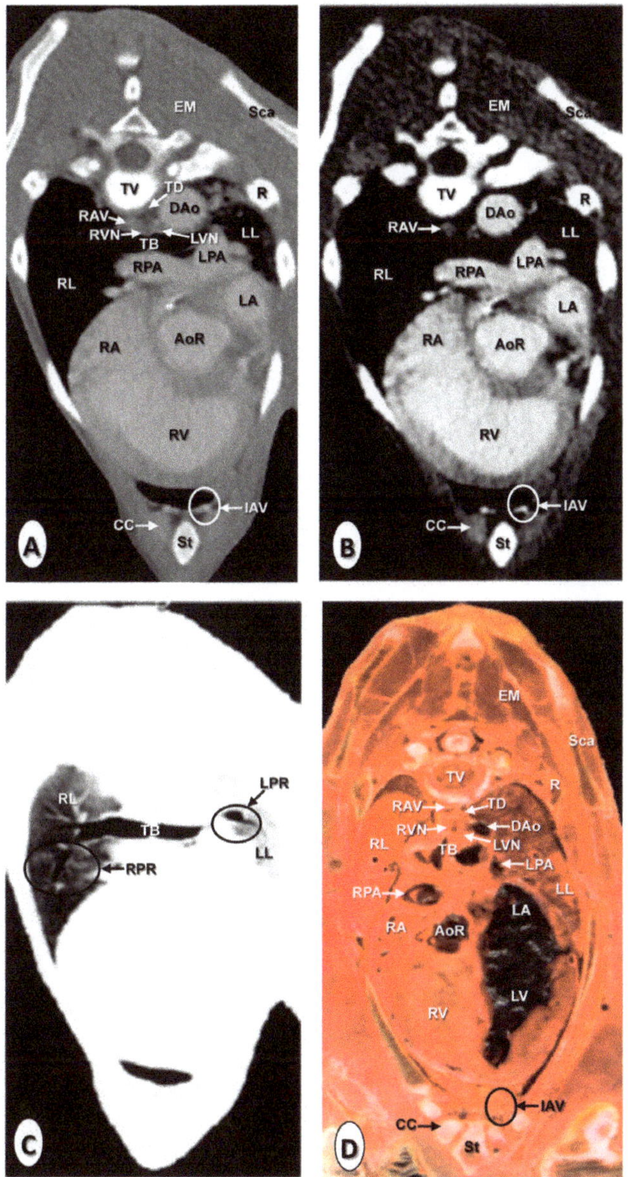

Figure 4. Transverse images at the level of the pulmonary arteries. (**A**) CT bone window; (**B**) CT mediastinal window; (**C**) CT lung window; and (**D**) anatomical section. These images are displayed so that the right side of the foal is to the viewer's left and the dorsal view is at the top. Aortic root (AoR); descending aorta (Dao); epaxial muscles (EM); internal thoracic artery and vein (IAV); left lung (LL); left pulmonary artery (LPA); left pulmonary root (LPR); left vagus nerve (LVN); rib: Costal bone (R); rib: Costal cartilage (CC); right atrium (RA); right azygos vein (RAV); right lung (RL); right pulmonary artery (RPA); right pulmonary root (RPR); right vagus nerve (RVN); right ventricle (RV); scapula (Sca); sternum (St); thoracic duct (TD); thoracic vertebra (TV); and tracheal bifurcation (TB).

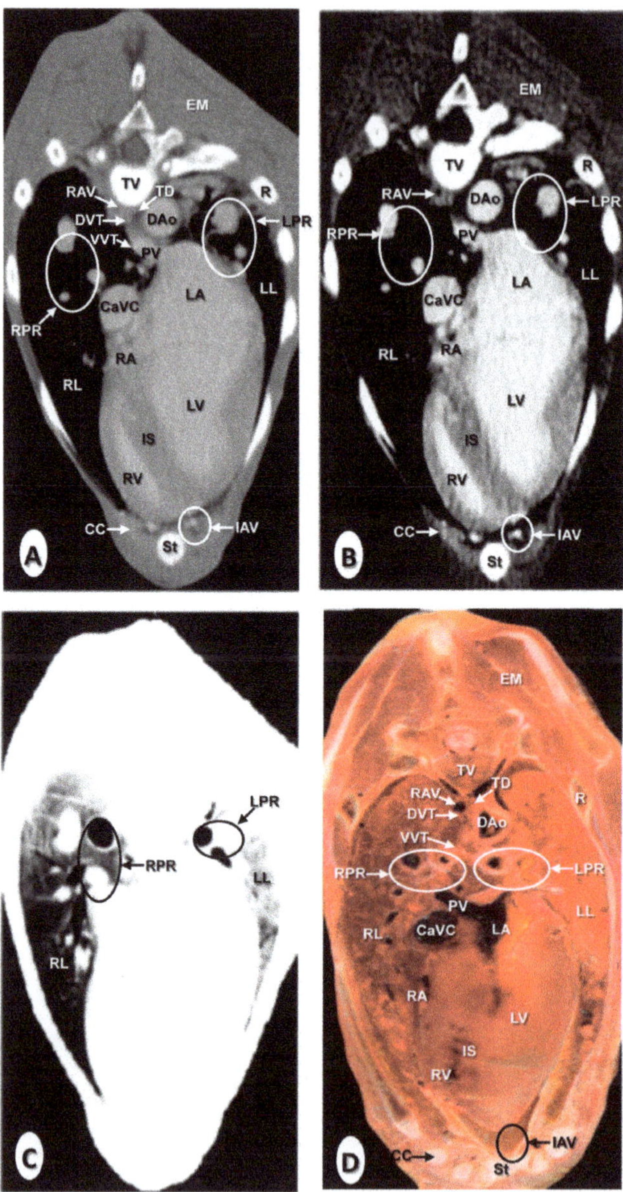

Figure 5. Transverse images at the level of the lung roots. (**A**) CT bone window; (**B**) CT mediastinal window; (**C**) CT lung window; and (**D**) anatomic section. These images are displayed so that the right side of the foal is to the viewer's left and the dorsal view is at the top. Caudal cava vein (CaCV); descending aorta (Dao); dorsal vagal trunk (DVT); epaxial muscles (EM); internal thoracic artery and vein (IAV); left atrium (LA); left lung (LL); left pulmonary root (LPR); pulmonary vein (PV); rib: Costal bone (R); rib: Costal cartilage (CC); right atrium (RA); right azygos vein (RAV); right lung (RL); right pulmonary root (RPR); right ventricle (RV); scapula (Sca); sternum (St); thoracic duct (TD); thoracic vertebra (TV); and ventral vagal trunk (VVT).

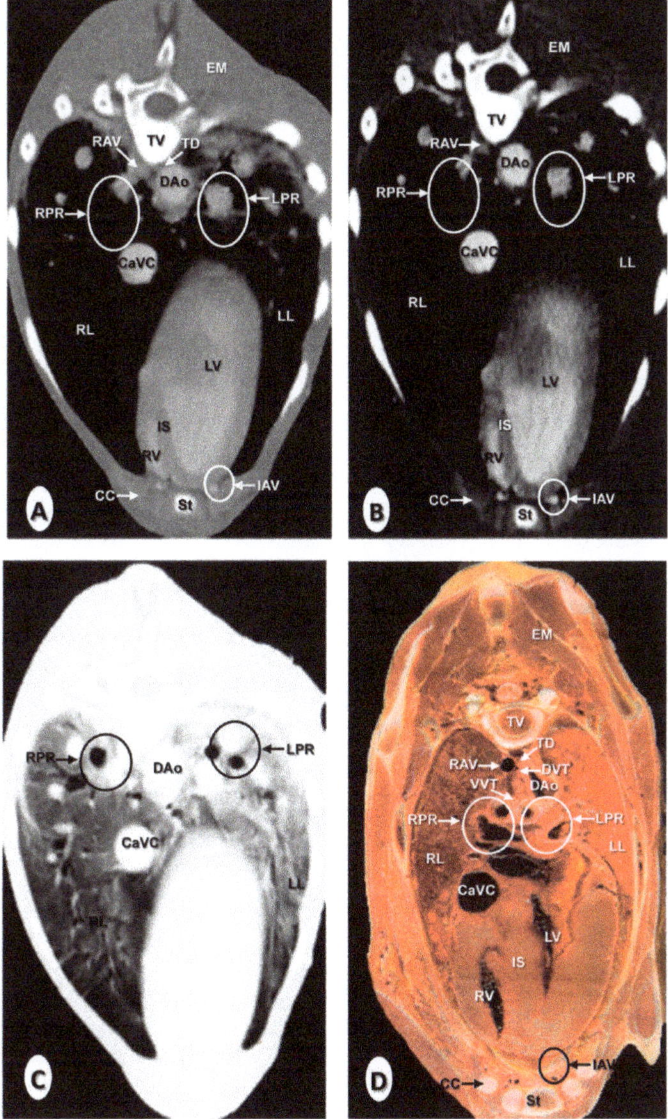

Figure 6. Transverse images at the level of the left ventricle. (**A**) CT bone window; (**B**) CT mediastinal window; (**C**) CT lung window; and (**D**) anatomical section. These images are displayed so that the right side of the foal is to the viewer's left and the dorsal view is at the top. Caudal vena cava (CaVC); descending aorta (Dao); dorsal vagal trunk (DVT); epaxial muscles (EM); internal thoracic artery and vein (IAV); interventricular septum (IS); left lung (LL); left pulmonary root (LPR); left ventricle (LV). rib: Costal bone (R); rib: Costal cartilage (CC); right azygos vein (RAV); right lung (RL); right pulmonary root (RPR); right ventricle (RV); sternum (St); thoracic duct (TD); thoracic vertebra (TV); and ventral vagal trunk.

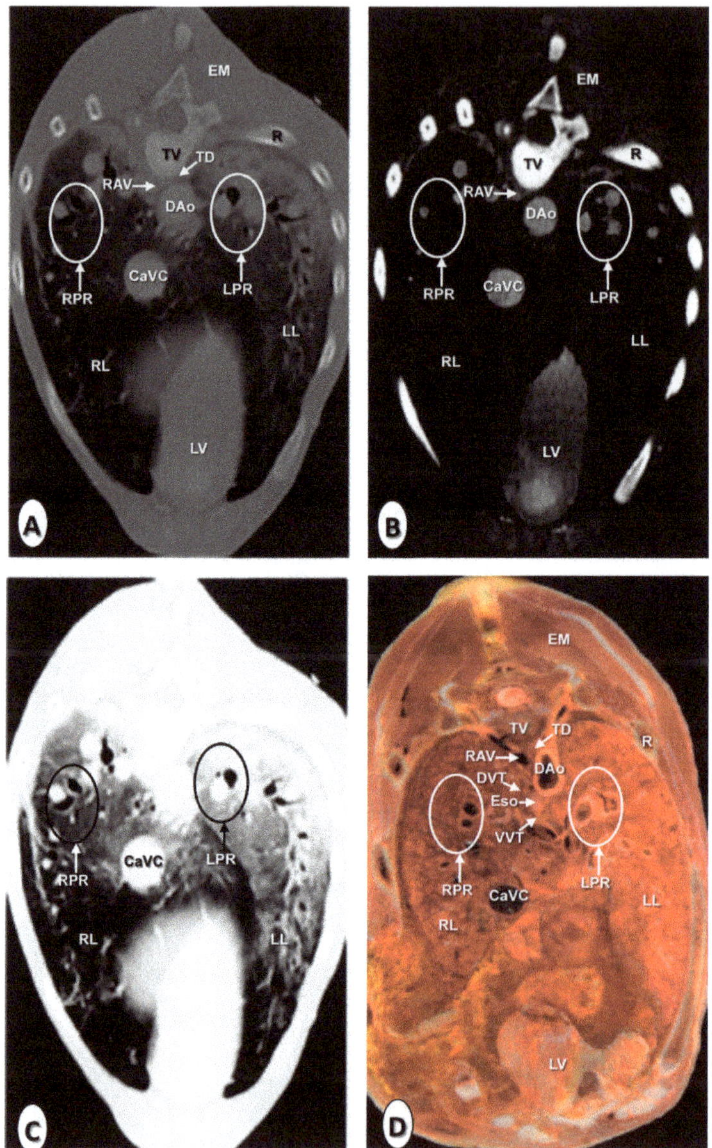

Figure 7. Transverse images at the level of the apex of the heart. (**A**) CT bone window; (**B**) CT mediastinal window; (**C**) CT lung window; and (**D**) anatomical section. These images are displayed so that the right side of the foal is to the viewer's left and the dorsal view is at the top. Caudal vena cava (CaVC); descending aorta (Dao); dorsal vagal trunk (DVT); epaxial muscles (EM); left lung (LL); left pulmonary root (LPR); left ventricle (LV); rib: Costal bone (R); right azygos vein (RAV); right lung (RL); right pulmonary root (RPR); thoracic duct (TD); thoracic vertebra (TV); and ventral vagal trunk (VVT).

3.3. *Gross Anatomical Dissections*

Figure 8 is a composition of two anatomical gross dissections at the level of the atrial (Figure 8A) and auricular (Figure 8B) surfaces of the heart. All chambers, grooves, and the main blood vessels were identified. Thus, the myocardial walls and the coronary groove were well visualised in both images (Figure 8A,B), as well as the subsinuosal interventricular (Figure 8A) and the paracoronary interventricular (Figure 8B) grooves. The location relative to the cranial and caudal vena cava leading into the right atrium could be clearly observed in all views of the heart (Figure 8A,B). In addition, the pulmonary veins were identified in the image corresponding to the atrial surface of the heart (Figure 8A), while the pulmonary trunk arising from the right ventricle was clearly visible in Figure 8B. In addition, the right and left pulmonary arteries were also well identified (Figure 8). The course of the ascending aorta arising from the left ventricle, and the main branches such as the brachiocephalic trunk, the right, and left subclavian arteries, and the descending aorta were also easily identified (Figure 8B).

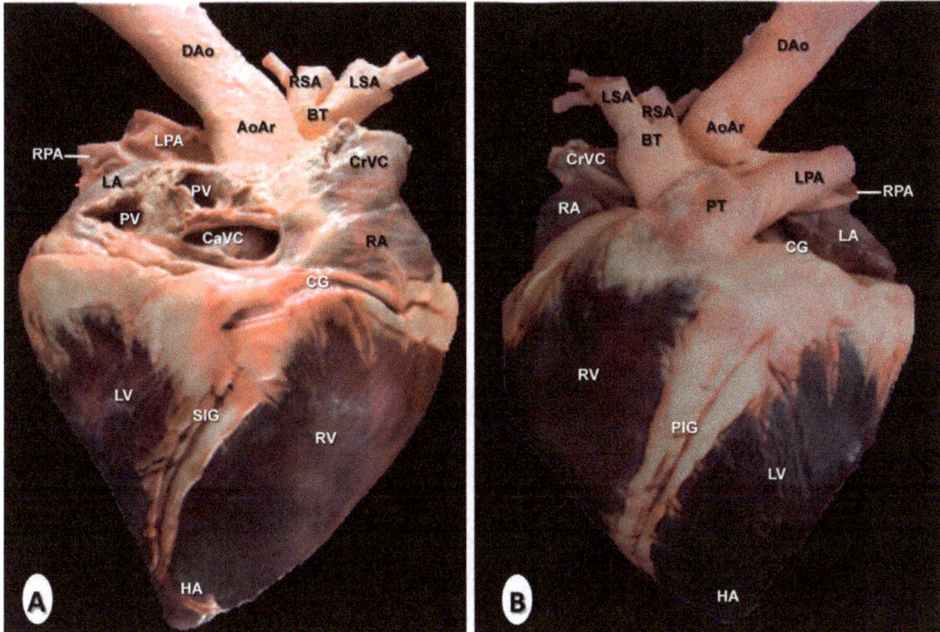

Figure 8. Anatomical dissections of the heart. (**A**) Atrial surface and (**B**) auricular surface. Aortic arch (AoAr); brachiocephalic trunk (BT); caudal vena cava (CaVC); cranial vena cava (CrVC); descending aorta (DA); heart apex (HA); left atrium (LA); left pulmonary artery (LA); left subclavian artery (LSA); left ventricle (LV); paracoronary interventricular groove (PIG); pulmonary trunk (PT); pulmonary vein (PV); right atrium (RA); right pulmonary artery (RPA); right subclavian artery (RSA); subsinuosal interventricular groove (SIG); and right ventricle (RV).

3.4. Cardiac Volume-Rendered Reconstructed CTA Images

Cardiac three-dimensional volume-rendered reconstructed images corresponding to right (Figure 9A) and left lateral surfaces (Figure 9B) and the base (Figure 10) of the heart are presented. Volume-rendered reconstructed CTA images provided a good visualisation of the heart and the major associated vessels. Thus, the cardiac chambers and the main associated blood vessels were identified in all CT reconstructed images (Figures 9 and 10). The location relative to the cranial and caudal vena cava leading into the right atrium could be clearly observed on all views of the heart (Figures 9 and 10). Other important vessels such as the right azygos and brachiocephalic veins joined to the cranial vena cava were seen (Figures 9 and 10). In addition to these observations, the junction of the pulmonary veins entering into the left atrium was identified in all volume reconstructed CT images (Figures 9 and 10). Figure 9 (panel A) shows the course of the pulmonary trunk originating from the right ventricle.

In contrast, the pulmonary artery bifurcation (right and left pulmonary arteries) was clearly identified on the dorsal aspect (Figure 10). The course of the ascending aorta arising from the left ventricle (Figures 9B and 10), and its main branches (such as the descending aorta and brachiocephalic trunk) were easily identified in Figures 9B and 10. The cranial branches of the brachiocephalic trunk such as the left and right subclavian arteries, and bicarotid trunk could be identified on all reconstructed CT images (Figures 9 and 10).

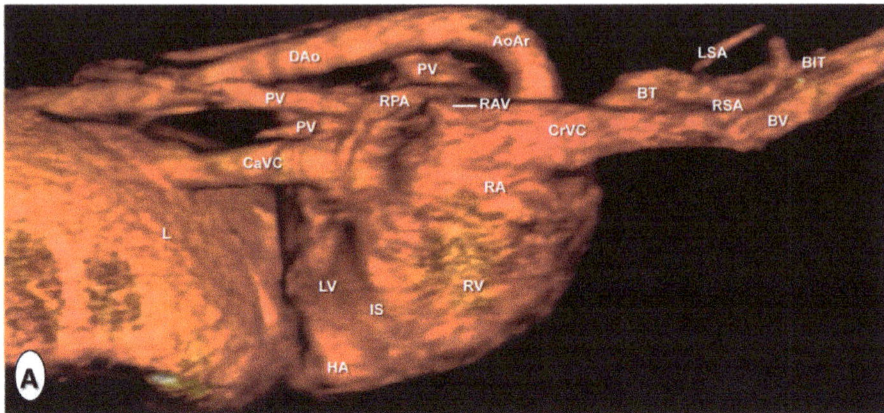

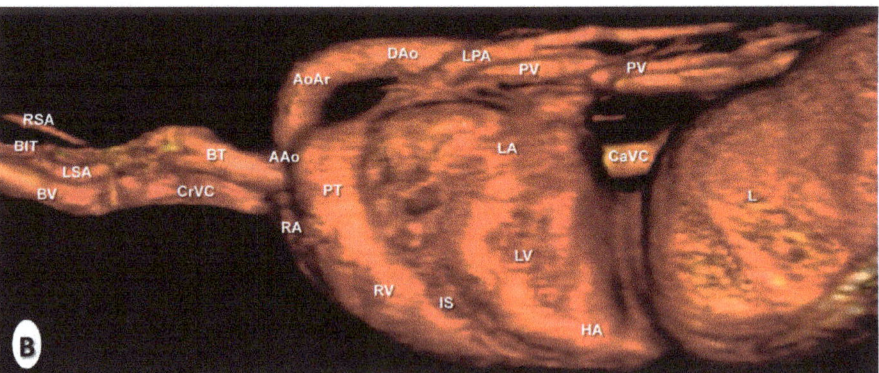

Figure 9. Three-dimensional volume-rendered reconstructed images of the normal neonatal foal heart and associated blood vessels. Parasagittal aspects. (**A**) Right lateral aspect and (**B**) left lateral aspect. Aortic arch (AoAr); ascending aorta (AAo); bicarotid trunk (BIT); brachiocephalic trunk (BT); brachiocephalic vein (BV); caudal vena cava (CaVC); cranial vena cava (CrVC); descending aorta (Dao); heart apex (HA); left atrium (LA); left pulmonary artery (LPA); left subclavian artery (LSA); left ventricle (LV); liver (L); pulmonary trunk (PT); pulmonary vein (PV); right atrium (RA); right azygos vein (RAV); right pulmonary artery (RPA); right subclavian artery (RSA); and right ventricle (RV).

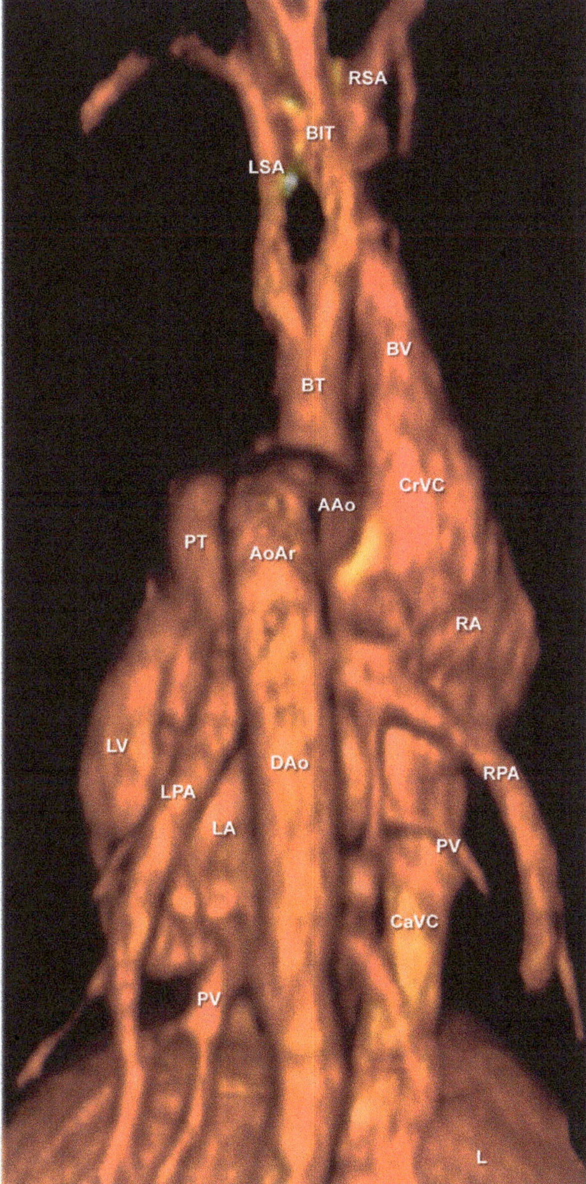

Figure 10. Three-dimensional volume-rendered reconstructed image of the normal neonatal foal heart and associated blood vessels. Dorsal aspect. Aortic arch (AoAr); ascending aorta (AAo); bicarotid trunk (BIT); brachiocephalic trunk (BT); brachiocephalic vein (BV); caudal vena cava (CaVC); cranial vena cava (CrVC); descending aorta (Dao); left atrium (LA); left pulmonary artery (LPA); left subclavian artery (LSA); left ventricle (LV); liver (L); pulmonary trunk (PT); pulmonary vein (PV); right atrium (RA); right azygos vein (RAV); right pulmonary artery (RPA); and right subclavian artery (RSA).

4. Discussion

In humans, advanced image-based diagnostic techniques, especially helical computed tomography angiography makes possible the evaluation of the cardiac and vascular thoracic structures due to its fast imaging acquisition, the acquisition of body sections from different tomographic planes, good anatomic resolution without superimposition, high contrast between different vascular structures, and excellent tissue-like differentiation [1–4]. In addition, the use of CTA allows the obtention of three-dimensional volume-rendered reconstructed images that provide excellent detail of the heart, and the arteries and veins of this region [1,5,6].

In veterinary medicine, the use of third or four generations of CT scanners has provided an excellent anatomic resolution of the thoracic structures [28,29]. In the present study, CTA images were obtained using a helical CT scanner that provided a qualitative overview of thoracic morphology, giving adequate information of midline thoracic vascular structures, a good depiction of the four chambers of the heart, as well as serving of a standard reference for the size and positions of the heart and main blood vessels. The use of a 16-slice configuration CT scanner and a similar protocol was reported in other studies performed in humans [1–6], neonatal foal [17–19], dog [30,31], cat [32,33], and goat [34].

Clinical evaluation of the equine thorax is laborious due to its anatomical complexity, which makes it difficult to diagnose diseases by physical examination and conventional diagnostic techniques. Nevertheless, advanced diagnostic techniques as CTA has shown considerable advantages over traditional imaging techniques since it gives an accurate anatomical detail of blood vessels, higher differentiation of tissue densities [30,31]. In addition, CTA is more sensitive in detecting diseases such as congenital abnormalities of the cardiovascular system including the heart, vascular malformations, injuries, tumors, aneurysms, vessels ruptures or tears, and pulmonary embolism [1–5,32].

In this research, an intravascular contrast medium administration was very helpful to identify the heart chambers, the main associated vessels, and the delineation of the adjacent non-vascular structures. In veterinary medicine, only a few studies have applied contrast-enhanced helical CT to perform anatomical or clinical studies of the thoracic cavity in dogs [31,35–37] and cats [32], as well as in other studies performed in the thorax [17–19] and abdomen [38] of foals. Nevertheless, to our knowledge, the use of intravenous contrast agents to describe the normal anatomy of the thorax in neonatal foals has not been reported before. In CT imaging, the use of an appropriate window width is a key to successful diagnosis [30,31]. In the present study, thoracic CTA images were evaluated by the use of bone, mediastinal, and lung window settings. The CT bone window provided some valuable anatomical information of the cortical and medullar marrow fat of the bones, whereas the CT mediastinal window provided an excellent detail of soft tissues, especially the heart and the major associated blood vessels. By contrast, the lung window setting gave a better definition of the respiratory tract and intrapulmonary vascular structures.

The images obtained by volume-rendered tomographic reconstruction are the most flexible 3D visualization tools [1,35–37]. In our study, the contrast CT volume-reconstructed images performed by the post-processing bone removal technique provided an excellent anatomical detail of the lateral and dorsal aspects of the heart and main associated vessels. Lateral CT reconstructed acquisitions were preferred for the evaluation of the anatomic relationships between the heart chambers and the main blood vessels, while the dorsal view was selected for identification of midline thoracic vascular structures because it yielded detailed information about the pulmonary vessels, and the main branches of the brachiocephalic trunk. Usually, motion artifacts make it difficult to identify various parts of the heart or the lungs on CT images [28]. In this study, the use of helical scanning equipment tomography on living foals minimized the artifacts generated by cardiovascular and respiratory movements. However, its use in equine medicine is currently limited because of its expense, availability, and complications of acquiring CT images in older foals and adult horses due to their physical size [17–19].

This CTA anatomic study has confirmed the valid use of cadavers to evaluate different anatomic patterns. The absence of blood flow in dead animals must be taken into account when compared with live specimens. Results from the current study showed that the use of frozen anatomical sections was helpful in the identification of different thoracic structures observed on transverse CTA images and guaranteed the matching accuracy. In addition, the identification of vascular structures of the foal thorax in the volume-rendered reconstructed CTA images were facilitated by gross anatomical dissections of the atrial and auricular surfaces of the heart. Therefore, the three foals used in this study showed cardiovascular anatomy similar to that described in the anatomical literature [26,27]. Thus, the main anatomical differences in the cardiovascular structure of equines compared to dogs such as the subclavian arteries and bicarotid trunk arising from the brachiocephalic trunk could be distinghished.

There are no previously published anatomic identifications as these reported in this study, which could be applied as an initial anatomic approximation to other CTA studies on foals. Therefore, the information provided could be used for the evaluation of CT images of foals with thoracic disease. In humans, new CT scanners have achieved improved diagnostic capabilities to evaluate a wide variety of congenital and acquired heart diseases [1–6]. With improvements in CT protocols and optimized scanners, CT angiography images will become an accurate method for evaluating the foal thorax [17–19], and in the diagnosis of several thoracic diseases described in equine medicine [20–25].

5. Conclusions

Helical CT images provided adequate detail of the thorax of normal neonatal foals and were a useful imaging modality for anatomical evaluation. This information could serve as an initial anatomic reference aid to clinicians for the diagnosis of suspected thorax-associated diseases in foals.

Author Contributions: Conceptualization, A.A. and J.R.J.; methodology, A.A., J.A.C., and M.L.D.-B.; formal analysis, A.A.; investigation, A.A., G.R., L.P., and M.M.; writing—original draft, A.A. and J.R.J.; supervision, A.A. All authors have read and agreed to the published version of the manuscript.

Funding: This research received no external funding and the University of Las Palmas de Gran Canaria funded the study.

Acknowledgments: The authors thank foals owners for the cession of the specimens for our study.

Conflicts of Interest: The authors declare no conflict of interest.

References

1. Lawler, L.P.; Fishman, E.K. Multi–detector row CT of thoracic disease with emphasis on 3D volume rendering and CT angiography. *Radiographics* **2001**, *21*, 1257–1273. [CrossRef] [PubMed]
2. Whiting, P.; Singatullina, N.; Rosser, J.H. Computed tomography of the chest: I. Basic principles. *BJA Educ.* **2015**, *15*, 299–304. [CrossRef]
3. Goo, H.W.; Park, I.S.; Ko, J.K.; Kim, Y.H.; Seo, D.M.; Yun, T.J.; Park, J.J.; Yoon, C.H. CT of congenital heart disease: Normal anatomy and typical pathologic conditions. *Radiographics* **2003**, *23*, 147–165. [CrossRef] [PubMed]
4. Kumamaru, K.K.; Hoppel, B.E.; Mather, R.T.; Rybicki, F.J. CT angiography: Current technology and clinical use. *Radiol. Clin. N. Am.* **2010**, *48*, 213–235. [CrossRef] [PubMed]
5. Kim, T.H.; Kim, Y.M.; Suh, C.H.; Cho, D.J.; Park, I.S.; Kim, W.H.; Lee, Y.T. Helical CT angiography and three-dimensional reconstruction of total anomalous pulmonary venous connections in neonates and infants. *Am. J. Roentgenol.* **2000**, *175*, 1381–1386. [CrossRef]
6. Ferretti, G.R.; Bricault, I.; Coulomb, M. Virtual tools for imaging of the thorax. *Eur. Respir. J.* **2001**, *18*, 381–392. [CrossRef] [PubMed]
7. Mazan, M.R.; Vin, R.; Hoffman, A.M. Radiographic scoring lacks predictive value in inflammatory airway disease. *Equine Vet. J.* **2005**, *37*, 541–545. [CrossRef]
8. Sprayberry, K.A. Ultrasonographic examination of the equine neonate: Thorax and abdomen. *Vet. Clin. N. Am. Equine Pract.* **2015**, *31*, 515–543. [CrossRef]
9. Marr, C.M. The equine neonatal cardiovascular system in health and disease. *Vet. Clin. N. Am. Equine Pract.* **2015**, *31*, 545–565. [CrossRef]

10. Rodríguez, M.J.; Latorre, R.; López-Albors, O.; Soler, M.; Aguirre, C.; Vázquez, J.M.; Querol, M.; Agut, A. Computed tomographic anatomy of the temporomandibular joint in the young horse. *Equine Vet. J.* **2008**, *40*, 566–571. [CrossRef]
11. Powell, S.E. Use of multi-detector computed tomographic angiography in the diagnosis of a parapharyngeal aneurysm in a 6-week-old foal. *Equine Vet. J.* **2010**, *42*, 270–273. [CrossRef] [PubMed]
12. Dahlberg, J.A.; Valdes-Martinez, A.; Boston, R.C.; Parente, E.J. Analysis of conformational variations of the cricoid cartilages in thoroughbred horses using computed tomography. *Equine Vet. J.* **2011**, *43*, 229–234. [CrossRef] [PubMed]
13. van Hoogmoed, L.; Yarbrough, T.B.; Lecouteur, R.A.; Hornof, W.J. Surgical repair of a thoracic meningocele in a foal. *Vet. Surg.* **1999**, *28*, 496–500. [CrossRef] [PubMed]
14. Puchalski, S.M.; Galuppo, L.D.; Drew, C.P.; Wisner, E.R. Use of contrast enhanced computed tomography to assess angiogenesis in deep digital flexor tendonopathy in a horse. *Vet. Radiol. Ultrasound* **2009**, *50*, 292–297. [CrossRef] [PubMed]
15. Vallance, S.A.; Bell, R.J.W.; Spriet, M.; Kass, P.H.; Puchalski, S.M. Comparisons of computed tomography, contrast enhanced computed tomography and standing low-field magnetic resonance imaging in horses with lameness localised to the foot. Part 1: Anatomic visualisation scores. *Equine Vet. J.* **2012**, *44*, 51–56. [CrossRef]
16. van Hamel, S.E.; Bergman, H.J.; Puchalski, S.M.; de Groot, M.W.; van Weeren, P.R. Contrast-enhanced computed tomographic evaluation of the deep digital flexor tendon in the equine foot compared to macroscopic and histological findings in 23 limbs. *Equine Vet. J.* **2014**, *46*, 300–305. [CrossRef]
17. Lascola, K.M.; O'Brien, R.T.; Wilkins, P.A.; Clark-Price, S.C.; Joslyn, S.K.; Hartman, S.K.; Mitchell, M.A. Qualitative and quantitative interpretation of computed tomography of the lungs in healthy neonatal foals. *Am. J. Vet. Res.* **2013**, *74*, 1239–1246. [CrossRef]
18. Lascola, K.M.; Joslyn, S. Diagnostic imaging of the lower respiratory tract in neonatal foals: Radiography and computed Tomography. *Vet. Clin. N. Am. Equine Pract.* **2015**, *31*, 497–514. [CrossRef]
19. Schliewert, E.C.; Lascola, K.M.; O'Brien, R.T.; Clark-Price, S.C.; Wilkins, P.A.; Foreman, J.H.; Mitchell, M.A.; Hartman, S.K.; Kline, K.H. Comparison of radiographic and computed tomographic images of the lungs in healthy neonatal foals. *Am. J. Vet. Res.* **2015**, *76*, 42–52. [CrossRef]
20. Marble, S.L.; Edens, L.M.; Shiroma, J.T.; Savage, C.J. Subcutaneous emphysema in a neonatal foal. *J. Am. Vet. Med. Assoc.* **1996**, *208*, 97–99.
21. Jean, D.; Laverty, S.; Halley, J.; Hannigan, D.; Léveillé, R. Thoracic trauma in newborn foals. *Equine Vet. J.* **1999**, *31*, 149–152. [CrossRef] [PubMed]
22. Maleski, K.; Magdesian, K.G.; LaFranco-Scheuch, L.; Pappagianis, D.; Carlson, G.P. Pulmonary coccidioidomycosis in a neonatal foal. *Vet. Rec.* **2002**, *151*, 505–508. [CrossRef] [PubMed]
23. Sanz, M.; Loynachan, A.; Sun, L.; Oliveira, A.; Breheny, P.; Horohov, D.W. The effect of bacterial dose and foal age at challenge on Rhodococcus equi infection. *Vet. Microbiol.* **2013**, *167*, 623–631. [CrossRef] [PubMed]
24. Hinchcliff, K.W.; Adams, W.M. Critical pulmonary stenosis in a newborn foal. *Equine Vet. J.* **1991**, *23*, 318–320. [CrossRef] [PubMed]
25. Chaffin, M.K.; Miller, M.W.; Morris, E.L. Double outlet right ventricle and other associated congenital cardiac anomalies in an American Miniature Horse foal. *Equine Vet. J.* **1992**, *24*, 402–406. [CrossRef] [PubMed]
26. World Association of Veterinary Anatomists. Arthrologia. In *Nomina Anatomica Veterinaria*, 6th ed.; International Committee on Veterinary Gross Anatomical Nomenclature: Hanover, NH, USA, 2005; pp. 73–75.
27. Schaller, O. Arthrologia. In *Illustrated Veterinary Anatomical Nomenclature*, 2nd ed.; Enke: Sttutgart, Germany; Ghent, Belgium; Columbia, MO, USA; Rio de Janeiro, Brazil, 2007; pp. 234–244.
28. Smallwood, J.E.; George, T.F. Anatomic atlas for computed tomography in the mesaticephalic dog: Thorax and cranial abdomen. *Vet. Radiol. Ultrasound* **1993**, *34*, 65–84. [CrossRef]
29. Samii, V.; Biller, D.; Koblik, P. Normal cross-sectional anatomy of the feline thorax and abdomen: Comparison of computed tomography and cadaver anatomy. *Vet. Radiol. Ultrasound* **1998**, *39*, 504–511. [CrossRef]
30. De Rycke, L.M.; Gielen, I.M.; Simoens, P.J.; van Bree, H. Computed tomography and cross-sectional anatomy of the thorax in clinically normal dogs. *Am. J. Vet. Res.* **2005**, *66*, 512–524. [CrossRef]

31. Cardoso, L.; Gil, F.; Ramírez, G.; Teixeira, M.A.; Agut, A.; Rivero, M.A.; Arencibia, A.; Vázquez, J.M. Computed tomography (CT) of the lungs of the dog using a helical CT scanner, intravenous iodine contrast medium and different CT windows. *Anat. Histol. Embryol.* **2007**, *36*, 328–331. [CrossRef]
32. Henninger, W. Use of computed tomography in the diseased feline thorax. *J. Small Anim. Pract.* **2003**, *44*, 56–64. [CrossRef]
33. Rodriguez, K.T.; O'Brien, M.A.; Hartman, S.K.; Mulherin, A.C.; McReynolds, C.J.; McMichael, M.; Rapoport, G.; O´Brien, R. Microdose computed tomographic cardiac angiography in normal cats. *J. Vet. Cardiol.* **2014**, *16*, 19–25. [CrossRef] [PubMed]
34. Ohlerth, S.; Becker-Birck, M.; Augsburger, H.; Jud, R.; Makara, M.; Braun, U. Computed tomography measurements of thoracic structures in 26 clinically normal goats. *Res. Vet. Sci.* **2013**, *92*, 7–12. [CrossRef] [PubMed]
35. Ohlerth, S.; Scharf, G. Computed tomography in small animals—Basic principles and state of the art applications. *Vet. J.* **2007**, *173*, 254–271. [CrossRef] [PubMed]
36. Bertolini, G.; Prokop, M. Multidetector-row computed tomography: Technical basics and preliminary clinical applications in small animals. *Vet. J.* **2011**, *189*, 15–26. [CrossRef]
37. Brewer, F.C.; Sydney Moise, N.; Kornreich, B.G.; Bezuidenhout, A.J. Use of computed tomography and silicon endocasts to identify pulmonary veins with echocardiography. *J. Vet. Cardiol.* **2012**, *14*, 293–300. [CrossRef]
38. Willems, D.S.; Kranenburg, L.C.; Ensink, J.M.; Kummeling, A.; Wijnberg, I.D.; Veraa, S. Computed tomography angiography of a congenital extrahepatic splenocaval shunt in a foal. *Acta Vet. Scand.* **2019**, *61*, 39. [CrossRef]

MDPI

Article

Morphometrical Study of the Lumbar Segment of the Internal Vertebral Venous Plexus in Dogs: A Contrast CT-Based Study

Valeria Ariete [1], Natalia Barnert [1], Marcelo Gómez [1,*], Marcelo Mieres [2], Bárbara Pérez [1] and Juan Claudio Gutierrez [3]

[1] Institute of Pharmacology and Morphophysiology, Austral University of Chile, Valdivia 5090000, Chile; valeria.ariete@gmail.com (V.A.); maildeatty@gmail.com (N.B.); barbaraperez@uach.cl (B.P.)
[2] Institute of Veterinary Clinical Sciences, Austral University of Chile, Valdivia 5090000, Chile; mmieres@uach.cl
[3] Department of Anatomy, Physiology and Cell Biology, School of Veterinary Medicine, University of California, Davis, CA 95616, USA; jcgutier@ucdavis.edu
* Correspondence: marcelogomez@uach.cl; Tel.: +56-632221216 or +56-984437732

Simple Summary: The internal vertebral venous plexus (IVVP) is a valveless venous network running inside the vertebral canal. The objective of this study was to morphometrically describe the IVVP, dural sac, epidural space and vertebral canal of the lumbar segment in dogs with enhanced computerized tomography. Six clinically healthy adult dogs were used for the study. Dorsal reconstructed computed tomography (CT) images showed a continuous rhomboidal morphological pattern for the IVVP. The dural sac was observed as an isodense structure with a rounded shape throughout the vertebral canal. The average percentage area occupied by the IVVP between L1 and L7 vertebrae ranged between 6.3% and 8.9% of the area of the vertebral canal, and the dural sac ranged between 13.8% and 72.2% of the vertebral canal. The epidural space accounted between 27.08% and 86.2% of the lumbar vertebral canal. CT venography is a safe technique that allows adequate visualization and evaluation of the lumbar IVVP and adjacent structures in dogs.

Abstract: The internal vertebral venous plexus (IVVP) is a thin-walled, valveless venous network that is located inside the vertebral canal, communicating with the cerebral venous sinuses. The objective of this study was to perform a morphometric analysis of the IVVP, dural sac, epidural space and vertebral canal between the L1 and L7 vertebrae with contrast-enhanced computed tomography (CT). Six clinically healthy adult dogs weighing between 12 kg to 28 kg were used in the study. The CT venographic protocol consisted of a manual injection of 880 mgI/kg of contrast agent (587 mgI/kg in a bolus and 293 mgI/mL by continuous infusion). In all CT images, the dimensions of the IVVP, dural sac, and vertebral canal were collected. Dorsal reconstruction CT images showed a continuous rhomboidal morphological pattern for the IVVP. The dural sac was observed as a rounded isodense structure throughout the vertebral canal. The average area of the IVVP ranged from 0.61 to 0.74 mm^2 between L1 and L7 vertebrae (6.3–8.9% of the vertebral canal), and the area of the dural sac was between 1.22 and 7.42 mm^2 (13.8–72.2% of the vertebral canal). The area of the epidural space between L1 and L7 ranged from 2.85 to 7.78 mm^2 (27.8–86.2% of the vertebral canal). This CT venography protocol is a safe method that allows adequate visualization and morphometric evaluation of the IVVP and adjacent structures.

Keywords: internal vertebral venous plexus; computed tomography; canine

Citation: Ariete, V.; Barnert, N.; Gómez, M.; Mieres, M.; Pérez, B.; Gutierrez, J.C. Morphometrical Study of the Lumbar Segment of the Internal Vertebral Venous Plexus in Dogs: A Contrast CT-Based Study. *Animals* **2021**, *11*, 1502. https://doi.org/10.3390/ani11061502

Academic Editors: Matilde Lombardero and Mar Yllera Fernández

Received: 18 March 2021
Accepted: 19 May 2021
Published: 22 May 2021

1. Introduction

The vertebral venous plexus (VVP) is a thin-walled, valveless venous network that surrounds the entire length of the vertebral column, terminating at the cephalad end in the cerebral venous sinuses [1,2]. According to its position inside or outside of the vertebral canal, the vertebral venous plexus can be divided into three intercommunicating divisions:

the internal vertebral venous plexus, external vertebral venous plexus, and basivertebral veins [1]. The internal vertebral venous plexus (*plexus vertebralis internus ventralis*) (IVVP) is also known as the longitudinal venous sinus, vertebral venous sinuses, epidural venous plexus, paravertebral veins, and meningorachidean plexus [3]. The IVVP lies within the vertebral canal, inside the epidural space, and along the dorsal surface of the vertebral bodies and intervertebral disks [1,2] (Figure 1). Cranially, the IVVP communicates with the basilar sinuses at the level of the foramen magnum, and caudally, it extends to the fourth or sixth caudal vertebra [1]. The VVP drains blood from the vertebral column, the paravertebral musculature, the spinal cord, the meninges, and the nerve roots of the spinal nerves [3]. In addition to its normal drainage function, this vascular network can also be a collateral pathway for blood return towards the heart in cases of occlusion or ligation of the caval venous system [4]. Clinically, in humans and animals, rupture of the IVVP has been associated with the etiology of spontaneous spinal epidural hematomas [5,6]. Doberman pinschers with a deficiency of von Willebrand factor (VIII-related antigen) can develop significant spinal epidural hemorrhage due to laceration of the IVVP, which results in progressive neurological deficits [7]. Following Hansen type I intervertebral disk herniation, the extruded nucleus pulposus can be expelled laterally and rupture the IVVP. This can result in bleeding, hematoma formation, and subsequent extradural spinal cord compression [8]. The continuity of the valveless IVVP throughout the length of the neuroaxis enables bacteria and tumor cells to travel from the thorax, abdomen, and pelvis to the head and vertebral column when intrathoracic or intra-abdominal pressure is increased [9]. This phenomenon has been termed paradoxical embolism, retrograde venous invasion, and Batson's phenomenon [9]. In dogs, metastasis of osteosarcomas and pheochromocytomas into the central nervous system is hypothesized to occur by retrograde venous spread through the VVP. Diskospondylitis, vertebral osteomyelitis, and infectious diskitis in dogs with primary sites of infection elsewhere in the body have been explained by this mechanism [10]. The IVVP may participate in the etiology of fibrocartilaginous embolism of the spinal cord vasculature (also known as embolic myelopathy) in dogs [11]. In humans, the IVVP also participates in the pathophysiology of other spinal cord vascular lesions, such as arteriovenous fistulas and venous malformations [12]. Recently, venous aneurism of the IVVP in a Scottish Deerhound was reported concurrently with severe dilatation of the venous sinuses [13]. Additionally, case reports in dogs have described abnormal enlargement of the IVVP in sighthounds by magnetic resonance imaging (MRI) with clinical signs of radiculopathies and myelopathy [14].

Understanding anatomical vertebral morphometry is an important factor when interpreting the pathological changes associated with several structural spinal diseases. The aims of the present study were to provide an in vivo morphometric description of the normal lumbar IVVP and surrounding vertebral structures using contrast-enhanced computed tomography (CT) in dogs.

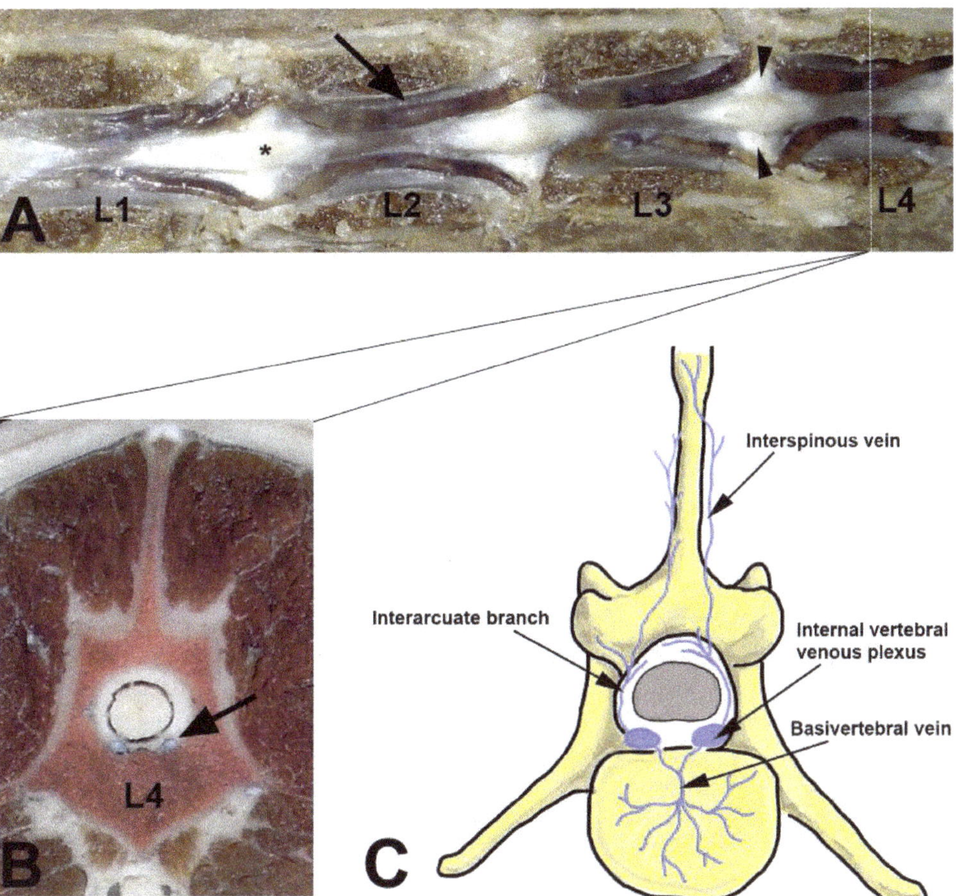

Figure 1. Lumbar internal vertebral venous plexus (IVVP) in a dog. (**A**) Dorsal anatomic view in which the vertebral arch and spinal cord was removed between L1 and L4. The IVVP is observed over vertebral bodies (arrow), intervertebral disk (arrow heads), and dorsolateral to the dorsal longitudinal ligament (asterix), with a typical rhomboidal appearance (arrow). (**B**) Transverse anatomic view of fourth lumbar vertebra where the IVVP (arrow) is located ventrally in the spinal epidural space and dorsal to the L4 vertebral body. (**C**) Schematic representation of some components of the vertebral venous plexus in the dog (Photographs correspond to previous dissections made by the author that do not correspond to the animals examined in this study).

2. Materials and Methods

2.1. Biological Material

For the study, 6 healthy adult dogs were used. The age range was between 2 and 6 years old, and the weight range between 12 and 28 kg, with an average weight of 19.5 kg. The dog breed included 1 Boxer, and 5 mixed-breed, with 3 males and 3 females. Owners provided written consent for research purposes via an authorization form. The experimental study was carried out according to the protocols proposed by the Bioethics Committee of the Austral University of Chile and all procedures were monitored throughout the study by a veterinarian

2.2. Anaesthesia Protocol

All CT studies were performed under general anesthesia with the patient positioned in dorsal recumbency. For the study, dogs were food-deprived for 12 h prior to the CT scan. Each dog was premedicated with 0.5 mg/kg of intramuscular xylazine (2% Xylazine, Xylavet 2% Lab. Alfasan, Worden, Holland). Then, animals were cannulated with a 20 G cannula in the cephalic vein to administer a contrast medium in a continuous infusion of saline (0.9% NaCl) in a volume of 10 mg/kg/h. Anesthesia was induced by intravenous administration of propofol (Propofol 1%, PropoFlo; Zoetis, NJ, USA) at a dose of 4 µg/kg and maintained with 2% isoflurane in oxygen.

2.3. CT Venography Protocol

For the study, a four-generation CT scanner (Picker PQ 6000, Picker International, Cleveland, OH, USA) was used. For imaging, dogs were positioned in dorsal recumbency with flexed hindlimbs over the CT table. Contiguous CT slices were obtained before and after the application of a nonionic iodinated contrast medium (Iohexol, Omnipaque® 300 mg of iodine/mL, Nycomed Inc, Princeton, NJ, USA) in a total dose of 880 mg/kg IV, applying 587 mg I/kg by bolus, and flow rate of 4 mL/s before the scan. Infusion of an additional 293 mg I/kg as continuous infusion during the scan was administered by means of a constant-rate infusion pump with a flow rate of 0.1 mL/min. Technical parameters included: Slice thickness of 2 mm and an interval of 2 mm and a scan field of view of 180 cm. Settings of the CT scanner included standard resolution, 100 mA and 130 kV per slice. The study area included from L1 to L7 vertebrae. The CT gantry was tilted such that the sections were parallel to intervertebral spaces. All dogs recovered routinely from anesthesia and were clinically normal at 24 and 48 h post CT examination.

2.4. Morphometric Analysis

The CT images were transferred to a computer for morphometric assessment. Measurements were obtained from CT transverse images using standardized soft tissue window settings (window width WW: 250–450; window level WL: 30–50). All measurements were performed between L1 to L7 at mid-vertebral body level. This segment was chosen due to the scarce morphometric information of components of the vertebral canal at this level in dogs. All image data were imported and recorded into the Osirix® DICOM viewer (Pixmeo Inc., Version 3.9.4., 32 Bit, Bernex, Switzerland), a software for morphometric analysis. Anatomical parameters were measured at the mid-level of the vertebral body of the lumbar segment (L1–L7). Measurements were taken in millimeters in the axial planes and were collected three times by the same observer. The following anatomical parameters were determined considering the anatomical structures shown in Figure 2:

- cross sectional area of the vertebral canal (mm^2);
- cross sectional area of the dural sac (mm^2);
- total area of the IVVP (area of the right and left IVVP component) (mm^2);
- area of the epidural space: vertebral canal area minus the area occupied by the dural sac (mm^2);
- percentage of the IVVP occupying the vertebral canal;
- percentage of the dural sac occupying the vertebral canal;
- percentage of the IVVP within the epidural space;

2.5. Statistical Analysis

Data are presented as the mean ± standard deviation and were analyzed using STATGRAPHICS (Centurion XVI version 16.1.12, 32 bits, StatPoint Inc., Rockville, MD, USA) software.

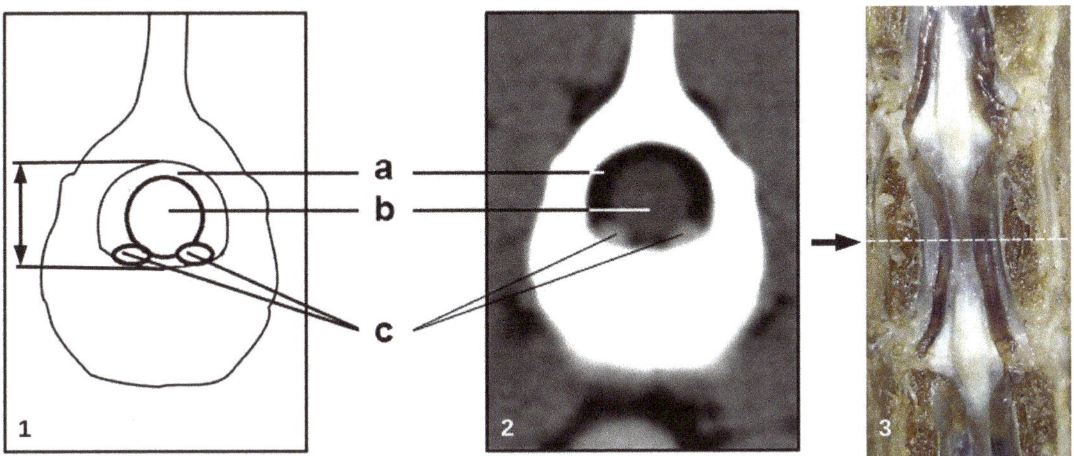

Figure 2. (**1**) Illustration and (**2**) transverse post-contrast CT image at (**3**) the level of L2 vertebra in an adult dog illustrating the measured anatomical parameters. The broken line in the photograph indicates the level of the transverse section where the measurements were performed in the L1-L7 vertebral segment. (**a**) Epidural space, (**b**) dural sac, (**c**) internal vertebral venous plexus.

3. Results

In all six dogs, appropriate opacification of the vertebral venous component was achieved in all of them. Nonselective CT venography developed in this study allowed adequate visualization and measurement of the IVVP and associated vertebral structures. The contrast injection protocol based on nonionic iodinated contrast medium in a total dose of 880 mg I/kg, 2/3 of which was delivered as an initial bolus before CT image acquisition and 1/3 of which was delivered in the form of a constant-rate infusion pump, resulted in an IVVP average density of 215 UH. Image acquisition time lasted between 25 and 35 min.

3.1. Internal Vertebral Venous Plexus

The internal vertebral venous plexus (IVVP) was observed as two symmetric longitudinal hyperdense channels, oval in shape, located in the epidural space, dorsal to the vertebral bodies, and ventral to the dural sac (Figure 2). In dorsal reconstruction CT images, it was possible to appreciate their rhomboidal distribution within the vertebral canal (Figure 3). Additionally, CT transverse images obtained at the level of the middle portion of the vertebral body allowed visualization of the basivertebral veins and tributaries of the IVVP (Figure 4). The IVVP between the L1 and L7 vertebrae had an average area of 0.66 mm^2, with a minimum value recorded of 0.43 mm^{2,} and the maximum of 1.01 mm^2. In vertebra L7, the IVVP presented the lowest average with a 0.61 mm^2 (±0.1) area, and L5 vertebrae presented the highest IVVP average area with a 0.74 mm^2 (±0.2) area (Table 1, Figure 5). The IVVP occupies an average of 7.67% of the L1–L7 vertebral canal. In the L1 vertebra, there was the highest percentage of the IVVP in relation to the vertebral canal, with 8.9%, while in the L4 vertebra, IVVP occupied the lower percentage, with 6.3% of the vertebral canal (Table 2).

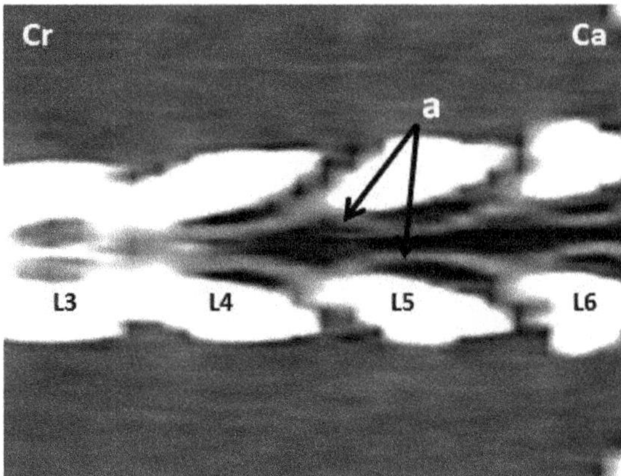

Figure 3. Dorsal reconstruction post-contrast CT images at the L3–L6 lumbar level in an adult medium size dog. (**a**) Intervertebral venous plexus, (**cr**) cranial, (**ca**) caudal.

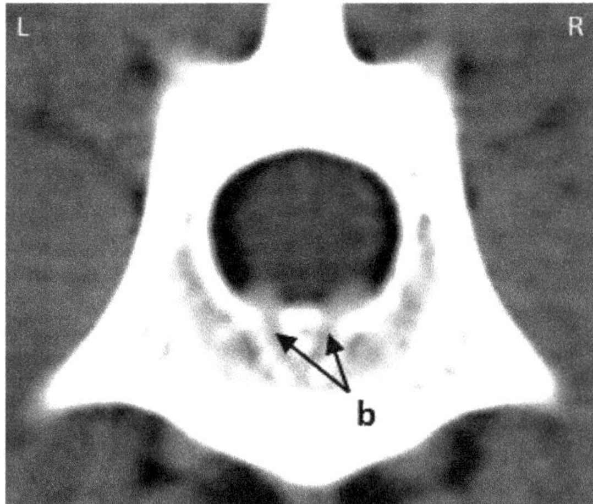

Figure 4. Transverse post-contrast CT image of the mid-portion of the L3 vertebral body from an adult medium size dog. (**b**) basivertebral veins, (**L**) left, (**R**) right.

Table 1. Average area (mm^2) and standard deviation of the IVVP, vertebral canal, and dural sac and epidural space between the L1 and L7 vertebral segments in six adult medium size dogs.

Morphometric Measures	Vertebral Segment						
	L1	L2	L3	L4	L5	L6	L7
Vertebral canal	7.87 ± 1.3	8.24 ± 1.8	9.21 ± 1.7	10.27 ± 1.9	9.77 ± 1.7	8.68 ± 1.2	8.28 ± 2
Dural sac	4.67 ± 1.1	5.04 ± 1.2	6.34 ± 1.3	7.42 ± 1.6	5.04 ± 0.9	1.78 ± 0.4	1.22 ± 0.1
IVVP	0.7 ± 0.2	0.66 ± 0.1	0.64 ± 0.1	0.63 ± 0.1	0.74 ± 0.2	0.66 ± 0.2	0.61 ± 0.1
Epidural space	3.2 ± 0.6	3.2 ± 0.9	2.87 ± 0.7	2.85 ± 0.8	4.72 ± 1.6	6.9 ± 0.8	7.78 ± 1.6

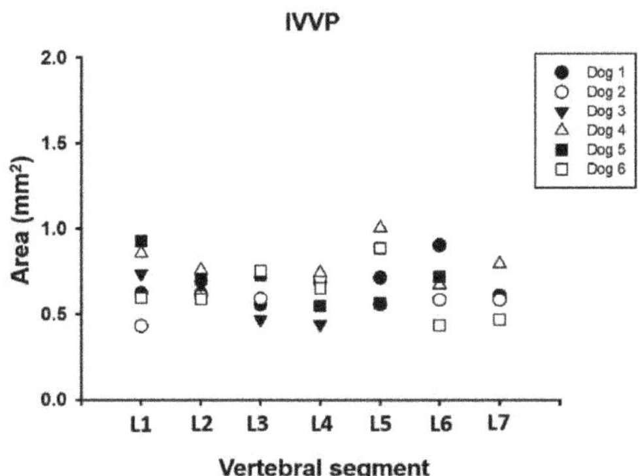

Figure 5. Area dimension of the internal vertebral venous plexus (IVVP) between L1 and L7 vertebral segment in six adult medium size dogs.

Table 2. Average percentage (%), minimum and maximum of the vertebral canal corresponding to the dural sac, internal vertebral venous plexus (IVVP), and epidural space between vertebrae L1 and L7 in six adult dogs.

Morphometric Measures	Vertebral Segment						
	L1	L2	L3	L4	L5	L6	L7
Dural sac	59 (47.7–67,4)	61.3 (49.1–69.4)	68.8 (58.8–76.6)	72.2 (62.6–77.8)	52.6 (46.8–75.3)	20.4 (17–21.2)	13.8 (11.3–15.7)
IVVP	8.9 (6.4–11.8)	8.3 (5.5–10.3)	7.1 (4.8–8.7)	6.3 (4.8–8.8)	7.4 (5.3–10.8)	7.7 (4.5–9.5)	7.9 (4.5–11.2)
Epidural space	41 (32.6–52.3)	38.7 (30.6–50.9)	31.2 (23.4–41.2)	27.8 (20.8–37.4)	47.4 (24.7–58.9)	79.6 (75.6–83)	86.2 (84.3–88.7)

3.2. Vertebral Canal

The vertebral canal measured an average of 8.94 mm^2 (±1.7) between the L1 and L7 vertebral segments, the minimum value recorded was 6.14 mm^2, and the maximum was 12.92 mm^2. Vertebra L1 had the lowest recorded value with a 7.87 mm^2 (±1.3) area, while the L4 vertebra had the highest value with a 10.27 mm^2 (±1.9) area (Table 1, Figure 6).

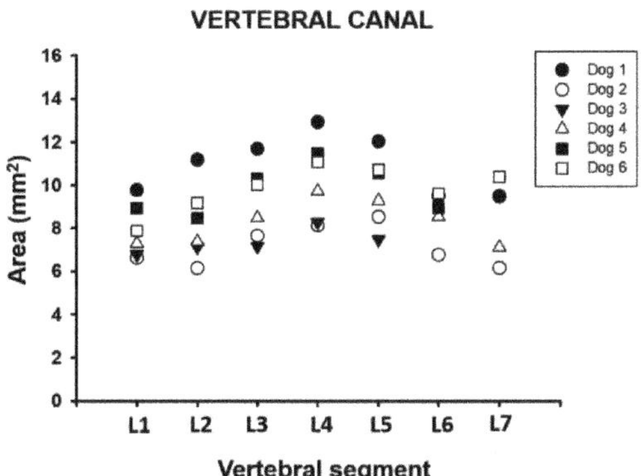

Figure 6. Area dimension of the vertebral canal between L1 and L7 vertebral segment in six adult medium size dogs.

3.3. Dural Sac

The dural sac had an average area of 4.83 mm^2 (±2.1) between vertebral segments L1 and L7. The minimum recorded value was 1.11 mm^2, and the maximum value was 10.05 mm^2. The dural sac in the L7 vertebra had the lowest value with a 1.22 mm^2 (±0.1) area, and the L4 vertebra had the highest average with a 7.42 mm^2 (±1.6) area (Table 1,

Figure 7). The dural sac occupied between 13.8% and 72.2% of the vertebral canal between the L1 and L7 vertebral segments in these dogs (Table 2). It is also possible to appreciate that there is an increase in the occupancy rate toward L4, where it reaches its maximum value, and then it decreases toward L7, where the lowest average value is recorded.

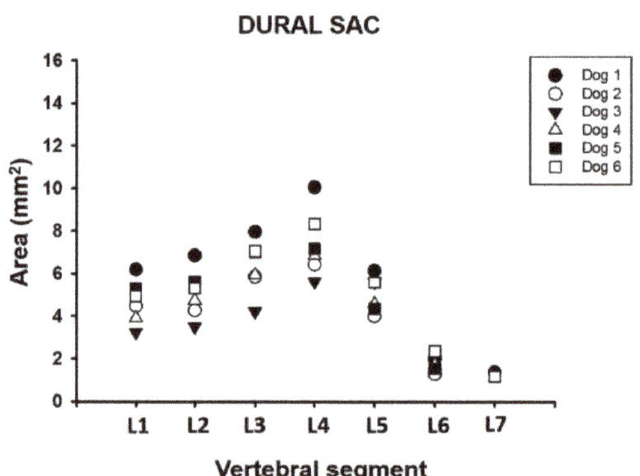

Figure 7. Area dimension of the dural sac between L1 and L7 vertebral segment in six adult medium size dogs.

3.4. Epidural Space

The average area of the epidural space between segments L1 and L7 ranged from 2.85 (±0.8) to 7.78 (±1.6) mm^2, corresponding to 27.8% to 86.2% of the vertebral canal (Tables 1 and 2, Figure 8).

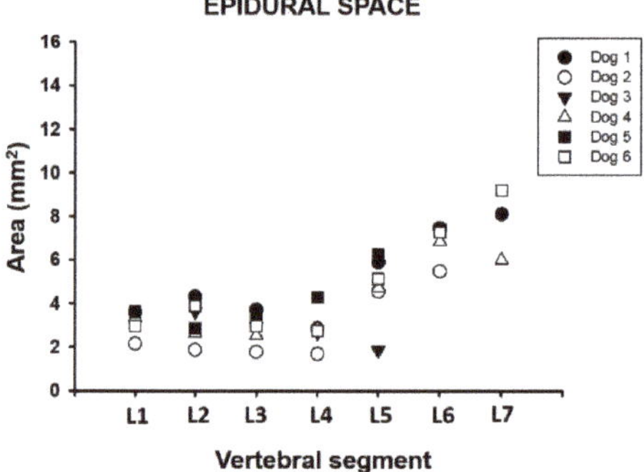

Figure 8. Area dimension of the epidural space between L1 and L7 vertebral segment in six adult medium size dogs.

4. Discussion

The nonselective CT venography technique developed in this study allowed adequate visualization and measurement of the lumbar IVVP and associated vertebral structures in a group of dogs. The contrast injection protocol based on nonionic iodinated contrast medium in a total dose of 880 mg I/kg, 2/3 of which was delivered as an initial bolus before CT image acquisition and 1/3 of which was delivered in the form of constant-rate infusion during CT acquisition, resulted in an IVVP density of 215 UH. According to different studies, for optimal vasculature viewing in CT examinations, it is necessary to use a window level ranging between 200 and 400 UH and a window width ranging between 200 and 2000 UH, depending on the vascular concentration of contrast medium [15]. There are various doses and administration protocols for contrast administration published for

CT venography in dogs. There have been protocols with values ranging from 720 mg I/kg to 814 mg I/kg in different studies, including the study of IVVP cervical level, diagnosis of portosystemic shunts, multiple vena cava anomalies, pancreatic TC analysis, and evaluation of pulmonary embolisms [16–21]. Although high doses of contrast medium were used in this CT venographic study, no adverse effects were observed in all animals examined. However, it is necessary to consider that large doses of contrast medium in angiographic studies may be associated with major risk of contrast agent hypersensitivity and renal insufficiency [22].

The IVVP appearance on the CT images consisted of two symmetrical oval hyperatenuated structures with defined margins located on the floor of the vertebral canal (Figure 2). The description observed in this study evidenced a rhomboidal distribution with a metameric organization of the venous system consistent with previous publications in dogs [1,17,18]. Regarding morphometry of the IVVP, the average area measurements for the lumbar segment (from 0.61 to 0.74 mm^2) were lower than values published previously for the cervical segment in dogs, where an average area of 1.33 mm^2 was recorded in similarly-sized dogs [18]. In addition, the progressive cranio-caudal decrease in the diameter of the IVVP described previously was also observed in previous studies [3,18]. The lumbar IVVP constituted, on average, 7.67% of the area of the vertebral canal and 19.69% of the epidural space, lower values than those reported for the cervical segment values, where these vessels occupy 12.4% of the vertebral canal and 30.61% of the epidural space [18]. These venous trunks tend to decrease in size gradually from vertebrae T4 to L7, away from each other at intervertebral spaces and converging in the center of the vertebral bodies. The size of the IVVP in the L5–L7 segment was smaller and variable, therefore values of the structures in this area should be taken with caution.

In humans, the IVVP is smaller in the cervical area, being bulkier and pronounced as it descends through the vertebral canal, reaching a maximum size at the level of the L4–L5 vertebrae [5]. This difference is most likely related to the postural difference between humans and canids [5]. Dissection studies in humans determined that the anterior IVVP, equivalent to the ventral IVVP in dogs, has a mean length of 103 mm (range 43–153 mm) and a maximum width of 5.8 mm. The maximum caliber of the anterior IVVP in humans is at the thoracolumbar segments (mean maximum width 7.2 $\pm$ 2.0 mm) [22].

Previous anatomical descriptions of the IVVP have been performed for humans, dogs, domestic cats, rodents, and even reptiles using various imaging modalities ranging from conventional radiography to more modern techniques, such as magnetic resonance imaging (MRI) and CT [17,18,23–26]. In vivo imaging studies are more appropriate to analyze the vascular anatomy in animals because of vein collapse in cadaver specimens.

In humans, epidural venous engorgement in the lumbar spine, including epidural varices, may cause pain in the lower back and radiculopathy secondary to nerve root impingement [27]. The diagnosis of IVVP engorgement is often mistaken for a herniated disc on radiologic interpretation, and the true diagnosis is finally made in the surgical intervention. In dogs, vertebral venous system abnormalities have been identified on MRI in 12% of all sighthounds, but the underlying cause is unknown [14]. Recently, severe dilation of the right IVVP causing significant compression of the spinal cord and nerve roots was identified in an adult Scottish Deerhound [13]. Bilateral vertebral venous sinus thrombosis and dilation causing cervical spinal cord compression were also recently detected in a 10-year-old male mixed dog [28]. In another study, the visibility of the IVVP was significantly ($p < 0.001$) different between Great Danes with and without clinical signs of cervical spondylomyelopathy [29]. Variation in the size of the IVVP in the central nervous system (CNS) can be explained by the Monro-Kellie doctrine, which establishes an inverse relationship between cerebrospinal fluid (CSF) volume and intracranial blood volume [30]. Hence, as CSF is removed from the intracranial compartment, more blood enters the intracranial compartment. Because the intracranial compartment and vertebral canal are nearly a closed system, the principles of the Monro-Kellie doctrine can be extended to the vertebral canal. As expected, the size of the dural sac increased with increased intracranial

CSF or blood volume and decreased with decreased intracranial CSF or blood volume [30]. In the brain, the dura mater is closely opposed by bone, whereas in the vertebral canal, the dura mater has the IVVP and fat separating it from the fixed bony vertebral canal. As the epidural fat is constant and likely has little movement in and out of the neural foramen, the IVVP is most likely to change in size with changes in dural sac volume [30].

In relation to the dural sac, the increase in the average area recorded at the L4 vertebra level is associated with medullary expansion of the L4-S3 segment, known as lumbar enlargement (*intumescentia lumbalis*) [1]. From this lumbar enlargement, the nervous roots of the lumbosacral plexus emerge, which can be seen in the transverse images from the L5 vertebra caudal. The dural sac area values recorded for each vertebral segment, with the exception of L4, are smaller than those published by Gómez et al. (2005) for the cervical segment in dogs, where an average area of 6.40 + 1.0 mm^2 was recorded. A CT study of the dural sac in humans indicated that between the L4 and L5 segments, the average area of the dural sac was >100 mm^2, which was higher than that recorded in the present study for the same vertebral segment in dogs [31]. In humans, a dural sac cross-sectional area < 100 mm^2 and dural sac anteroposterior diameter <10 mm are frequently considered to assess the severity of spinal canal stenosis [31,32]. The percentage of the cross-sectional area of the dural sac found in our study was between 50 and 72% of the vertebral canal between L1 and L4. These values were similar to those reported by Lim et al. (2018) [33], with values of 64% in a group of 12 normal Beagle dogs. In this study, a significant reduction in the ratio cross-sectional area of the dural sac (40%) in patients with different spinal disorders (IVDD, spinal tumors, hematomas, etc.) to that of the vertebral canal was observed [33].

The vertebral canal area measured 8.94 mm^2 between the L1 and L7 vertebral segments on average. Similar results were observed in a morphometric CT study of the thoracic spine performed in 13 German Shepherd dogs for the T2–T13 vertebral level (8.36 $\pm$ 4.03 mm^2) [34]. A decrease in vertebral canal diameter and/or area that results in compression of spinal cord and/or nerve roots is termed absolute stenosis, whereas a diameter that is less than normal but does not cause compression of neural elements is termed relative stenosis [35,36]. Relative vertebral canal stenosis results in decreased available space for the spinal cord to compensate for extradural space-occupying conditions. Relative vertebral canal stenosis therefore predisposes animals to develop clinical signs when relatively mild space-occupying pathologies, such as age-related intervertebral disc protrusion or ligamentous hypertrophy, occur [35,36].

The percentage of the vertebral canal of the spinal epidural space was minimal between L1 and L4 (27–41%), indicating that these segments are more subject to clinically significant epidural compressive lesions and that L5–L7 had a greater amount of vertebral canal available for the dural sac (47–87%). The large size of the epidural space between L5 and L7 also provides a secure space for epidural instruments or devices (i.e., catheters) in this region. The spinal epidural space surrounds the dural sac and is limited dorsally by the epidural fat, ligamentum flavum and periosteum, ventrally by the dorsal longitudinal ligament, IVVP and vertebral bodies, and laterally bordered by the vertebral pedicles and intervertebral foramina [1]. In MRI and CT imaging studies, severe stenosis is usually associated with subjective signs of the absence of epidural fat [37–39]. There are no quantitative studies using CT images that evaluated the dimensions or proportion of the lumbar epidural space in dogs.

Limitations of the present study were associated with the small number of dogs included and the variety of dogs of different sizes evaluated. Further studies to increase the study population and standardize various breeds to allow analysis by breed (small, medium, large, and giant) size are necessary. These data should thus be interpreted with caution given that measurements were only acquired at the mid-vertebral level and did not include the intervertebral cross-sectional area parameters.

5. Conclusions

This preliminary study provides reference values of the lumbar IVVP and adjacent structures in a group of dogs. The IVVP constitutes a complex vascular network of the vertebral column that has recently received attention in veterinary medicine. The understanding of its anatomy and morphometry is necessary to adequately diagnose new pathological entities that involve this venous plexus and in preoperative imaging evaluation of the venous morphometry may be useful to avoid complications related with vascular structures.

Author Contributions: Conceptualization, V.A., N.B., M.G., and M.M.; methodology, V.A., N.B., M.G., and M.M.; investigation, V.A., M.G., B.P., and J.C.G.; resources, M.G., M.M., B.P., and J.C.G.; writing—original draft preparation, M.G., B.P., and J.C.G.; and writing—review and editing, V.A., N.B., M.G., B.P., and J.C.G. All authors have read and agreed to the published version of the manuscript.

Funding: This research received no external funding.

Institutional Review Board Statement: The study was conducted according to the guidelines of the Declaration of Helsinki, and approved by the Comité de Bioética "Uso de Animales en la Investigación" from Universidad Austral de Chile (UACh), code 22-2011.

Informed Consent Statement: Informed consent was obtained from all dog owners.

Conflicts of Interest: The authors declare no conflict of interest.

References

1. Evans, H. *Miller's Anatomy of the Dog*, 3rd ed.; WB Saunders Company: Philadelphia, PA, USA, 1993; pp. 713–715.
2. International Committee on Veterinary Gross Anatomical Nomenclature; World Association of Veterinary Anatomists (W.A.V.A.). *Nomina Anatomica Veterinaria (2012)*; Editorial Committee: Hannover, Germany; Columbia, MO, USA; Ghent, Belgium; Sapporo, Japan, 2012.
3. Gómez, M.; Freeman, L. Revisión del plexo venosos vertebral en el perro. *Int. J. Morphol.* **2003**, *21*, 237–244.
4. Parke, W. *Applied Anatomy of the Spine*, 3rd ed.; WB Saunders Company: Philadelphia, PA, USA, 1992; pp. 35–88.
5. Groen, R.J.; Groenewegen, H.J.; van Alphen, H.A.; Hoogland, P.V. Morphology of the human internal vertebral venous plexus: A cadaver study after intravenous Araldite CY 221 injection. *Anat. Rec.* **1997**, *249*, 285–294. [CrossRef]
6. Kreppel, D.A.G.; Seeling, W. Spinal hematoma: A literatura survey with meta-analysis of 613 patients. *Neurosurg. Rev.* **2003**, *26*, 1–49. [CrossRef] [PubMed]
7. Applewhite, A.A.; Wilkens, B.E.; McDonald, D.E.; Radasch, R.M.; Barstad, R.D. Potential central nervous system complications of von Willebrand's disease. *J. Am. Anim. Hosp. Assoc.* **1999**, *35*, 423–429. [CrossRef] [PubMed]
8. Amsellem, P.; Toombs, J.; Laverty, P.; Breur, G. Loss of deep pain sensation following thoracolumbar intervertebral disk herniation in dogs: Pathophysiology. *Compend. Contin. Educ. Pract. Vet.* **2003**, *25*, 256–264.
9. King, A.S. *Physiological and Clinical Anatomy of the Domestic Mammals*; Oxford University Press: Oxford, UK, 1987; pp. 24–30.
10. Thomas, W.B. Diskospondylitis and other vertebral infections. *Vet. Clin. N. Am. Small Anim. Pract.* **2000**, *30*, 169–182. [CrossRef]
11. Penwick, R. Fibrocartilaginous embolism and ischemic myelopathy. *Compend. Contin. Educ. Pract. Vet.* **1989**, *11*, 287–297.
12. Spetzler, R.F.; Detwiler, P.W.; Riina, H.A.; Porter, R.W. Modified classification of spinal cord vascular lesions. *J. Neurosurg.* **2002**, *96* (Suppl. 2), 145–156. [CrossRef]
13. Vicens, L.; Tapia, R.; Bergman, W.; Bergknut, N. Venous Sinus Aneurysm in a Scottish Deerhound. In Proceedings of the 31st Annual Symposium of the ESVN-ECVN, Kopenhagen, Denmark, 20–22 September 2018.
14. Vernon, J.C.; Durand, A.; Guevar, J.; José-López, R.; Hammond, G.; Stalin, C.; Gutierrez-Quintana, R. Vertebral venous system abnormalities identified with magnetic resonance imaging in sighthounds. *Vet. Radiol. Ultrasound* **2017**, *58*, 399–410. [CrossRef]
15. d'Anjou, M.; Schwarz, T. Hearts and vessels. In *Veterinary Computed Tomography*, 1st ed.; Schwarz, T., Saunders, J., Eds.; Wiley-Blackwell: Oxford, UK, 2011; pp. 229–242.
16. Thompson, M.; Graham, J.; Mariani, C. Diagnosis of a porto-azygous shunt using helical computed tomograpy angiography. *Vet. Radiol Ultrasound* **2003**, *44*, 287–291. [CrossRef] [PubMed]
17. Gomez, M.; Freeman, L.; Jones, J.; Lanz, O.; Arnold, P. Computed tomographic anatomy of the canine cervical vertebral venous system. *Vet. Radiol. Ultrasound* **2004**, *45*, 29–37. [CrossRef] [PubMed]
18. Gomez, M.A.; Jones, J.C.; Broadstone, R.V.; Inzana, K.D.; Freeman, L.E. Evaluation of the internal vertebral venous plexus, vertebral canal, dural sac, and vertebral body via nonselective computed tomographic venography in the cervical vertebral column in healthy dogs. *Am. J. Vet. Res.* **2005**, *66*, 2039–2045. [CrossRef] [PubMed]
19. Caceres, A.; Zwingenberger, A.; Hardam, E.; Lucena, J.; Schwarz, T. Helical computed tomography of the normal canine páncreas. *Vet. Radiol. Ultasound* **2006**, *47*, 270–278. [CrossRef]

20. Jung, J.; Chang, J.; Sunkyoung, O.; Junghee, Y.; Mincheol, C. Computed tomography for evaluation of pulmonary embolism in a experimental model and heartworm infested dogs. *Vet. Radiol. Ultrasound* **2010**, *51*, 288–293. [CrossRef] [PubMed]
21. <named-content content-type="background:white">Oui, H.; Kim, J.; Bae, Y.; Oh, J.; Park, S.; Lee, G.; Jeon, S.; Choi, J. Computed tomography angiography of situs inversus, portosystemic shunt and multiple vena cava anomalies in a dog. *J. Vet. Med. Sci.* **2013**, *75*, 1525–1528. [CrossRef]
22. Andreucci, M.; Faga, T.; Pisani, A.; Sabbatini, M.; Michael, A. Acute kidney injury by radiographic contrast media: Pathogenesis and prevention. *Biomed. Res. Int.* **2014**, *2014*, 362725. [CrossRef]
23. Pollard, R.; Puchalski, S. CT contrast media an applications. In *Veterinary Computed Tomography*; Schwarz, T., Saunders, J., Eds.; Wiley-Blackwell: Oxford, UK, 2011; pp. 57–66.
24. Stringer, M.D.; Restieaux, M.; Fisher, A.L.; Crosado, B. The vertebral venous plexuses: The internal veins are muscular and external veins have valves. *Clin. Anat.* **2012**, *25*, 609–618. [CrossRef]
25. Demondion, X.; Delfaut, E.M.; Drizenko, A.; Boutry, N.; Francke, J.P.; Cotten, A. Radio-anatomic demonstration of the vertebral lumbar venous plexuses: An MRI experimental study. *Surg. Radiol. Anat.* **2000**, *22*, 151–156. [CrossRef]
26. Zippel, K.C.; Lillywhite, H.B.; Mladinich, C.R. New vascular system in reptiles: Anatomy and postural hemodynamics of the vertebral venous plexus in snakes. *J. Morphol.* **2001**, *250*, 173–184. [CrossRef]
27. Zippel, K.C.; Lillywhite, H.B.; Mladinich, C.R. Anatomy of the crocodilian spinal vein. *J. Morphol.* **2003**, *258*, 327–335. [CrossRef]
28. Kotzé, S.; Vorster, J.; Hoogland, P. A morphological study of the vertebral venous plexus and its connections in the Cape dune mole-rat, *Bathyergus suillus* (Bathyergidae). *J. Morphol.* **2010**, *272*, 280–286. [CrossRef] [PubMed]
29. Fredrickson, V.L.; Patel, A.; Pham, M.H.; Strickland, B.A.; Ohiorhenuan, I.; Chen, T. Spine Surgery Complicated by an Engorged Lumbar Epidural Venous Plexus from Cerebrospinal Fluid Overshunting: A Case Report and Review of the Literature. *World Neurosurg.* **2018**, *111*, 68–72. [CrossRef] [PubMed]
30. Rhue, K.E.; Taylor, A.R.; Cole, R.C.; Winter, R.L. Bilateral Vertebral Venous Sinus Thrombosis Causing Cervical Spinal Cord Compression in a Dog. *Front. Vet. Sci.* **2017**, *4*, 8. [CrossRef] [PubMed]
31. Martin-Vaquero, P.; da Costa, R.C. Magnetic resonance imaging features of Great Danes with and without clinical signs of cervical spondylomyelopathy. *J. Am. Vet. Med. Assoc.* **2014**, *245*, 393–400. [CrossRef] [PubMed]
32. Mokri, B. The Monro-Kellie hypothesis: Applications in CSF volume depletion. *Neurology* **2001**, *56*, 1746–1748. [CrossRef]
33. Mannion, A.F.; Fekete, T.F.; Pacifico, D.; O'Riordan, D.; Nauer, S.; von Büren, M.; Schizas, C. Dural sac cross-sectional area and morphological grade show significant associations with patient-rated outcome of surgery for lumbar central spinal stenosis. *Eur. Spine J.* **2017**, *26*, 2552–2564. [CrossRef]
34. Zhou, Z.; Jin, Z.; Zhang, P.; Shan, B.; Zhou, Z.; Zhang, Y.; Deng, Y.; Zhou, X. Correlation Between Dural Sac Size in Dynamic Magnetic Resonance Imaging and Clinical Symptoms in Patients with Lumbar Spinal Stenosis. *World Neurosurg.* **2020**, *134*, e866–e873. [CrossRef]
35. Lim, J.; Yoon, Y.; Hwang, T.; Lee, H.C. Novel vertebral computed tomography indices in normal and spinal disorder dogs. *J. Vet. Sci.* **2018**, *19*, 296–300. [CrossRef]
36. Dabanoglu, I.; Kara, M.E.; Turan, E.; Ocal, M.K. Morphometry of the thoracic spine in German shepherd dog: A computed tomographic study. *Anat. Histol. Embryol.* **2004**, *33*, 53–58. [CrossRef]
37. Sturges, B.K.; Westworth, D.R. Congenital spinal malformations in small animals. *Vet. Clin. N. Am. Small Anim. Pract.* **2010**, *40*, 951–981.
38. Bailey, C.S.; Morgan, J.P. Congenital spinal malformations. *Vet. Clin. N. Am. Small Anim. Pract.* **1992**, *22*, 985–1015. [CrossRef]
39. Thrall, D. *Textbook of Veterinary Diagnostic Radiology*, 7th ed.; Elsevier: St. Louis, MO, USA, 2018.

MDPI

Review

The Cat Mandible (I): Anatomical Basis to Avoid Iatrogenic Damage in Veterinary Clinical Practice

Matilde Lombardero [1,*], Diana Alonso-Peñarando [2,3] and María del Mar Yllera [1]

[1] Unit of Veterinary Anatomy and Embryology, Department of Anatomy, Animal Production and Clinical Veterinary Sciences, Faculty of Veterinary Sciences, University of Santiago de Compostela–Campus of Lugo, 27002 Lugo, Spain; mar.yllera@usc.es
[2] Department of Animal Pathology, Faculty of Veterinary Sciences, University of Santiago de Compostela–Campus of Lugo, 27002 Lugo, Spain; diana.alonso.penarando@rai.usc.es
[3] DVM at Veterinary Clinic Villaluenga, Villaluenga de la Sagra, 45520 Toledo, Spain
* Correspondence: matilde.lombardero@usc.es; Tel.: +34-982-822-333

Simple Summary: Nowadays, cats are one of the most common companion animals. They differ from dogs in some important aspects. However, most of the veterinary clinics are oriented towards the care and treatment of dogs, where the cat patient is clinically treated like a small dog. The cat mandible and related structures have some particularities that should be taken into account, when treating a cat, to avoid any unintended medical (iatrogenic) damage. The feline mandible has fewer teeth than a dog's one, but tooth roots and the neurovascular supply account for up 70% of the volume of the mandibular body. This fact makes mandibular fracture repair challenging. In addition, the cat mandible has a prominent angular process that, when the cat is under anesthesia and his mouth is wide open (during oral or transoral manipulation), compresses the maxillary artery (that supplies blood to the brain) inducing temporal or permanent blindness and/or deafness. Other particularities of the cat jaw are also addressed to get a comprehensive knowledge of its functional anatomy, essential to an effective feline clinical practice.

Abstract: Cats are one of our favourite pets in the home. They differ considerably from dogs but are usually treated clinically as small dogs, despite some anatomical and physiological dissimilarities. Their mandible is small and has some peculiarities relative to the dentition (only three incisors, a prominent canine, two premolars and one molar); a conical and horizontally oriented condyle, and a protudent angular process in its ventrocaudal part. Most of the body of the mandible is occupied by the mandibular dental roots and the mandibular canal that protects the neurovascular supply: the inferior alveolar artery and vein, and the inferior alveolar nerve that exits the mandible rostrally as the mental nerves. They irrigate and innervate all the teeth and associated structures such as the lips and gingiva. Tooth roots and the mandibular canal account for up to 70% of the volume of the mandibular body. Consequently, when fractured it is difficult to repair without invading the dental roots or vascular structures. Gaining a comprehensive anatomical knowledge and good clinical practice (such as image diagnosis before and post-surgery) will help in the awareness and avoidance of iatrogenic complications in day-to-day feline clinical practice.

Keywords: anatomy; feline; lower jaw; neurovascular supply; temporomandibular joint; tooth

Citation: Lombardero, M.; Alonso-Peñarando, D.; Yllera, M.d.M. The Cat Mandible (I): Anatomical Basis to Avoid Iatrogenic Damage in Veterinary Clinical Practice. *Animals* **2021**, *11*, 405. https://doi.org/10.3390/ani11020405

Academic Editor: Ralf Einspanier
Received: 21 December 2020
Accepted: 29 January 2021
Published: 5 February 2021

Publisher's Note: MDPI stays neutral with regard to jurisdictional claims in published maps and institutional affiliations.

1. Introduction

Cats have been fully appreciated animals for a long time. Ancient Egyptians mummified cats, as pets were buried with their owner, or they were used as votive offerings that depicted the gods [1]. Animals were revered by associating them with deities. In fact, the goddess Bastet was symbolised as a cat or even a woman with a feline head [2].

Nowadays, cats are one of the most numerous pets at home, sharing our company and friendship. According to the American Pet Products Association's (APPA) 2019–2020

National Pets Owners Survey [3], it is estimated that in the USA, 67% of households own a pet; that is 84.9 million homes with an estimated 63.4 million dogs and 42.7 million cats, among other pets (such as freshwater/saltwater fish, birds, small animals and reptiles, accounting all together for 137 million animals being kept at home). However, compared with 2019 figures, in the USA, an estimated 89.7 million dogs and 94.2 million cats is reported (made in base of multiplying the number of households that own the pet by the average number of pets owned by household), which is a large decrease in both species, but no explanation was given, unless maybe it was a poor rough calculation. In Europe, there were at least 85 million households (38% of all households) owning a minimum of one pet animal [4]: 106.4 million cats and 77.4 million dogs. Hence, in Europe, the numbers of cats largely overtake those of dogs as pets; as they do not take much space, they do not need to go outdoors during the day, and their ownership is more affordable, among other reasons. However, the majority of pet clinics are more oriented towards dog management and treatment. Concerning cat size, they are usually small and consequently their surgery is more difficult. In addition, their lifespan is longer than the dog's, with an average life span of 12 years and ages of 20 years or more not uncommon. The current longevity record cited by Beaver in 2003 [5] was 36 years. Regarding clinical surgery solutions, up to now, little has been is done by companies to provide surgical prostheses and other products adapted to such a small animal, as they are more oriented towards different dog breeds. Hence, unfortunately, and in accordance with Little [6], cats are still the 'poor stepchild' in companion animal medicine, receiving less attention in research into common medical problems or improved diagnostic and treatment approaches than is given to their canine counterparts.

Cats should not be considered or treated as small dogs as there are many differences. This review will be focused specifically on the cat mandible and other related aspects. Taking into account the key relevance of anatomy in high-quality medicine, the treatment of the feline mandible has many important features, most of them due to the singularity of its anatomy when compared to that of the dog. Knowledge of the functional anatomy of the cat mandible will give us clues as to appropriate treatment of the feline patient, minimising the occurrence of potential iatrogenic injuries. Mandibular injuries are varied: from those related to tooth extraction complications to mandibular fractures, or from temporomandibular joint pathologies to neural complications due to an excessive opening of the mouth (usually during oral or transoral manipulation). Consequently, we propose to review the cat mandible and the anatomically related structures from the point of view of functional anatomy with a clinical orientation, as it is of crucial importance in face morphology and for feeding and grooming. Thus, any alteration of the mandible would have important repercussions on the general appearance of the feline patient, affecting both food intake and grooming, as house cats can spend up to 50% of their waking time grooming their coat or performing related behaviour [5]. The various grooming behaviours are important to a normal healthy cat. In the absence of grooming, excess debris can tangle fur, causing painful tugging of the skin and even infection.

2. The Mandible

Starting with the anatomical description of the feline mandible, and according to the Veterinary Anatomical Nomenclature [7,8], the cat jaw has two halves (*mandibula*) joined rostrally by an *articulatio intermandibularis,* known as the mandibular symphysis (in carnivores, partially conformed by a *synchondrosis* and a *sutura intermandibulares*). Each mandibular half has a horizontal part (body, or *corpus mandibulae*) and a vertical part placed caudally (*ramus mandibulae*; Figure 1). The mandibular ramus has a dorsal part, oriented slightly caudally, with a coronoid process (*processus coronoideus*) on the top (which attaches the temporal muscle), and two faces with both *fossae* for muscle fixation: a masseteric fossa (very deep, laterally—*fossa masseterica*) and a pterigoid fossa (medially—*fossa pterygoidea*). More or less at the same level as the dorsal margin of the body, completely caudally, there is the condylar process (*processus condylaris*), a cylinder oriented horizontally that articulates with the mandibular fossa of the temporal bone and participates in the temporomandibular joint (TMJ) (*articulatio temporomandibularis*). On the caudal part and ventrally, there is an angular process (*processus angularis*), oriented caudally.

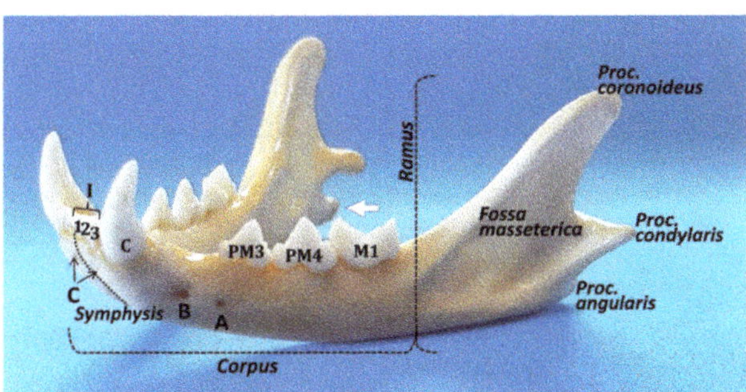

Figure 1. Rostro-lateral view of the cat mandible. The most relevant details are shown. A: posterior mental foramen; B: main or middle mental foramen; and C: rostral mental foramen. The prominence of the *processus angularis* is more manifest in a medial view (white arrow).

The mandibular body has two faces—medial or lingual (*facies lingualis*), and lateral or buccal and labial (*facies buccalis* and *labialis*)—separated by two margins - the dorsal or alveolar (*margo alveolaris*), and the ventral—(*margo ventralis*). The ventral margin is smooth, whereas the alveolar margin is very irregular with deep pits where the root teeth are fitted (*alveoli dentales*). The body has two parts, a rostral or incisive part (*pars incisiva*) and the molar part (*pars molaris*). The incisive part supports the incisive and canine teeth, whereas the molar part contains the premolar and molar teeth. All the lower teeth form the *arcus mandibularis inferior*. The buccal face of the mandible is uneventful, except for the two mental foramina (*foramina mentalia)*: the main mental foramen and the posterior mental foramen (or three, the third, minute, one below the incisors, similar to a *foramen nutricium*, but through which the anterior mental nerves exit to innervate the area below the incisors) [9]. Lobprise & Dodd [10] also mention three foramina: the rostral foramen in the incisive part, the middle mental foramen at the level of the labial frenulum, and the caudal mental foramen between the two roots of the third premolar. In contrast, the lingual face has only a mandibular foramen (*foramen mandibulae*) in the rostral part of the ramus for the neurovascular supply of the mandible, which enters into the mandibular canal and provides the sole blood supply (the mandibular artery (*A. alveolaris inferior*-ramus of the maxillary artery (*A. maxillaris*)—to the alveolar bone and teeth, and the mandibular vein (*V. alveolaris inferior*)) and the inferior alveolar (sensory) nerve (*N. alveolaris inferior*). According to Starkie & Stewart [11], a tough fibrous sheath surrounds the main nerve and its larger branches, and only the cat (in contrast to the rabbit and sheep) presents a sheath of compact bone surrounding the neurovascular bundle [11], a fact that in theory would help to localise the feline mandibular canal by radiography.

Pitakarnnop et al. [12] analysed 44 parameters on dried feline bones and demonstrated that there are only three hallmarks for sex identification in cats. One of them is the coronoid process of the mandibular ramus (with an accuracy rate up to 88.2%): 'looking at a lateral view of the mandible, the coronoid process in females was more curved than in males'. The other two were described in the *os coxae* [12]. Hence, the mandible could give us additional and useful information for some scientific domains, such as forensic, developmental and evolutionary sciences, and also for zooarchaeological studies.

To the outer surface of the mandible is attached the Mm. *masseter* and *buccinator* (*partes buccalis* and *molaris*); the inner surface to the Mm. *pterygoidei* (*lateralis* and *medialis*), *mylohyoideus*, *geniohyoideus*, and *geniohyoglossus*; both surfaces at the lower border to the M. *digastricus*, and the upper process (*processus coronoideus*) to the M. *temporalis* [7,8].

When the jaw is closed, the lower incisors normally strike immediately caudal to the upper incisors. The lower canine occludes between the lateral upper incisor and the upper canine. Hence, this arrangement provides a shearing action, particularly between the cheek teeth [13]. The upper and lower teeth do not touch when the jaws move in a sagittal plane. However, when the cat chews on one side of the mouth, the lower jaw must be brought to that side, so the buccal surface of the lower teeth may shear upwards and forwards against the occlusal surface of the upper teeth [13].

Bite force is generated by the interaction of the masticatory muscles, the mandibles and maxillae, the TMJs, and the teeth. The main factors affecting the bite forces in dogs and cats are body weight and the skull's morphology and size [14]. In cats, biting forces of 20–23.25 kg at the canine and up to 28 kg at the carnassial teeth have been reported [15] or, according to Kim et al. [14], an average of 73.3 Newtons (N) and 118.1 N, respectively. However, if converted to kg-force (kgf), it is obvious that these measurements do not match with [15], as they result in 7.47 and 12.04 kgf, respectively.

3. Temporomandibular Joint

It is worth pointing out that the temporomandibular joint (TMJ) is a cardinal feature that defines the class Mammalia and separates mammals from other vertebrates. Despite its status as a mammalian identifier, the TMJ shows remarkable morphological and functional variations in different species, reflecting the great adaptive diversification of mammals in feeding mechanisms [16]. During evolution, the common features of the TMJ (such as modified hinge joint, fibrocartilaginous articular surfaces, and two synovial joint compartments separated by an articular disc) persisted mostly invariable, except for a few species [17]. The simple components of the TMJ present adaptations, both in form and function, to satisfy the needs of the species, such as feeding and communication [17]. The evolutionary variants include adaptations in the orientation of the joint cavity from parasagittal (many rodents) to transverse (many carnivores), among other features [16]. However, function still remains a problem, because muscles, movements, and joint loads are to a great extent species-dependent [16].

The feline TMJ, which works as a hinge, is a synovial condylar joint formed between the condyloid process of the *ramus mandibulae* and the *fossa mandibularis* of the *pars squamosa* of the temporal bone. Caudally, the deep gutter of the temporomandibular articulation is bounded by a prominent retroarticular process placed behind the mandibular fossa [18]. Medial dissection shows a close relationship of the medial aspect of the articular capsule with the mandibular nerve, the tympanic cord, and maxillary artery. This area is particularly sensitive due to the exaggerated opening of the mouth when using different instruments or clinical exploration or dentistry [19].

It is the unique joint with a whole articular disc (*discus articularis*) that has developed owing to a slight articular surface incongruity [20]. However, in the cat, the temporomandibular joint is highly congruent [17] and the articular disc is very thin and poorly developed. According to Arredondo et al. [21], it is attached around its entire periphery to the capsule, dividing the synovial cavity into two separate spaces—dorsal and ventral. The periphery of the disc is irrigated by small branches from the articular temporomandibular artery [21]. In the cat, the condyloid process formed by the *caput mandibulae* consists of a slender, transverse roller 13–15 mm wide with a diameter of 2–3 mm (Figure 2). The axis of the articulated roller is oriented transversely at the line of the occlusal plane [17]. Caudally oriented, it presents a highly curved convexity medial to the mandibular body that narrows laterally and the outer end is often pointed [22].

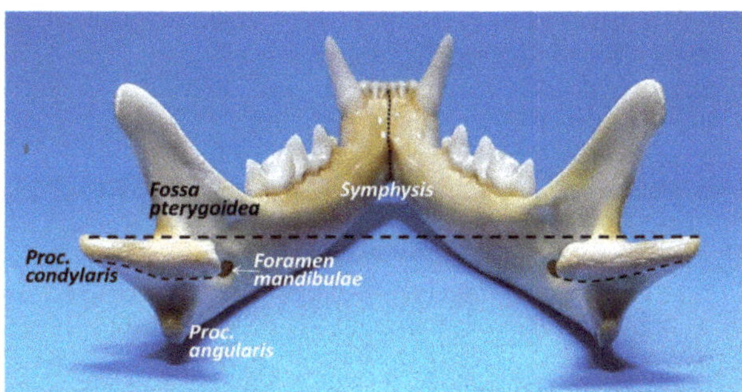

Figure 2. Caudal view of the cat mandible. Both cone-shaped condylar processes are horizontally oriented.

The mandibular fossa, which lies under the base of the zygomatic arch and is 12–15 mm wide, has a concave *facies articularis* placed between the retroarticular process (*processus retoarticularis*), which is a caudoventral extension (medially placed) of the mandibular fossa, and a rounded and pronounced articular eminence (*tuberculum articularis*) rostrolateral to the mandibular fossa [22] (Figure 3). Therefore, in cats, the mandibular head is completely surrounded by bony structures of the temporal bone, with both eminences acting like two stops to limiting anteroposterior movements, but allowing mandibular motion to the sagittal plane of the cutting edge of the molar/premolar border P4/M1 (just opening and closing the mouth), with very limited lateral [23,24]. According to Crompton et al. [25], the retoarticular process resists the posteriorly directed force of the temporal muscle and the articular tubercle resists the anteriorly directed force of the superficial masseter muscle. When the mouth is closed, the mandibular dental arch fits into the dental arch of the maxilla and leaves no gaps to allow lateral or transverse movement. However, according to Knospe [22], the sideways movement of the lower jaw in the transverse plane occurs only when the oral cavity is slightly to strongly open, providing 2–3 mm to the right or left, with a total of approximately 5 mm, allowing crushing shears in the cat's P4/M1, splitting the jaw pressure into a vertical cutting and a transverse pressure component.

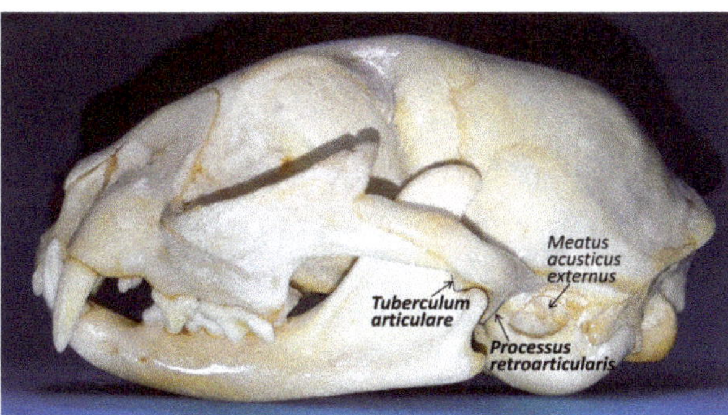

Figure 3. Lateral view of the cat head. The temporomandibular joint has two 'bumpers' (one rostrally and other caudally) to keep in place the condylar process in the *fossa mandibularis* of the temporal bone.

Problems or perception of pain in opening or closing the mouth should always include a complete evaluation of the bilateral TMJ. Imaging with radiographs can be challenging due to superimposition of maxillofacial/cranial structures, requiring some rotation in either the lateral (10–30°) or long axis (10–30°) to isolate the individual structures [10].

In addition, Gracis and Zini [23] stated that the evaluation of the vertical mandibular range of motion or range of mandibular abduction (the distance between the maxillary and mandibular incisor teeth at maximum mandibular extension) should be incorporated into every diagnostic examination, as it may be valuable in showing changes over time for a single patient (concurrently with the patient under general anaesthesia and the musculature is relaxed). Consequently, early detection of a reduction in joint mobility allows a prompt diagnosis of these limiting pathologies or conditions. Those conditions may affect intra-articular or extra-articular TMJ structures, such as ankylosis secondary to fracture, joint luxation, dysplasia and osteoarthritis, which are relatively common TMJ lesions in cats [21], or fracture, osteomyelitis, bone neoplasia, retrobulbar masses, neuromuscular diseases, and trismus [23]. It should be taken into account that in cats there is a positive correlation between the vertical mandibular range of motion and (1) body weight and (2) age. In addition, male cats can open their mouth wider than can females [23].

4. Mandibular Teeth

The cat has double dentition: deciduous and later permanent teeth. It is edentulous at birth but develops a set of deciduous teeth that start erupting between 2 and 8 weeks after birth [13]. At 60 days, the deciduous dentition is complete [14]. Between 3 and 6 months of age, the deciduous teeth are shed as the permanent teeth erupt, and their full crown height should be achieved by 10–12 months of age [13]. There are four types of teeth depending on their shape and function: incisors (I), canines (C), premolars (PM), and molars (M). The permanent teeth number 30 in all. The formula of the permanent teeth includes I 3/3, C 1/1, PM 3/2, M1/1. The deciduous formula is similar but lacks molar teeth (26 teeth in total). The incisors and canine teeth of the cat are all single-rooted, as those of the dog. The mandibular first and second premolars are normally absent. The mandibular third and fourth premolars and the single mandibular molar each have two roots. The roots of the premolar are nearly equal in size, but the mesial root of the mandibular molar is approximately three times the width of the distal root [10]. The mandibular first molar tooth is considered a carnassial tooth (as well as the maxillary fourth premolar tooth).

Regarding the simple or brachydont tooth structure, as a general overview, the crown of the tooth is covered with enamel, whereas the roots are covered with cementum. Both hard-tissue layers meet at the cement–enamel junction, near the cervical portion of the tooth. Dentine constitutes the major part of the mature tooth. The difference between enamel and dentine microhardness is a result of the percentage of mineralisation they present. Dentine is synthesised by the odontoblasts lying at the pulp's periphery. Since dentine is produced throughout life in a vital tooth, the permanent teeth of old cats have thicker dentineal walls and narrower pulp cavity compared to those of young cats [26]. The enamel thickness is reported to be 0.1–0.3 mm in cats and 0.1–0.6 mm in dogs [10] and, additionally, cat enamel is less hard than that of dogs, according to Hayashi & Hideo [27]. They also reported that enamel microhardness is higher in the outer layer than in the central or inner layer (since calcium ions in saliva infiltrate from the tooth surface), and there is an age-related increase in the microhardness of enamel [27]. In the premolars, the enamel hardness is higher at the top and middle of the crown, and it decreases in the cervical portion. Similarly, the dentine is harder in the cusp than in the rest of the tooth and the dentine microhardness decreases from the outer to the inner part. They also reported that the comparative microhardness enamel/dentine ratio varied from 3–9/1 [27]. Hence, data support that cat teeth are far more fragile compared to dog teeth; that being so, special care should be taken when manipulating.

The dental root is inserted in the dental alveolus and kept in place by the periodontium, which is made up of the gingiva, periodontal ligament, cementum, and alveolar bone. The space between the tooth and the free gingiva is the gingival sulcus, which should be no deeper than 0.5 mm in cats. The periodontal ligament attaches the root to the alveolar bone [26]. The alveolar bone appears with tooth eruption and disappears with tooth loss [14]. It surrounds the alveolar socket with an extension of cortical bone into the alveolus that outlines a radiopaque lamina dura in radiographic images [26]. The dental sac contacting with the cementum forms fibroblasts that produce collagen fibres while the other components of the periodontal ligament are developing. These are blood vessels, lymphatics, nerves, and various types of connective tissue cells. The nerves of the periodontal ligament are important as they provide additional senses to the tooth. They harbour pain fibres (similar to the pulp), but also pressure, heat, and cold fibres (not present in the pulp) [10].

However, as the aim of this review is not feline dentistry, specialised textbooks should be consulted for further information regarding teeth. Nonetheless, the more frequent dental pathologies seen in cats should be mentioned: periodontal disease and tooth resorption. Periodontal disease is perhaps the most common oral disease seen in dogs and cats and involves the periodontium and is the major cause of tooth loss. In general terms, periodontal disease is caused essentially by the accumulation of plaque on the tooth surfaces, and the severity correlates directly with the quantity of such deposits, producing gingivitis and forming a periodontal pocket. These changes lead to destruction of the gingival tissue with recession of the gums and retraction of the periodontal membrane. Ultimately there is infection of the dental root, destruction of alveolar bone, and dental loss. It is clear that a soft diet correlates positively with periodontal disease [13]. Cats suffer severely with progression of periodontitis and can be seriously anorexic. Gingival recession is common with marked tooth loss as a result of external tooth resorption, where the inflammation at the cement–enamel junction leads to collapse of the tooth crown with retention of the roots. These roots may become chronically inflamed or cystic and be very painful with periodontal abscessation. Treatment consists of removing degenerated teeth and any retained roots [13].

Dental disease is more common in older cats and can lead to other health problems, so maintaining oral health is important. To prevent dental disease, the single most effective method is to brush the cat's teeth daily with a pet-specific toothpaste or powder, although it can sometimes take some time to train them to allow their teeth to be brushed.

It must be taken into account that the morphology of the mandible is conditioned by food habits [28]. As a carnivore, the cat has a mastication pattern consisting of an up and down or hinge movement of the mandible. Their TMJ lies on or close to the same plane as that of the lower dentition, with molars designed for crushing and slicing [25]. In contrast, in the herbivore group, which includes the Ungulates, the main action consists of a grinding movement of the mandible [28], as their TMJ lies above the occlusal plane of the maxillary dentition. In carnivores, the combination of a tall coronoid process and extension of the skull posterior to the TMJ permits both a large gape and a powerful bite. Another characteristic feature of this group is that the temporal muscle is considerably stronger than the masseter muscle [25].

5. Mandibular Neurovascular Supply

The first vessel to leave the maxillary artery in carnivores is the a. *articularis temporomandibularis* destined to the mandibular joint. The a. *alveolaris mandibularis* arises in a rostrolateral direction from the first part of the maxillary artery and runs towards the mandibular foramen through which it enters into the mandibular canal and provides blood supply to the mandible. Within this canal, it gives off the *rami dentales* to the molar and premolar teeth. Other branches pass through the alveolar canal to the canine and incisor teeth [29]. It exits at the caudal, middle, and rostral mental foramina to supply the lower lips [6]. In carnivores, the *rami mentales* leave the mandibular canal trough the mental foramina and ramify in the region of the *margo interalveolaris* in the gingiva of the incisor teeth and in the lower lip [29]. Veins often exist concurrently with arteries and empty by way of the maxillary and linguofacial veins into the external jugular vein [6].

In the mandibular canal, when examined in cross-section, the mandibular nerve is located in the dorsolateral portion of the canal with the vein in the ventromedial portion and the artery in the middle [10], although according to Davis & Story [30] the a. *alveolaris inferior* is situated lateral to the inferior alveolar nerve as the two enter the foramen.

The maxillary and mandibular branches (*nervi maxillaris* and *mandibularis*) of the trigeminal nerve (nervus trigeminus) are sensory, but the mandibular branch also supplies motor function to the masticatory musculature (temporal muscle, masseter muscle and the lateral and medial pterigoid muscles, which close the mandible) and other muscles. The digastric muscle is the only one responsible for opening the jaws; its rostral belly is stimulated by the mandibular branch of the trigeminal nerve, as the caudal belly is innervated by the facial (VII) nerve [10,31]. The facial nerve also provides motor function to many cutaneous facial muscles and is responsible for taste in the rostral two-thirds of the tongue [6]. The inferior alveolar nerve contains afferent fibres from the ipsilateral lower lip, areas of oral mucous membrane and mandibular teeth. It supplies sensory innervation of the lower teeth and, after exiting the mental foramina as the mental nerves, it innervates the lower lip. According to Robinson [9], fibres supplying the teeth are found in all branches, except the mental ones. At the mandibular foramen, the inferior alveolar nerve is a single bundle with the mandibular artery and vein lying inferior to it. Within the mandible, the nerve divides into several branches, which conform to a basic pattern with some individual variation. Three branches, splitting dorsally the main trunk, supply the alveolar processes (the posterior, middle, and anterior alveolar branches); another supplies the canine and incisor regions (canine/incisor branch), and there are four mental branches (posterior, main and two anterior) that leave through the various mental foramina. Interconnecting fibres are often seen between these principal branches [9]. There is no apparent bilateral symmetry and all the branches contain fibres from at least two adjacent tissues, including afferences from the pulp, periodontal ligament, mucous membrane, and skin. In addition, the nerves supplying one tooth come from different branches because they do not travel in a unique branch of the main trunk (Figure 4). The most proximal branch splitting from the superior aspect of the main trunk is the posterior alveolar branch. It contains afferents from the molar, fourth premolar, and occasionally third premolar teeth [9]. The next alveolar branch is the middle alveolar branch, leaving dorsally the main trunk beneath the distal root of the molar tooth to supply the third premolar, although in 50% of specimens it also supplies the canine teeth and the third incisor. The last alveolar branch is the anterior alveolar nerve, which supplies the canine and third incisor teeth in addition to mechanoreceptor afferents from the mucous membrane and skin adjacent to these teeth. Occasionally it also carries fibres from the second and first incisors and the third premolar teeth. Other branches splitting laterally from the main trunk, such as the posterior and main mental, leave the mandible through the posterior and main mental foramina, respectively, to supply the buccal gingival margin, the mucous membrane on the labial side of the alveolar process and the skin of the chin and lip from the anterior to caudal part (just up to the rostral edge of the molar tooth). The anterior mental branch divides in two terminal branches: the canine/incisor nerve and the anterior mental nerve. The latter does not contain any pulpal or periodontal afference. Lastly, the canine/incisor branch carries fibres from the canines and all three incisors.

Relative to the existence of transmedian innervation, Robinson [9] said that there is no evidence of transmedian innervation of tooth pulps; nonetheless, the cutaneous innervation in the anterior mental nerve crosses the midline for 1–2 mm. In contrast, Anderson & Pearl [32] reported the existence of an extensive transmedian innervation of the teeth in the cat; an innervation which is particularly dense in the canine teeth and extends at least as far laterally as the third premolar teeth. Wilson et al. [33], using the horseradish peroxidase technique, reported that, in addition to the inferior alveolar nerve, the nerve to mylohyoid and possibly other accessory neural pathways is involved in incisor innervation in cats.

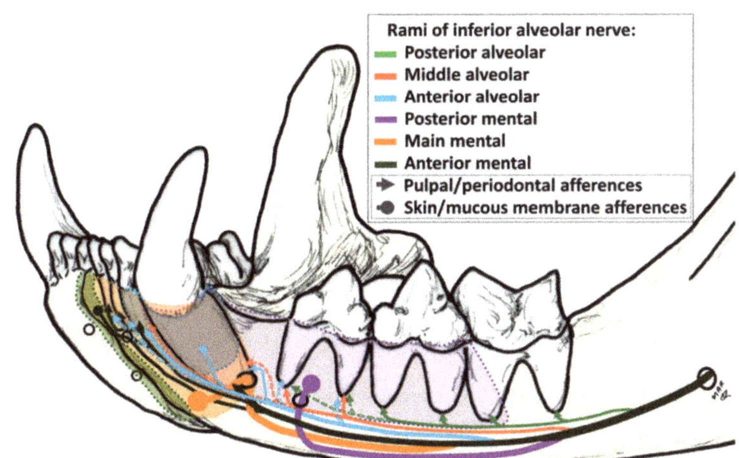

Figure 4. Innervation of the cat mandible modified from Robinson [8]. Diagram showing the inferior alveolar nerve through the mandibular canal and its division in *rami* to innervate different structures. Dashed lines indicate inconstant afferences. The thickness of the line when detaching from the inferior alveolar nerve is proportional to its real thickness, according to Robinson [8].

According to Izumi et al. [34], ligation or cutting of the inferior alveolar nerve always elicits an increase in gingival blood flow. They also reported that blood flow in the cat gingiva and periodontal ligament is controlled by sympathetic α-adrenergic fibres for vasoconstriction; regarding vasodilatation, sensory fibres are involved besides the mast cells of the gingiva. According to Skerrit [31], a unilateral deficit of the mandibular nerve results in weakness of the chewing muscles when biting and atrophy of the temporal and masseter muscles. When the deficit is bilateral, it leads to drooping of the mandible and inability to close the mouth [31].

Mental nerve block, when needed, should be done at the main or middle mental foramen and anaesthetises the buccal soft tissues and the mandibular incisors and canine on the side injected [35]. The mandibular nerve can be anaesthetised by intraoral or extraoral techniques. Anaesthesia of the nerve results in desensitisation of the mandibular body, the lower portion of the mandibular ramus, all mandibular teeth on the same side, the labial/buccal surfaces of the mandible, and the mucosa and skin of the lower lip and chin [35].

The mandibular canal is not a medullary canal, and treating fractures of the body via an intramedullary pin through this canal will damage the associated neurovascular bundle. In many fractures or tumoural reparative surgery of the mandible, the inferior alveolar nerve is damaged and then resected. However, some experiments provide evidence that peripheral nerve fibres are important, not only in normal bone homeostasis and skeletal growth, but also in their influence on the repair mechanism of bone fracture. Many experiments suggest that sensory and sympathetic nerve fibres do have a role in bone remodelling and osteogenic differentiation of precursor cells during skeletal growth. Hence, the loss of sensory nerves could result in a decrease in the quality of new bone, as reported by Cao et al. [36] on the mandibular distraction osteogenesis in rabbits, stating that the peripheral sensory nervous system plays an important role in bone regeneration. Sensory nerves also play an important role in regulating bone resorptive activity, as shown by Yamashiro et al. [37] during experimental tooth movement in rats. In bone, the areas with the highest metabolic activity receive the richest sensory and sympathetic innervation, which has an effect on the activity of both osteoblasts and osteoclasts [37]. As seen in many articles, the inferior alveolar nerve could also be affected by reparative osteosynthesis, and as in the feline mandible there is very little space to safely place the screws required to fix the metal plates to resolve the fracture. Regarding the integrity of the inferior alveolar nerve, we consider a conservative option should be prevalent over inferior alveolar nerve resection so as to not impair the outcome of bone repair and sensitivity. Hence, we propose that surgeons should be as conservative as possible and try to leave the nerve intact, with no attempt to pull the nerve as this causes a lesion, and being sure where to place the screw in order to not affect the neurovascular bundle or any dental root.

6. Mandibular Radiographic Images

Dental radiography requires general anaesthesia to get accurate projections and avoid any trauma or damage to the equipment.

Cats have two mandibular halves that connect rostrally through a mandibular symphysis. The symphysis is represented radiographically as a radiolucent border between the two mandibles. The portions of the mandibles associated with the symphysis are roughly parallel [38].

The lateral projection of the feline mandible (Figure 5) is bordered ventrally by the ventral cortex and dorsally by the cusps of the premolars and molar. The area corresponding to the location of the mandibular canal, containing the neurovascular bundle (the alveolar mandibular nerve, artery and vein) appears as a radiolucent area just dorsal to the ventral cortex and ventral to the dental roots. When evaluating any tooth, the following are assessed: crown (and enamel) and pulp chamber, root and root canal, periodontal ligament space and alveolar bone. The relative radiolucent line outlining the roots (*lamina lucida*) is the periodontal ligament space. It is wider early in age and is typically widest at the coronal and apical one-third of the root [38]. Adjacent to this ligament space is a radiodense line (*lamina dura*), which is the cortical bone of the alveolus. Contiguous to it is the trabecular bone of the alveolus. The crown is covered by a more radiodense margin, which is the enamel, and the bulk is dentine, which is not as radiodense as enamel but is radiodense compared to bone. Because the cementum has nearly the same radiodensity as bone, it is not obvious radiographically. In multirooted teeth, bone should be present up to the apex of the furcation. The centre of the root is the radiolucent root canal, which houses the radicular portion of the pulp.

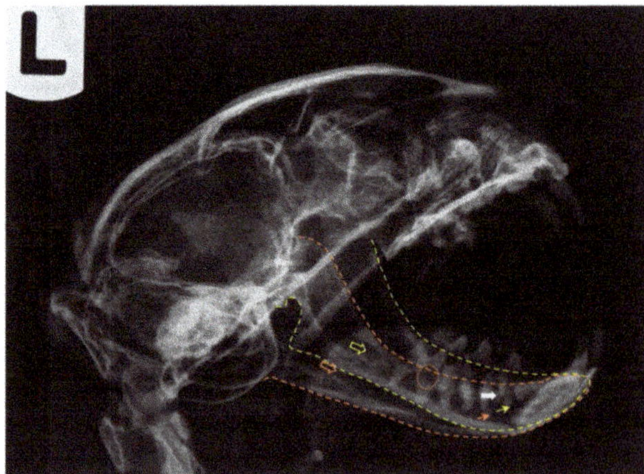

Figure 5. Postmortem radiography of a cat head, laterolateral projection (left lateral recumbency). The yellow outline corresponds to the left mandible and the orange one to the right jaw. The respective coloured empty arrows point the mandibular foramina, and the small arrows indicate the main mental foramina. The orange circle shows signs of tooth root resorption and the white arrow points evidences of a dental root fracture. Note that a plastic needle cap was used to keep the mouth open. As plastic is radiolucent, it does not interfere with the radiologic image.

According to Milella & Smithson [39] ten radiographic films should be taken to assess accurately each tooth in the cat's mouth (upper incisors, upper left canine (anterior-posterior oblique and lateral), upper right canine (anterior-posterior oblique and lateral), upper left maxillary premolars and molar, upper right premolars and molar, lower canines and incisors, lower right mandibular teeth, lower left mandibular premolars and molar). However, in older cats, full mouth radiographs are more to be recommended.

Hoffman & Ridinger [40] pointed out the value of obtaining annual dental radiographies, as well as the importance of interpreting dental radiographic findings in the light of the patient's systemic health and oral examination findings. This information will be essential when planning treatment in differentiating between primary oral disease and those secondary to systemic disease.

7. Topographical Considerations

The arteries of significance in clinical dentistry and oral surgery have the same origin in the common carotid artery (*arteria carotis comunis*), which branches into the internal and external carotid arteries (*arteriae carotis externa* and *interna*) [10]. The internal carotid artery blood supply is insignificant in the cat. Consequently, the external carotid artery, continuing as the maxillary artery, provides the majority of cerebral blood flow in the cat. The maxillary artery lies medial to the angular process of the mandible and branches into the maxillary rete before entering the skull through the orbital fissure [10]. Exclusively in the cat, the maxillary artery forms a network known as *rete mirabile a. maxillaris*, which is extracranial, near the *foramen ovale*. This network extends dorsally and laterally to the apex of the periorbital region. In the cat, the arteries for the eye and accessory structures arise from this network as do also the *rami retis*, which pass through the *fissura orbitalis* to connect with the *circulus arteriosus cerebri*. At the same time, the maxillary artery itself, being traceable through this network as a stronger vessel, leads the *a. infraorbitalis* [28].

According to Skerritt [41], there is an inverse relationship between the degree of development of the internal carotid and that of the anastomosing ramus of the maxillary artery: when one is large, the other is small. In no species are both of these channels fully developed. In cat and sheep, the lumen of the internal carotid artery becomes obliterated in the weeks or months after birth (although at birth it is fully functional). As a result, the whole of the adult brain is supplied by maxillary blood via the anastomosing ramus of the maxillary artery. In addition, a *rete mirabile a. maxillaris* occurs on the anastomosing ramus of the maxillary artery in all species in which the supply from the maxillary artery is well developed [41].

The significance of the *rete mirabile* has long been discussed. It was thought that it might eliminate pulsation before the blood reaches the brain itself. However, more recent observations indicate that the rete is involved in thermoregulation, as reported by Baker & Hayward [42]; using different anatomical nomenclature, they stated that the plexus of arteries that make up the carotid rete in the cat seems to be able to modify the temperature of central arterial blood as it enters the cranial cavity. In this short paper (only three pages) published in Nature, they described that in the cat, the carotid rete lies outside the cranial cavity, near the apex of the orbit, and is termed an 'extracranial rete', while in artiodactyls the rete is intracranial, lying in the cavernous sinus at the base of the skull. The extracranial rete lies within a venous lake, and the large surface area of the interlacing network of vessels may permit a 'countercurrent' exchange of heat between the arterial blood of the *rete* and the venous return from various regions of the head, having a profound effect on brain temperature that may have significant thermoregulatory consequences, allowing large, rapid changes in the temperature throughout the cranium (0.1–0.7 °C). The presence of the extracranial carotid rete is the most important cerebrovascular difference between the cat and the other species studied by Baker & Hayward [42].

Consequently, in the cat, maxillary arterial blood is distributed to all of the brain and any disturbance of the blood flow may have dramatic consequences. Regarding this aspect, it is currently proven that overextension of the mandibles of the cat (during oral and transoral procedures, such as intubation) can lead to compression of the rete and/or compression of the maxillary artery by the angular process of the mandible (Figure 6), leading to cerebral ischaemia and resulting in temporary or permanent cortical blindness, loss of hearing, or possibly death [43,44].

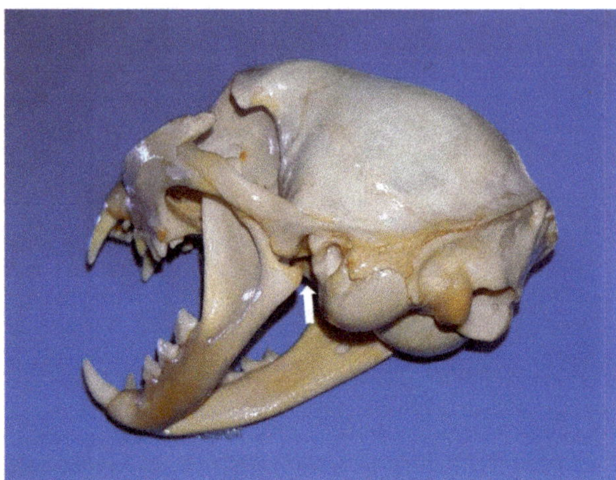

Figure 6. Caudolateral view of the cat head with the mouth open. The white arrow points to the channel narrowed by the *processus angularis*, through which the *a. maxillaris* runs to lead the *rete mirabile*.

A comprehensive understanding of the functional anatomy of all structures associated with the caudal angle of the cat mandible may explain why keeping a cat's mouth wide open for a prolonged period of time may result in temporary or permanent neurological deficits, unilaterally or bilaterally, post-anaesthesia [26]. Cats and dogs have the maxillary artery running through this area, but what gives only cats an increased risk of cerebral ischaemia when the mouth is wide open is that the mandibular angular process, presses against the area through which the maxillary artery passes. This reduces to some extent the maxillary artery blood flow. However, an additional feature is decisive to this occurrence: the internal carotid artery (leading the main blood supply to the brain, retina and inner ear) is functionally absent in cats, so all the blood to the brain is supplied exclusively by the bilateral maxillary arteries. Consequently, the longer the duration of the pressure, the higher the risk of onset of cerebral ischaemia and/or blindness and/or deafness. De Miguel García et al. [44] suggested (where the use of a mouth gag is essential for surgery) to reduce the size of the gag or to close the cat's mouth every few minutes (throughout the procedure) to enable restoration of the blood supply. However, the reason why cats are unequally affected is still unknown, although it could be due to collateral or altered blood flow through the basilar arteries (although, according to Skerritt [41], they only carry blood away from the arterial circle).

Thus, the use of spring-loaded mouth gags is no longer recommended in feline patients [10,26] as they apply continual force to keep the mouth open to such an abnormal degree [45]. It is also reported that the vascular flow is more compromised on the side ipsilateral to the mouth gag, perhaps because the distance between the angular process of the mandible and tympanic bulla is smaller on the ipsilateral side [46]. Another drawback would be that the tighter the lips and cheek become when the mouth is wide open, the more difficult it is to retract them, to perform surgery or a thorough oral examination [26]. In contrast, custom-made plastic mouth props (made from a syringe cap) have many benefits: (1) they are gentler on the jaws and appear to induce fewer alterations in blood flow; and (2) they are radiolucent, hence they do not interfere with diagnostic radiological imaging [26], among others. However, if a mouth gag must be used in a feline patient, the smallest possible gag (ideally between 20 and 30 mm), should be chosen to retract the upper and lower lips without difficulty and the duration of use minimised [26,47].

The Nomina Anatomica Veterinaria [7] does not clarify which species present molar salivary glands (*glandulae molares*). Okuda et al. [48] described the bulge lingual to the lower molar tooth to be a small salivary gland corresponding to the lingual molar gland. Interestingly, there is no equivalent salivary gland in the dog. Previously, Orsini & Hennet [14] stated in their review entitled 'Anatomy of the mouth and teeth of the cat' that 'just medial to the lower first molar on the floor of the oral cavity is a mass-like flap of oral mucosa with no known function; its prominent appearance leads to an incorrect identification as an abnormal finding'. This major salivary gland in the cat has two parts (Figure 7): the buccal and lingual molar gland. The secretory portion of the buccal molar gland is close to the commissure (between the M. *orbicularis oris* and the mucous membrane of the lower lip at the angle of the mouth) and it empties into the buccal cavity by several small ducts; and the lingual molar gland located within a membranous bulge caudolingual to the mandibular molar tooth (constituting the molar pad) [10]. Sometimes, the membranous molar pad enlarges and may be traumatised (when chewing), being important not to be mistaken for a tumour or polyp [49].

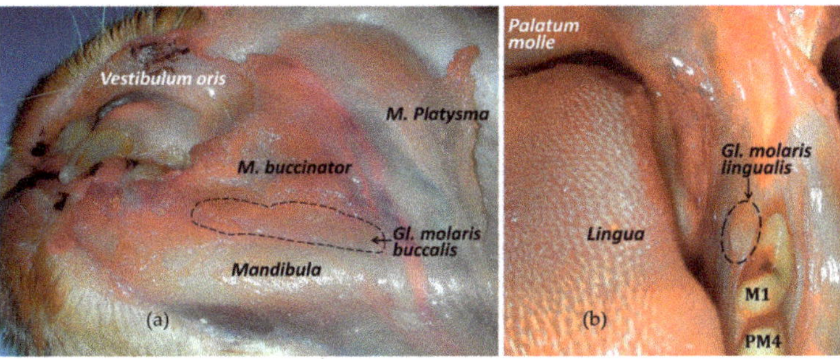

Figure 7. The cat molar salivary glands. (**a**) Ventrolateral view of the buccal molar gland. (**b**) Frontal view of the lingual molar gland, caudomedially to the first molar tooth.

8. Conclusions

Clinicians must have a deep knowledge of the functional anatomy of cats (taking into account that they present some important differences with respect to the dog) to achieve an effective and high-quality cat medicine.

A good knowledge of the anatomy of the mandible and the TMJ, and their relation to other important structures, such as blood vessels and nerves, is essential for an accurate interpretation of radiographic images and tomographic diagnostic techniques, in order to make a diagnosis and achieve good results in the management of different conditions.

Clinicians and surgeons are increasingly aware of animal welfare to avoid or, at least, minimise any suffering due to iatrogenic complications. Such as those related to a temporary or permanent blindness and/or deafness following general anesthesia when the mouth is held open with a gag (to the compression of the angular process on the maxillary artery, disrupting the blood flow to the maxillar *rete mirabile* and the brain); the resection of the lingual molar pad (that includes the lingual molar salivary gland), or those complications secondary to dental extraction.

Author Contributions: Conceptualization, M.L. and M.d.M.Y.; Writing—Original Draft Preparation, M.L., D.A.-P. and M.d.M.Y.; Writing—Review & Editing, M.L., D.A.-P. and M.d.M.Y. All authors have read and agreed to the published version of the manuscript.

Funding: This research received no external funding.

Institutional Review Board Statement: The Institutional Review Board statement was not required in this study as it has been carried out with cat cadavers, previously euthanised due to other pathologies. An oral consent was given by the owners allowing to use their dead pets for research purposes.

Data Availability Statement: Data sharing is not applicable to this article as no new data were created or analyzed in this study.

Conflicts of Interest: The authors declare no conflict of interest.

References

1. Johnston, R.; Thomas, R.; Jones, R.; Graves-Brown, C.; Goodridge, W.; North, L. Evidence of diet, deification, and death within ancient Egyptian mummified animals. *Sci. Rep.* **2020**, *10*, 1–14. [CrossRef] [PubMed]
2. Málek, J. *The Cat in Ancient Egypt*, 2nd ed.; British Museum Press: London, UK, 2006; ISBN 9780714119700.
3. American Pet Products Association's (APPA) 2019–2020 National Pets Owners Survey. Published 2020. Available online: https://www.americanpetproducts.org/press_industrytrends.asp (accessed on 2 July 2020).
4. The European Pet Food Industry (FEDIAF). Annual Report 2020. Published 2020. Available online: http://www.fediaf.org/annual-report.html (accessed on 2 July 2020).
5. Beaver, B.V. *Feline Behavior*, 2nd ed.; Elsevier Sciences: St. Louis, MO, USA, 2003.
6. Little, S. Dental and oral diseases. In *The Cat: Medicine and Management*, 1st ed.; Elsevier-Saunders: St. Louis, MO, USA, 2012; pp. 329–370, ISBN 978-1-4377-0660-4.
7. Nomina Anatomica Veterinaria, 6th ed.; World Association of Veterinary Anatomists (WAVA), 2017. Available online: http://www.wava-amav.org (accessed on 9 January 2020).
8. Schaller, O. *Illustrated Veterinary Anatomical Nomenclature*, 2nd ed.; Enke Verlag: Stuttgart, Germany, 2007; ISBN 978-3-8304-1069-0.
9. Robinson, P.P. The course, relations and distribution of the inferior alveolar nerve and its branches in the cat. *Anat. Rec.* **1979**, *195*, 265–271. [CrossRef] [PubMed]
10. Lobprise, H.; Dodd, J.R. *Wiggs's Veterinary Dentistry*, 2nd ed.; Willey-Blackwell: Hoboken, NJ, USA, 2019; ISBN 97811188116165.
11. Starkie, C.; Stewart, D. The Intra-Mandibular Course of the Inferior Dental Nerve. *J. Anat.* **1931**, *65*, 319–323.
12. Pitakarnnop, T.; Buddhachat, K.; Euppayo, T.; Kriangwanich, W.; Nganvongpanit, K. Feline (*Felis catus*) Skull and Pelvic Morphology and Morphometry: Gender-Related Difference? *J. Vet. Med. Ser. C Anat. Histol. Embryol.* **2017**, *46*, 294–303. [CrossRef] [PubMed]
13. Simpson, J.W.; Else, R.W. Disease of the oral cavity and pharynx. In *Digestive Disease in the Dog and Cat-Library of Veterinary Practice Series-BSAVA*, 1st ed.; Blackwell Scientific Publications: Oxford, UK, 1991; pp. 1–38. ISBN 0-632-02931-5.
14. Kim, S.E.; Arzi, B.; Garcia, T.C.; Verstraete, F.J.M. Bite forces and their measurement in dogs and cats. *Front. Vet. Sci.* **2018**, *5*, 1–6. [CrossRef]
15. Orsini, P.; Hennet, P. Anatomy of the mouth and teeth of the cat. *Vet. Clin. N. Am. Small Anim. Pr.* **1992**, *22*, 1265–1277. [CrossRef]
16. Herring, S.W. TMJ anatomy and animal models. *J. Musculoskelet. Neuronal. Interact.* **2003**, *3*, 391–407.
17. Arzi, B.; Staszyk, C. The Temporomandibular Joint through the Lens of Comparative Anatomy. In *Contemporary Management of Temporomandibular Disorders*; Springer: Cham, Switzerland, 2019; pp. 41–50. [CrossRef]
18. Dyce, K.M.; Sack, W.O.; Wensing, C.J.G. *Textbook of Veterinary Anatomy*, 4th ed.; Saunders-Elsevier: St. Louis, MO, USA, 2010; ISBN 978-1-4160-6607-1.
19. Latorre, R.; Arredondo, J.; Rodríguez, M.J.; Agut, A.; Vazquez-Autón, J.M.; López-Albors, O. Clinical anatomy of the relation between the temporomandibular joint and the mandibular nerve in the cat. Abstracts Presented at the Joint Meeting of the American Association of Clinical Anatomists (27th Annual Meeting) and the International Society for Plastination (15th Annual Meeting) in Honolulu, Hawaii, July 20–23, 2010. *Clin. Anat.* **2010**, *23*, 1005–1040. [CrossRef]
20. König, H.E.; Liebich, H.-G. *Anatomía de los Animales Domésticos. Tomo 1*, 2nd ed.; Editorial Médica Panamericana: Buenos Aires, Argentina, 2005; ISBN 84-7903-746-6.
21. Arredondo, J.; Agut, A.; Rodríguez, M.J.; Sarriá, R.; Latorre, R. Anatomy of the temporomandibular joint in the cat: A study by microdissection, cryosection and vascular injection. *J. Feline Med. Surg.* **2013**, *15*, 111–116. [CrossRef]
22. Knospe, C.; Roos, H. Zur Articulatio temporomandibularis der Hauskatze (Felis silvestris catus). *Anat. Histol. Embryol.* **1994**, *23*, 148–153. [CrossRef]
23. Gracis, M.; Zini, E. Vertical mandibular range of motion in anesthetized dogs and cats. *Front. Vet. Sci.* **2016**, *3*, 1–8. [CrossRef]
24. Watson, P.J.; Fitton, L.C.; Meloro, C.; Fagan, M.J.; Gröning, F. Mechanical adaptation of trabecular bone morphology in the mammalian mandible. *Sci. Rep.* **2018**, *8*, 1–12. [CrossRef] [PubMed]
25. Crompton, A.W.; Lieberman, D.E.; Aboelela, S. Tooth Orientation during Occlusion and the Functional Significance of Condylar Translation in Primates and Herbivores. In *Amniote Paleobiology: Perspectives on the Evolution of Mammals, Birds, and Reptiles, Carrano, M.T., Gaudin, T.J., Blob, R.W., Wible, J.R., Eds.*; University of Chicago Press: Chicago, IL, USA, 2006; pp. 367–388.
26. Reiter, A.M.; Soltero-Rivera, M.M. Applied Feline Oral Anatomy and Tooth Extraction Techniques: An illustrated guide. *J. Feline Med. Surg.* **2014**, *16*, 900–913. [CrossRef]
27. Hayashi, K.; Kiba, H. Microhardness of enamel and dentine of cat premolar teeth. *Jpn. J. Vet. Sci.* **1989**, *51*, 1033–1035. [CrossRef] [PubMed]

28. Hoshi, H. Comparative morphology of the mammalian mandible in relation to food habit. *Okajimas Folia Anat. Jpn.* **1971**, *48*, 333–345. [CrossRef] [PubMed]
29. Nickel, R.; Schummer, A.; Seiferle, E. Arteries from the head and neck. In *The Anatomy of the Domestic Animals: The Circulatory System, the Skin and the Cutaneous Organs of the Domestic Mammals*; Springer: Berlin, Germany, 1981; Volume 3, pp. 99–119, ISBN 978-0-387-91193-9.
30. Davis, D.D.; Story, H.E.; Story, H.E. *The Carotid Circulation in the Domestic Cat*; Field Museum of Natural History: Chicago, IL, USA, 1943. [CrossRef]
31. Skerritt, G. Clinical Neurology. In *King's Applied Anatomy of the Central Nervous System of Domestic Mammals*, 2nd ed.; Willey-Blackwell: Oxford, UK, 2008; pp. 374–415, ISBN 9781118401064.
32. Anderson, K.V.; Pearl, G. Transmedian innervation of canine tooth pulp in cats. *Exp. Neurol.* **1974**, *44*, 35–40. [CrossRef]
33. Wilson, S.; Johns, P.I.; Fuller, P.M. Accessory innervation of mandibular anterior teeth in cats: A horseradish peroxidase study. *Brain Res.* **1984**, *298*, 392–396. [CrossRef]
34. Izumi, H.; Kuriwada, S.; Karita, K.; Sasano, T.; Sanjo, D. The nervous control of gingival blood flow in cats. *Microvasc. Res.* **1990**, *39*, 94–104. [CrossRef]
35. Bellows, J. *Small Animal Dental Equipment, Materials, and Techniques*, 2nd ed.; Willey-Blackwell: Hoboken, NJ, USA, 2019; ISBN 9781118986622.
36. Cao, J.; Zhang, S.; Gupta, A.; Du, Z.; Lei, D.; Wang, L.; Wang, X. Sensory nerves affect bone regeneration in rabbit mandibular distraction osteogenesis. *Int. J. Med. Sci.* **2019**, *16*, 831–837. [CrossRef]
37. Yamashiro, T.; Fujiyama, K.; Fujiyoshi, Y.; Inaguma, N.; Takano-Yamamoto, T. Inferior alveolar nerve transection inhibits increase in osteoclast appearance during experimental tooth movement. *Bone* **2000**, *26*, 663–669. [CrossRef]
38. Lemmons, M. Clinical Feline Dental Radiography. *Vet. Clin. N. Am. Small Anim. Pract.* **2013**, *43*, 533–554. [CrossRef]
39. Milella, L.; Smithson, A. Radiology: Dental. Felis. Available online: www.vetstream.com/treat/felis/freeform/radiology-dental (accessed on 13 November 2020).
40. Hoffman, S.L.; Ridinger, S.K. Diagnostic Imaging in Veterinary Dental Practice. *J. Am. Vet. Med. Assoc.* **2020**, *256*, 1105–1109. [CrossRef] [PubMed]
41. Skerritt, G. Arterial supply to the central nervous system. In *King's Applied Anatomy of the Central Nervous System of Domestic Mammals*, 2nd ed.; Willey-Blackwell: Oxford, UK, 2018; pp. 56–72. ISBN 9781118401064.
42. Baker, M.A.; Hayward, J.N. Carotid Rete and Brain Temperature of Cat. *Nature* **1967**, *216*, 139–141. [CrossRef]
43. Barton-Lamb, A.L.; Martin-Flores, M.; Scrivani, P.V.; Bezuidenhout, A.J.; Loew, E.; Erb, H.N.; Ludders, J.W. Evaluation of maxillary arterial blood flow in anesthetized cats with the mouth closed and open. *Vet. J.* **2013**, *196*, 325–331. [CrossRef]
44. De Miguel Garcia, C.; Whiting, M.; Alibhai, H. Cerebral hypoxia in a cat following pharyngoscopy involving use of a mouth gag. *Vet. Anaesth. Analg.* **2013**, *40*, 106–108. [CrossRef]
45. Fraser, J.; Olin, S.J.; Saim, D. Feline Complications from Mouth Gags. *Cliniciansbrief* **2018**, *5*, 65–69.
46. Martin-Flores, M.; Scrivani, P.V.; Loew, E.; Gleed, C.A.; Ludders, J.W. Maximal and submaximal mouth opening with mouth gags in cats: Implications for maxillary artery blood flow. *Vet. J.* **2014**, *200*, 60–64. [CrossRef] [PubMed]
47. Reiter, A.M. Open wide: Blindness in cats after the use of mouth gags. *Vet. J.* **2014**, *201*, 5–6. [CrossRef] [PubMed]
48. Okuda, A.; Inouc, E.; Asari, M. The membranous bulge lingual to the mandibular molar tooth of a cat contains a small salivary gland. *J. Vet. Dent.* **1996**, *13*, 61–64. [CrossRef]
49. Clarke, D.E.; Caiafa, A. Oral Examination in the Cat: A systematic approach. *J. Feline Med. Surg.* **2014**, *16*, 873–886. [CrossRef]

Review

The Cat Mandible (II): Manipulation of the Jaw, with a New Prosthesis Proposal, to Avoid Iatrogenic Complications

Matilde Lombardero [1,*,†], Mario López-Lombardero [2,†], Diana Alonso-Peñarando [3,4] and María del Mar Yllera [1]

1 Unit of Veterinary Anatomy and Embryology, Department of Anatomy, Animal Production and Clinical Veterinary Sciences, Faculty of Veterinary Sciences, Campus of Lugo—University of Santiago de Compostela, 27002 Lugo, Spain; mar.yllera@usc.es
2 Engineering Polytechnic School of Gijón, University of Oviedo, 33203 Gijón, Spain; uo254292@uniovi.es
3 Department of Animal Pathology, Faculty of Veterinary Sciences, Campus of Lugo—University of Santiago de Compostela, 27002 Lugo, Spain; diana.alonso.penarando@rai.usc.es
4 Veterinary Clinic Villaluenga, calle Centro n° 2, Villaluenga de la Sagra, 45520 Toledo, Spain
* Correspondence: matilde.lombardero@usc.es; Tel.: +34-982-822-333
† Both authors contributed equally to this manuscript.

Simple Summary: The small size of the feline mandible makes its manipulation difficult when fixing dislocations of the temporomandibular joint or mandibular fractures. In both cases, non-invasive techniques should be considered first. When not possible, fracture repair with internal fixation using bone plates would be the best option. Simple jaw fractures should be repaired first, and caudal to rostral. In addition, a ventral approach makes the bone fragments exposure and its manipulation easier. However, the cat mandible has little space to safely place the bone plate screws without damaging the tooth roots and/or the mandibular blood and nervous supply. As a consequence, we propose a conceptual model of a mandibular prosthesis that would provide biomechanical stabilization, avoiding any unintended (iatrogenic) damage to those structures. The improvement of imaging techniques and a patient-specific prosthesis made of full biocompatible material are part of the future trends to improve patients' recovery.

Abstract: The cat mandible is relatively small, and its manipulation implies the use of fixing methods and different repair techniques according to its small size to keep its biomechanical functionality intact. Attempts to fix dislocations of the temporomandibular joint should be primarily performed by non-invasive techniques (repositioning the bones and immobilisation), although when this is not possible, a surgical method should be used. Regarding mandibular fractures, these are usually concurrent with other traumatic injuries that, if serious, should be treated first. A non-invasive approach should also first be considered to fix mandibular fractures. When this is impractical, internal rigid fixation methods, such as osteosynthesis plates, should be used. However, it should be taken into account that in the cat mandible, dental roots and the mandibular canal structures occupy most of the volume of the mandibular body, a fact that makes it challenging to apply a plate with fixed screw positions without invading dental roots or neurovascular structures. Therefore, we propose a new prosthesis design that will provide acceptable rigid biomechanical stabilisation, but avoid dental root and neurovascular damage, when fixing simple mandibular body fractures. Future trends will include the use of better diagnostic imaging techniques, a patient-specific prosthesis design and the use of more biocompatible materials to minimise the patient's recovery period and suffering.

Keywords: anatomy; feline; lower jaw; mandibular fracture; neurovascular supply; temporomandibular joint; tooth

Citation: Lombardero, M.; López-Lombardero, M.; Alonso-Peñarando, D.; Yllera, M.d.M. The Cat Mandible (II): Manipulation of the Jaw, with a New Prosthesis Proposal, to Avoid Iatrogenic Complications. *Animals* **2021**, *11*, 683. https://doi.org/10.3390/ani11030683

Academic Editor: Ralf Einspanier

Received: 21 December 2020
Accepted: 26 February 2021
Published: 4 March 2021

1. Introduction

As seen in the first part (The cat mandible (I) [1]), a deep knowledge of the functional anatomy of the feline mandible will be the basis for interpreting the diagnostic images that, in combination with symptoms, will help to achieve an accurate diagnosis about the current pathology. Afterwards, the clinician should come to a decision and a treatment has to be prescribed. A meticulous therapeutic decision-making process is essential to choose among alternatives and analyse consequences, as some treatment procedures include surgery.

Hence, the present review will be focused on how to solve different pathological situations of the feline mandible and temporomandibular joint, such as luxation or fractures. In addition, we will insist on the importance of keeping the integrity of the tooth roots and the mandibular canal neurovascular bundle, as they could be damaged during fracture repair.

2. Temporomandibular Joint Luxation

The temporomandibular joint (TMJ) is characterised by the presence of a whole intra-articular disc, although is very thin in the cat and in the dog. Many elements are related to this synovial condylar joint and gaining a comprehensive knowledge of this joint is required to correctly interpret its diagnostic images. This ability is essential in order to make a diagnosis and to achieve good results in the management of different conditions, minimising the incidence of surgical iatrogenic lesions. While TMJ caudal luxation might occur with fractures in the retroarticular process, mandibular fossa or mandibular head, rostral mandible movement is more common [2]. Unilateral rostrodorsal luxation of the mandibular condyle causes the lower jaw to shift laterorostrally to the luxation-opposite side and the inability of the cat to close its mouth fully due to tooth-to-tooth contact. This distinguishes it from open-mouth jaw locking, where the mouth is held wide open with no contact between mandibular and maxillary teeth [3]. The unilateral rostrodorsal luxation of the mandibular condyle could easily be reduced by placing a pencil between the maxillary fourth premolar (PM4) and the mandibular first molar (M1) on the affected side only and closing the mandible against the pencil while simultaneously relieving the jaw caudally (easing the condyle from the articular eminence). However, this treatment is contraindicated in animals with open-mouth jaw locking [3].

Open-mouth jaw locking is characterised by an inability to close the mouth that usually results from fixed mandibular coronoid process displacement lateral to the ipsilateral zygomatic arch and abnormal contact pressure between these two structures [2]. The history is important in the diagnosis, as it is usually observed after animals have yawned, groomed or vocalised [2]. Physical findings of a wide-open mouth and palpable coronoid process superficial to the zygomatic arch help to distinguish open-mouth jaw locking from TMJ dislocation [2]. Manual reduction is the first-line treatment method in open-mouth jaw locking, secondary to coronoid process zygomatic arch interlocking and temporomandibular dislocation. Resolution may be spontaneous or require manual correction, but recurrence is possible [2]. However, and according to Reiter and Lewis [3], the treatment consists of two phases: an acute treatment (under sedation) consisting of opening the jaw even further to release the coronoid process from the lateral aspect of the zygomatic arch, and then closing the mouth. A tape muzzle should be placed until definitive surgery is performed. This consists of a partial resection of the coronoid process, partial resection of the zygomatic arch, or a combination of both [3].

3. Mandibular Fractures

3.1. General Considerations

Mandibular fractures are commonly seen in practice, comprising up to 6% of all fractures in dogs, and 11–23% of all fractures in cats, according to Glyde and Lidbetter [4]. They occur more frequently than maxillary fractures [3]. The majority of mandibular fractures in cats may result from road traffic accidents, fighting injuries or falls from heights [4–6] and, unfortunately, also due to human abuse [7]. A traumatic aetiology commonly involves serious concurrent injuries requiring prompt clinical attention, mainly to the brain, maxilla and chest [8]. Management of life-threatening injuries and normalisation of patient physiology is required before surgical stabilisation of mandibular fractures [8].

Less commonly, pathological mandibular fractures may occur secondary to periodontal, oral neoplasia or metabolic disease, and iatrogenic fractures can also occur during dental treatment [3,4]. Although various types of injuries and trauma are typically responsible for fractures of the upper (maxilla) and lower jaw, certain risk factors may predispose a cat to fractures, including oral infections (e.g., periodontal disease, osteomyelitis), certain metabolic diseases (e.g., hypoparathyroidism) and congenital or hereditary factors resulting in a weakened or deformed jaw [9].

Independently of the aetiology, in cats, and according to Umphlet and Johnson [10], mandibular fractures accounted for 14.5% of all fractures seen in a total of more than 500 cat specimens (n = 517). Symphyseal fractures were the most common (73.3%), followed by fractures of the body (16%), condyle (6.7%) and coronoid process (4%). Complications developed more commonly in cats with multiple or open fractures. Clinical union occurred after an average of 6 weeks (range 3–12 weeks) for symphyseal fractures, 10 weeks (range 8–16 weeks) for body fractures, 6 weeks for coronoid fractures and 6 weeks (range 4–8 weeks) for condylar fractures [10]. In contrast, in dogs, fractures in the premolar region are significantly more frequent than in other regions [11]. Umphlet and Johnson [11] also reported that fractures in the rostral portion of the mandible had shorter average time to clinical union than did other mandibular fractures. However, the average time to clinical union for fractures in the caudal portions of the mandible was longer than that currently reported [11]. Nonetheless, overall prognosis depends on type, extent, location of trauma, quality of home care and selection of treatment modality [9].

Accordingly, Little [12] reported that mandibular fractures in cats are typically located in the area of mandibular symphysis or the mandibular ramus (fractures of the condylar process or coronoid process). The midportion of the mandibular body is less frequently fractured in cats.

A non-invasive approach should receive primary consideration, and an invasive option is employed only if non-invasive treatment is insufficient or impractical [13]. Tape muzzling is a non-invasive and inexpensive treatment option for mid-body and caudal mandibular fractures and for TMJ luxation and open-mouth jaw locking after manual correction [13]. Tape muzzles could be used as temporary, definitive or adjunctive therapy. This is a good method in cases of pathological fractures or where the bone is very porous and will not support a fixative device. Where fractures are stable, this is also a good technique [14]. It may be curative for mandibular fractures in immature, adolescent and young adult animals with good bone healing capacities. In addition, muzzling allows some TMJ movement, thus reducing the risk of ankylosis between fractured bones in that area [13].

In the cat, when the jaw is immobilised to allow healing of the fracture, the mouth should be kept open no less than 5 mm and no more than 10 mm (as measured between the incisal edges of the maxillary and mandibular incisors) to allow for the tongue to protrude and lap water and a slurry diet. If the mouth is open too far, it will result in difficulty in swallowing [13]. This immobilisation of the mandible, to limit oral opening, could be done with a tape muzzle, fixation with composite spanning the ipsilateral canine teeth, or through labial buttons placed with suture material [15].

Southerden et al. [16] reported that there is a low level of asymmetry between contralateral mandibles in cats, allowing the use of a mirror image of an intact mandible for planning and evaluating the accuracy of fixation of a contralateral mandible. The most consistent measurement among 27 specimens was the lateral ramus inclination angle. However, the least consistent measurements were ramus height and jaw width at the mental foramen [16]. This type of study may facilitate the development of standardised pre-contoured locking plates for cat mandible repair.

Regarding mandibular biomechanics, as reported by Spodnick and Boudrieau [17], a continuity of tensile to compressive stresses exists from one side of the bone to the other during bending stress. Maximal tensile stresses are present at the oral (alveolar) surface and maximal compressive stresses at the aboral (ventral) surface; therefore, distraction is created at the oral margin. These bending moments increase from caudal to rostral; furthermore, shear forces are maximal at the ramus, while rotational forces are most prominent rostral to the canine teeth and maximal at the mandibular symphysis [17]. Taking these biomechanics into account would be of great help when fixing a mandibular fracture. Consequently, invasive jaw fracture repair techniques (osseous wiring, external skeletal fixation and bone plating) should be carefully planned and be accompanied by dental radiography, both intra- and post-operatively [16].

3.2. Symphyseal Fractures

Mandibular symphysis fractures are the simplest and are best treated with cerclage wire [4]. Palpable instability of the symphysis is not pathognomonic for traumatic symphyseal separation, as instability may result from periodontal disease, laxity of the ligamentous attachment, neoplasia, or fracture of the mandible [18]. Relative to the placement of a circumferential wire for mandibular symphyseal fracture repair, and according to Glyde and Lidbetter [4], it is necessary to place a hypodermic needle to act as support, making a hole from the oral cavity (from the caudolateral edge of the canine tooth) and leave the wire end at the level of a skin incision in the intermandibular space; however, on the other side, the needle is introduced from the bottom to the oral cavity to allow removing the needle once the wire is in place. This is a logical sequence, taking into account that the wire should make a loop to press the two mandibles together and reassure the mandibular symphysis. Make sure that the incisor teeth remain in alignment; otherwise, step defects can be generated [15]. The wire should be removed once union has been achieved [4].

Parasymphyseal fracture is a common iatrogenic injury during extraction of the mandibular canine tooth in cats [15]. The fracture may occur due to pre-existing periodontal or endodontic disease, insufficient preparation prior to extraction, or excessive force used by the operator, or a combination [15,19]. It turns into an important pathology that remains undetected if postoperative radiographs are not obtained, as the fracture often is non-displaced [19]. As a recommendation of good practice, Hoffman [20], a diplomate of the American Veterinary Dental College and board-certified in veterinary dentistry, advised that two dental X-rays should be always taken: (1) before extractions (this will allow the veterinary dentist to assess the health of the bone and the anatomy of each tooth, including its roots, taking into account that advanced dental disease contributes to bone loss and increased risk of iatrogenic trauma), and (2) after an extraction to ensure that the entire dental root has indeed been removed. In addition, the client should be informed upfront that a jaw fracture is a possibility, to avoid difficulties regarding iatrogenic jaw fractures secondary to tooth extraction. Furthermore, her advice is that once you are faced with an iatrogenic mandibular fracture, the case should be always referred to a specialist [20]. Parasymphyseal fractures could be treated with circumferential wire [15].

3.3. Body Fractures

When assessing the mandibular body for fractures, the direction of the fracture and the location of the fracture in relation to the dental roots should be evaluated. From a biomechanical point of view, fractures could be simple (in which the fracture line is perpendicular to the long axis of the mandible), or oblique fractures, that may be described as favourable or unfavourable according to the difficulty of immobilisation. This distinction results from the forces that the muscles of mastication place on the mandible as they either compress (favourable) or distract (unfavourable) the fracture segments [15,21] (Figure 1). Hence, a fracture that travels dorsocaudal to ventrorostral is considered favourable, whereas a fracture that travels dorsorostral to ventrocaudal and distracts the fracture fragments is considered unfavourable [15,21,22]. Favourable fractures compress because the upward pulling of the masseter and temporalis muscles and the downward and caudal pulling of the digastricus will hold the fracture segments in apposition, to a large extent. They are relatively stable, and stabilisation of the tension surface may be all that is required for bone healing [21]. In contrast, in unfavourable fractures, the alveolar crestal bone is considered the tension surface, while the ventral cortex is considered the compression surface [21], and the muscular forces will lead to considerable displacement of the fracture segments [3].

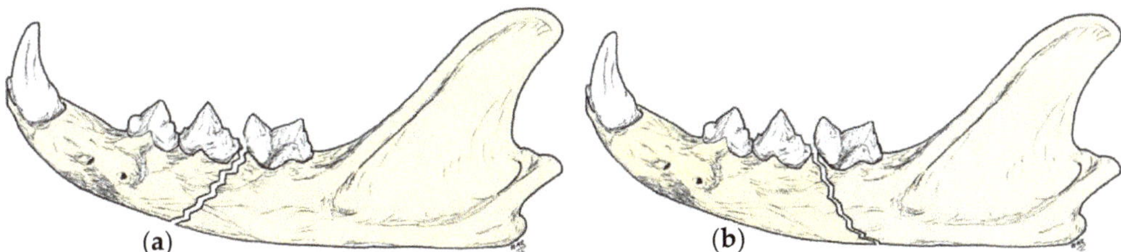

Figure 1. Drawings depicting the lateral view of the feline left mandible with a favourable fracture (**a**), as it compresses the fracture fragments, and an unfavourable fracture (**b**) in which the fracture segments are distracted.

As far as fracture repair is concerned, it is extremely important to take notice of a large percentage of the mandibular body being occupied by deep dental roots—between 50% and 70% of its dorsomedial depth [4]—and the neurovascular structures (along the mandibular canal), facts that largely limit the areas for safe implant/screw placement (i.e., for plate fixation). This is particularly true in cats, where only a small amount of 'free' bone exists rostrally to the first premolar and caudally to the molar tooth [4]. Radiologically, the mandibular canal can be placed parallel and just coronal to the ventral cortex (the radiopaque bone in the ventral mandibular body margin) as a thick, horizontal radiolucent line in close apposition to the mandibular premolar and molar dental root apexes [18,23]. In that sense, Bellows [21] advises "Unless you've had advanced training, avoid plating jaw fractures for fear of compromising dental roots. Also avoid placing intramedullary pins into the mandibular canal. The mandibular canal carries the neurovascular structures—it is not an intramedullary canal". Accordingly, one myth stated by Hoffman et al. [24], to dispel common misconceptions relative to "intramedullary pins in the mandibular canal are an option for treatment of fractures of the mandibular body", said that experimentally, inserting pins into the body of the mandible results in delayed healing and considerable damage to many dental roots, whereas clinically, malalignment is a common complication.

As stated by Lantz [25], the principles of facial repair include (1) restoration of occlusion and anatomic reduction, (2) stable fixation to neutralise detrimental forces on the fracture line(s), (3) preservation of blood supply, (4) preservation of soft tissue attachment to bone fragments by gentle tissue manipulations and minimal tissue elevation, (5) avoidance of iatrogenic dental trauma, not injuring the dental roots, and (6) extraction of diseased teeth at the line(s). The selected method of repair should provide occlusion and, ideally, rigid stability of all major fragments. In addition, the device should allow immediate return to oral function, be lightweight, economical and not cumbersome for the patient. However, Spodnick and Boudrieau [17] reported that removal of teeth may increase complications due to disruption of the blood supply and iatrogenic trauma to the adjacent tissues, including further displacement of the bone fragments, elimination of occlusal landmarks useful in realigning bone segments to allow functional occlusion, elimination of available structures for use in the fixation of bone fragments and creation of a large bony defect adding to the difficulty of reduction and stabilisation. Preservation of teeth involved within a line in a mandibular fracture has also been reported to have a favourable prognosis if optimal reduction and stabilisation of the jaw has been achieved. Therefore, removal of teeth is not advised unless the teeth involved are fractured (even here universal removal is not recommended if the tooth contributes to stabilisation and if the fracture of the tooth does not involve the root). Nonetheless, advanced periodontitis or periapical abscessation are situations in which in-line teeth should be removed since they have contributed to pathological fracture of the mandible [17].

Mandibular body fractures may be treated in various ways. Favourable body fractures could heal using conservative methods to get stabilization of the tension surface [15,21]. Internal fixation with interfragmentary wires is indicated for simple mandibular body fractures, such as unfavourable fractures. The primary function of osseous wiring is to reduce fractures and prevent their displacement by the passive function of the muscles of mastication. The first rule of osseous wiring is to do no harm or to attempt to avoid injury to other structures within the jaws, such as the mandibular canal, roots of teeth and the periodontal ligament. The second rule is always to attempt to re-establish normal functional occlusion [15], as the mouth must be able to shut after surgery. To repair an unfavourable fracture, two intra-osseus wires are needed (one dorsally and the other ventrally) between the two fragments [15], or a triangular method could be used instead, consisting in one hole rostrally to fix two wires in a perpendicular arrangement coming from two holes placed caudally to the fracture. The dorsal wire should have a horizontal disposition and the second wire provides additional stability and prevents shear or rotation of the fragments around the primary wire [8]. Additionally, interdental wiring could also be used to increase the fracture stability [15].

Another option to treat unfavourable fractures is by using internal fixation with conventional bone miniplates and screws. The advantages of internal rigid fixation in the treatment of mandibular fractures are the accurate restoration of normal anatomy and occlusion and rapid return to normal function [16]. Nevertheless, every effort should be made to avoid injury to the roots and periodontal ligament of the teeth and the mandibular canal during placement of the anchorage holes through the bone [15].

The mandible is not an easy bone in which to use plating techniques for stabilisation of fractures. According to Higgins [8], at least two plates are required in the mandible. One is placed along the alveolar surface to resist bending and act as a tension band. A secondary plate is required on the ventral surface to resist shear and rotation. However, the anatomy of the mandible in the cat precludes placement in a biomechanically advantageous position of the minimum two pins in each fragment without the risk of compromising either the dental roots or the mandibular canal [4,26]. It should be noted that dental damage may result in pain, infection, tooth death, periapical lesions and ultimately failure of fixation because of persistent infection and inflammation at the fixation site [4]. The application of other metal implants should be undertaken noting the following: large teeth occupy 70% of the depth of the bone, damage to vessels and nerves within the mandibular canal should be avoided and feeding tubes should be considered [27]. Thus, bone plates can be placed in the caudal toothless part of the mandible at the junction of the body and the ramus [4].

The article by Greiner et al. [28] biomechanically evaluates two plating configurations (using one or two internal fixation plates) to fix a simple transverse caudal mandibular fracture, in particular, shown in Figure 1 (and onwards), in which the legend states that the ideal region for miniplate application (where the bone is able to support internal fixation) in the cat mandible is depicted in blue. However, the blue colour coincides all along with the path of the mandibular neurovascular bundle, the main blood and sensory supply. They considered the nonideal region for miniplate application to be the area occupied by the dental roots, even though other figures show superimposition of the screw holes with the dental roots. Biomechanically, it could be an acceptable model, but it must be extremely painful for the cat patient. This happens because, usually, the mandible is considered like any other long bone, without taking into account its particularities: (1) the jaw has specialized structures, the teeth, whose roots are inserted in the alveoli and kept in position thanks to the periodontal ligaments, and (2) it does not have a medullary canal, but an inner canal that runs along the jaw (from caudal to rostral) providing blood supply and sensitive innervation, not only to the bone, but also the teeth and other related structures, such as mucosa and gingiva. Thus, if any of these structures become damaged, it does not matter that the prosthesis would be acceptable from a biomechanical point of view, because it is functionally unsatisfactory. Hence, it would be advisable that researchers, clinicians and surgeons be aware of the importance of maintaining the integrity of the neurovascular supply and dental roots in order to achieve a favourable outcome.

Plate placement on the ventral (compression) border of the mandibular body, which avoids the neurovascular structures of the mandibular canal, increases the load on the plate. Mono-cortical application of bone plates to the mid-buccal surface of the mandible has been recommended to reduce the risk of iatrogenic damage [4]. Consequently, standard bone plates in accordance with the feline mandibular size must be used. Nowadays, there are many companies producing osteosynthesis systems with titanium microplates (up to a minimum thickness of 0.6 mm to fix with 1 mm diameter screws). These sizes, and similar, are recommended for the repair of feline mandibular fractures. The plate should therefore be placed on the ventral surface to avoid these structures, although the tension side is the oral side. Plating should only be considered for simple fractures that can be very accurately reduced. Any malalignment may lead to dental malocclusion, which would be difficult to correct. Locking plates may be more useful than non-locking implants as, if plate contouring is not perfect, they should not distort the fragments. The most common complications of surgical repair are malocclusion and osteomyelitis [15,21].

In the case of comminuted or open fractures, in which soft tissue wounds prevent the use of internal fixation, an external fixation with pins and a mandibular bumper bar is appropriate for their reduction [26].

According to Spodnick and Boudrieau [17], the management of simple fractures (large fragments without comminution) could be done with routine induction and endotracheal intubation per os for anaesthetic maintenance and surgery. In these instances, anatomical re-alignment and reduction of the fragments, rather than dental occlusion, is used to determine the accuracy of surgical reduction. Alternatively, in cases of severely comminuted fractures or those with bone loss, dental occlusion must be used to access the accuracy of the surgical reduction. In these cases, the endotracheal tube impedes the assessment of occlusion by preventing full closure of the mouth and must be replaced to bypass the mouth (endotracheal intubation via pharyngotomy). Hence, occlusion is used to determine the accuracy of the reduction when comminution or gaps in the bone are present. Simpler fractures may be reconstructed anatomically. As usually one side of the head/face is more severely injured, a reasonable approach is to repair the side with the simpler fractures first. It is highly recommended to repair the mandible from caudal to rostral, with symphyseal separations secured as the final step [17]. Interestingly, the ventral approach to each mandible facilitates exposure and bone fragment manipulation, including the ability to perform an accurate reduction and stabilisation. Consequently, the patient should be placed in dorsal recumbency to get the head in the ventral approach, fixing it by taping the maxilla to the table [17].

3.4. Fractures of the Ramus

Fractures of the mandibular ramus are relatively stable because the surrounding muscles usually prevent gross displacement of the fracture segments [3]. Condylar process fractures often heal as pain-free and functional non-union without surgery, but comminuted fractures could result in a TMJ ankylosis in immature and young cats [3]. Ramus fractures in dogs and humans are frequently stabilised using internal rigid fixation with plates and screws, providing accurate reduction and good construct stability [16]. But in cats, this is complicated because of their small size, the need for greater contouring of implants, difficulty in positioning small fragile fragments of bone during the application of implants and the very small cross-sectional surface area of bone at the fracture sites, making accurate anatomical reduction challenging [16]. However, Southerden et al. [16] proposed the development of a small range of standard pre-contoured locking plates for the fixation of caudal mandibular fractures in cats due to the small variation in shape and size of mandibles between animals (among 38 specimens).

3.5. Impairment of the Nervous Supply

It should be taken into account that more than the upper two-thirds of the body of the mandible is occupied by dental roots. The ventral third includes the mandibular canal, containing the inferior alveolar nerve and associated blood vessels, and the inferior alveolar artery and vein. The inferior alveolar nerve provides sensory innervation for the teeth and leaves the bone through three mental foramina as the mental nerves. These nerves are sensory to the soft tissues of the rostral part of the mandible. The blood vessels in the mandibular canal supply all the teeth by the way of small dental branches (*rami dentales*) entering the apical foramina and the bone itself [17].

In humans, according to Misch and Resnik [29], and regarding nerve injuries after dental implant procedures, traumatic and iatrogenic nerve complications may involve total or partial nerve resection, crushing, stretching or entrapment injuries. As a consequence, the resulting sensory deficits may range from nonpainful minor loss of sensation to a permanent and severe debilitating pain dysfunction. Regarding oral and maxillofacial tumours in cats, Little [12] shows an example of mandibulectomy (removing left total and right partial mandibles), in which "the inferior alveolar artery and vein entering and exiting the mandibular canal through the mandibular foramen at the medial aspect of the mandible are ligated and transected". But nothing is said about the alveolar mandibular nerve, as it is supposed that its resection might become very painful (even post-resection) as it has a sensory component. The point is that humans can express and describe their loss of sensation or pain, but what about cat patients after mandibular surgery? It is reasonable to assume that they can also suffer from the same type of impairment, but they cannot describe their sensations and could be suffering from severe debilitating pain dysfunctions or neuropathic pain and not even be willing to eat normally by themselves again.

In mandibular fractures, besides mandibular reconstruction, the integrity of the mandibular nerve is a fundamental aspect to take into consideration, since it has a sensitive, in addition to the motor, component. Consequently, it is not recommended to proceed to cut the mandibular nerve just to remove it, given that it can induce trigeminal ganglion degeneration, as reported by Gobel and Binck [30] following pulp removal in cats, inducing degenerative changes in primary trigeminal axons and in neurons in the *nucleus caudalis*.

Mandibular fracture types, their incidence and treatment methods are summarized in Table 1.

Table 1. Mandibular fracture types. Their incidence, possible treatment methods and some recommendations are compiled.

	Incidence (%)	Cause	Treatment Methods	Pay Special Attention to …
Mandibular fractures in general	11–23 [4] 14,5 [10]	→ Road traffic accidents, fighting injuries, falls from heights, human abuse [4–7]. → Secondary to neoplasia, metabolic disease, and dental treatment [8].	→ A non-invasive treatment should be considered first [13]. → In multiple fractures, repair the mandible from caudal to rostral [17].	→ The naturally contaminated environment of the oral cavity. → Not damaging teeth and tooth roots. → Keep the integrity of the neurovascular supply in the mandibular canal. → Multiple or open fractures cause more complications [10]. → Infection and persistent periodontal disease can lead to osteomyelitis and non-union of fracture fragments [15].
Mandibular Fracture Types	**Incidence (%) [10]**	**Treatment Methods**	**Clinical Union (Weeks) [10]**	**Pay Special Attention to … Recommendations**
Symphyseal	73.3	→ Cerclage wire [4,15,26].	6 (3–12)	→ Be sure that the incisor teeth remain in alignment; otherwise, step defects can be generated [15].
Parasymphyseal		→ Osseous circumferential wiring [15].		→ Fracture is often non-displaced [19]. → Possibility of iatrogenic fracture after canine extraction when pre-existing periodontal disease, insufficient preparation prior to extraction or use of excessive force, or a combination [12,15,19]. → Do two radiographies: before and after dental extraction [20]. → Inform the client in advance that iatrogenic fracture is a possibility after canine extraction [20].
Body	16.0	→ Simple fractures (fracture line is perpendicular to the long axis of the mandible): internal fixation with interfragmentary wires [26]. → Oblique fractures: ◆ Favourable (mastication muscles compress the bone fragments, relatively stable): Conservative methods are enough to get stabilization of the tension surface [15,21]. ◆ Unfavourable (mastication muscles distract bone fragments): (A) Two intraosseus wires (one dorsally and the other ventrally) between the two fragments [15]; (B) Triangular method (two wires caudally to the fracture in a 90° angle to reach the same rostral hole [15]; (C) Internal fixation with conventional bone miniplates and screws [8,16]. → Open or comminuted fractures: external fixation with pins and mandibular bumper bar [26].	10 (8–16)	→ Keep the integrity of the neurovascular supply in the mandibular canal. → Avoid damaging the roots and periodontal ligament of the teeth [4,15]. → In-line tooth removal is not advised unless the teeth involved are fractured [8,17]. → Do postoperative radiographs, especially when severe periodontitis or other debilitating mandibular bone pathologies [19]. → The wiring must be placed so that its acts as a tension band to create interfragmentary compression [8]. → The most common complications of surgical repair are malocclusion and osteomyelitis [8]. → Ventral approach facilitates exposure, bone fragment reduction and stabilization [17]. → Use bone plates and screws in accordance with the cat mandibular size.
Ramus:				
Condylar process	6.7	→ Simple fractures heal by themselves as a functional and painless nonunion [3].	6 (4–8)	→ Comminuted fractures could generate TMJ ankylosis in young cats [3].
Coronoid process	4.0	→ A non-invasive treatment should be considered first [13].	6	

4. Prosthesis Proposal to Fix a Simple Fracture of the Mandibular Body

Taking into account all the concerns previously revealed, we propose a fixation method to repair simple fractures of the mandibular body that will provide acceptable rigid biomechanical stabilisation yet avoiding dental root and neurovascular damage.

The current prosthesis design proposal was made by using the design program Solidworks®. This computer-aided design (CAD) model was built starting from the three-dimensional (3D) model of a real jaw. A cloud of points, attained after laser scanning a real jaw, has been post-processed to obtain the 3D model of the cat mandible used. Meshlab and Cloud Compare software were used to post-process the data obtained by laser scanning.

The proposed prosthesis (Figures 2–5) is a conceptual design that, if referred to the technological maturity level scale known as Technology Readiness Levels (TRL), would be equivalent to a TRL2, which corresponds to the formulation of a conceptual solution without yet testing experimentally. Hence, seven other TRL levels would lay ahead, including research in the laboratory environment (up to TRL4), then, the simulation environment (TRL5–6), and finally, the real environment (TRL7–9).

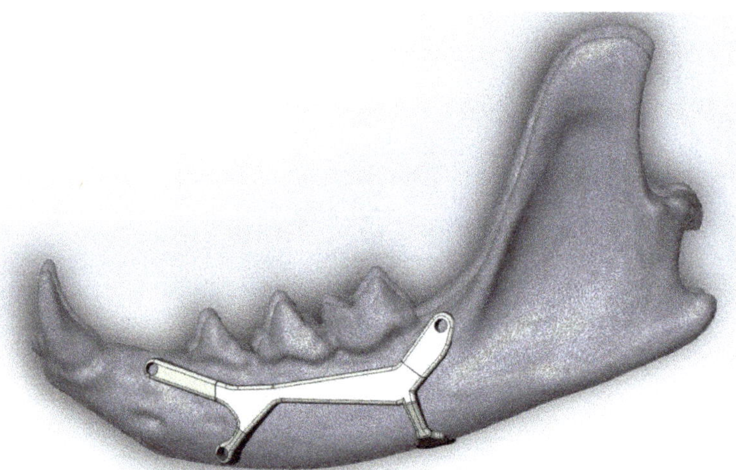

Figure 2. Lateral view of the mandibular prosthesis conceptual model in place, as the upper screws should be fixed where there are no dental roots. This model would be useful to repair body fractures between the third premolar and the first molar.

We think it could achieve good results in further TRL levels as it has three fixation points with small screws and the fourth is like a folded tab for fastening the ventral edge, therefore avoiding drilling the mandibular canal. The design with a two-sided 'Y' was chosen in a way that the screw positions do not imply any risk of perforating dental roots or damaging the neurovascular support. This shape also allows keeping safe the mental nerve that goes through the main mental foramen and also avoiding the caudal mental foramen.

The suitable material to implement this prosthesis, given its high requirements of resistance and biocompatibility, is titanium. The optimal manufacturing method would be 'additive manufacturing' (a transformative approach to industrial production that enables the creation of lighter, stronger parts and systems), i.e., metal 3D printing, since the prosthesis is a unique piece of small size, and up to now, this is the only method that makes production economically viable. In addition, given that its geometric complexity does not pose problems for its manufacture, this method is considered the best option.

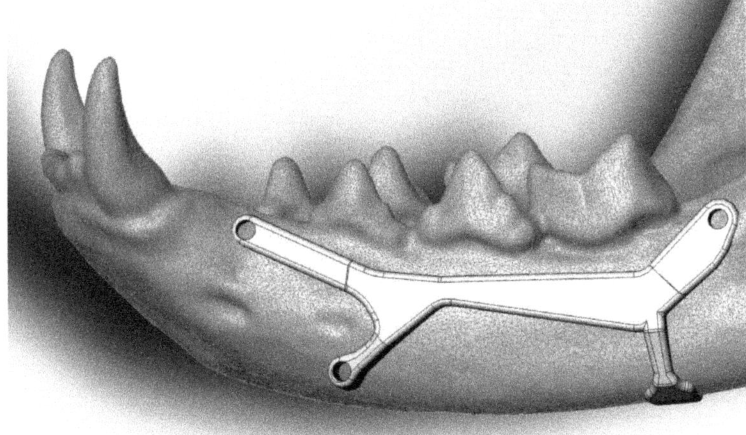

Figure 3. Magnification of Figure 2. The asymmetric shape of the anterior horizontal "Y" avoids damaging the nerves that come out through the main and mental foramina.

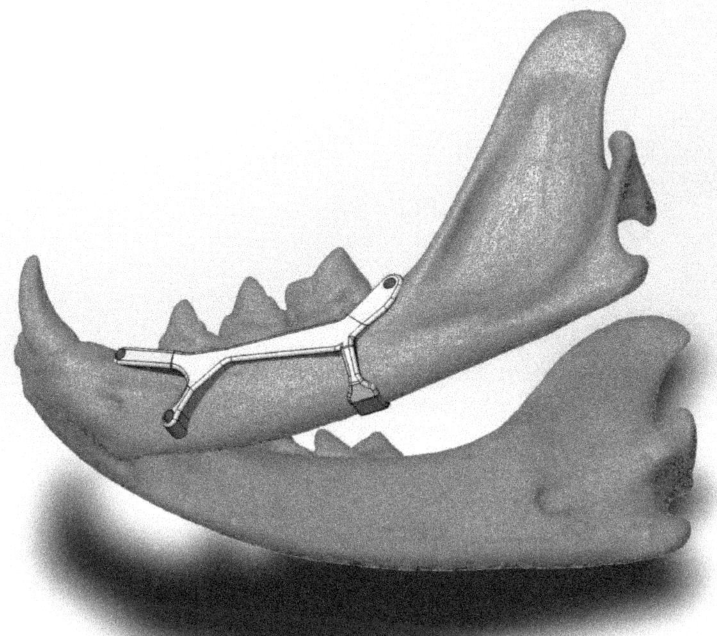

Figure 4. A ventrolateral view of the left jaw displaying the prosthesis conceptual model to show the fourth fixing point with no screw holes. This part consists of a flat hook-like device that surrounds and embraces the mandibular ventral margin to avoid damaging the neurovascular supply when drilling the mandibular canal. As this prosthesis is custom-designed, the flat hook size (thickness, width and length) will be variable, depending on the patient.

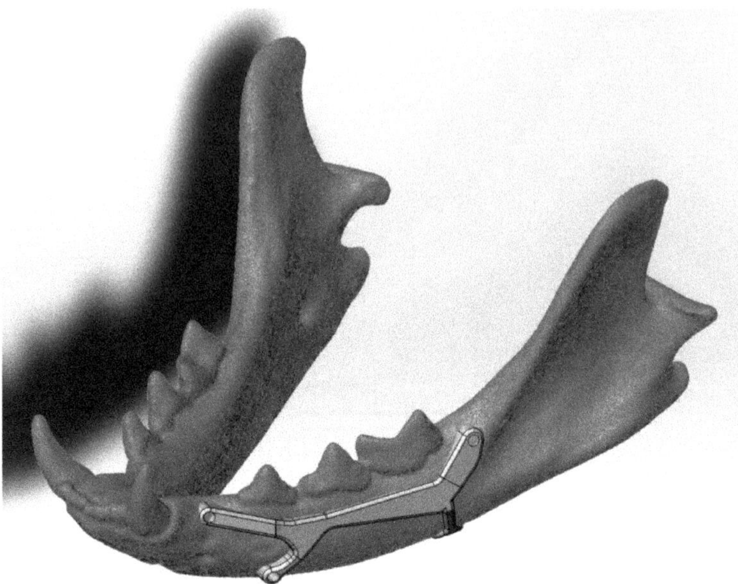

Figure 5. A frontolateral view of the left mandible with the proposed prosthesis. Note the prosthesis thickness is variable depending on the mandibular area in order to fully adapt to its contour. Prosthesis thickness does not exceed 1.2 mm at any point.

In order to promote good adhesion of the prosthesis to the bone, and trying to avoid damage to the bone, the surface of the prosthesis should have a high roughness (undulations, geometric patterns with protrusions or some geometry with an equivalent effect). Thus, the effective area of contact with the bone (i.e., the actual contact area) would be minimised while increasing the friction force of the prosthesis with respect to the bone, which greatly reduces the possible relative slipping that may exist between the elements.

4.1. *Calculations*

Considering that at the time of application of the prosthesis, the jaw is immobilised, and that once the bone has healed the stresses of the prosthesis are greatly reduced, the function of the prosthesis at the beginning consists fundamentally in ensuring that the bone fragments are correctly and firmly positioned. A calculation of resistance based on a bite force of 10 kg is considered conservative.

4.2. *Flexural Strength*

Considering this force and that the central part of the prosthesis is the weakest part (its most critical section), which is placed 2 cm away from the canines, it is calculated that it supports a moment (M) of:

$$M = F \times d = 50 \text{ N} \times 25 \text{ mm} = 1250 \text{ N}\cdot\text{mm}$$

where:

- F is the biting force (N)
- d is the distance between the point of application of the force and the point of evaluation (the maximum distance to obtain the most critical effort) (mm)

Note that the force considered is 5 kg, since it is assumed that half of the effort is carried by the other jaw.

Taking into account the moment calculated previously, a first approximation of flexure resistance could be obtained dividing the moment by the bending moment resistance (Wb), as is shown in the next equation [31], with a considered thickness of 1.5 mm and width of about 3 mm, there is a maximum bending stress of:

$$\sigma^{max} = \frac{M}{\frac{b \cdot h^2}{6}} = \frac{1250\ \mathrm{N{\cdot}mm}}{\frac{1.5\ \mathrm{mm} \times (3\ \mathrm{mm})^2}{6}} = 555.56\ \mathrm{MPa} \tag{1}$$

Equation (1): Maximum traction tension considering only flexion efforts.
where:

- b corresponds to thickness value (mm)
- h corresponds to height value (mm)
- M corresponds to flexure moment value (N·mm)
- σ corresponds to tension value (MPa)

The value of the bending moment resistance is obtained considering a squared section of flexure; if the design of the prosthesis changes, as a result of an irregular section chosen, the finite element method (FEM) would be needed to calculate the maximum bending tension.

As it can be seen, the stress obtained is considerably lower than the elastic limit of titanium. Thus, it could be considered a valid measurement. However, a balance should be struck between material resistance and physiology. Regarding the field of material resistance, it is advisable to make the prosthesis with a width as large as possible to achieve greater rigidity. However, this is inadvisable from the physiological point of view. Therefore, it is a question of achieving a balance between good rigidity and comfort.

4.3. Shear Resistance of Screws

Screws with a diameter of 1 mm have been considered, and taking into account the previous data and the conservative hypothesis that all the force is supported by one of them, the following shear stress is obtained [31]:

$$\tau^{max} = \frac{4}{3} \times \frac{F}{\pi r^2} = 85\ \mathrm{MPa} \tag{2}$$

Equation (2): Maximum shearing tension considering bite force
where:

- τ corresponds to shear tension in screws (N/mm)
- F corresponds to bite force (F)
- r corresponds to screw radius (mm)

The resulting stress is considerably lower than the shear strength of titanium, which means that the diameter of the screws can be even smaller.

Depending on the shape or position of the screws, a deeper calculation based in the FEM would be needed.

As previously indicated, the prosthesis shown is purely conceptual, and although an idea of the thickness and width that might be needed has been given, the calculations must be reviewed in practical clinical cases. Moreover, it would be convenient to undertake a more detailed study based on calculation software using the finite element method to obtain a more precise notion of the stresses to which the prosthesis is subjected. A topological optimisation of the geometry using suitable software could also be useful.

The current proposal would be equivalent to a TRL2 prototype according to the well-known TRL scale applied to technology, whose value (from 1–9) gives an indication of the level of maturity of the product, which implies that it is the formulation of a concept that has not yet been validated in the laboratory. Therefore, our proposal considers the mandibular prosthesis concept like an embryo, without practical validity, although its correct evolution and development could mean important advances in the field of intervention for mandibular fractures, not only in cats, but also in other species, by avoiding complications caused by damaging the integrity of the tooth roots and neurovascular supply when placing the screws to fix the jaw prosthesis.

This prosthesis model aims to visually reveal the proposals and ideas expressed throughout this review, without delving into the details about main dimensions, materials, manufacturing process and other aspects that may affect its implementation. Prostheses of these characteristics should be designed specifically for each specimen, adapted in a 3D model of the cat patient's jaw. In this way, not only will the size be adjusted, but the surfaces will also adapt much better, and thus a better grip will be ensured with all the advantages that this entails for fracture healing.

5. Future Trends

On one hand, the use of virtual surgical planning (VSP) and CAD/CAM (Computer-Aided Design/Computer-Aided Manufacturing) technology has added a new dimension to surgical planning, especially in the areas of craniomaxillofacial trauma, orthognatic surgery and reconstructive maxillofacial surgery [32]. This technology allows increased accuracy of reconstruction, decreased operative time, decreased flap ischaemic times, ease of use, improved predictability of outcomes, improved patient satisfaction and decreased complications [30]. In addition, with the aid of the surgical cutting guide, the positioning and fixation of the prosthesis would be accurate and almost flawless, thereby greatly reducing the operating time [32].

On the other hand, a comparison of three cone-beam computed tomography (CBCT) methods with dental radiography was analysed by Heney et al. [33]. They found that CBCT methods were better suited than dental radiography to the identification of anatomical structures in the full mouth of cats [15,21]. Cone-beam CT may prove to be the next major advancement in veterinary dentoalveolar and maxillofacial imaging because of its ability to provide 3D imaging at a lower cost than conventional CT, and with a lower radiation risk. The use of rapid scan technology, which allows faster image acquisition than conventional CT, and the ability to post-process the volumetric data into various two-dimensional (2D) and 3D reconstructions, makes CBCT an attractive imaging modality [33].

Liptak et al. [34] were one of the first to report a partial reconstruction of the mandibular body with a customised 3D-printed titanium prosthesis in a cat after removal of a mandibular osteosarcoma, as mandibulectomy in cats is associated with a high complication rate [35], including short-term (<4 weeks) and long-term (>4 weeks) adverse effects such as dysphagia or inappetence, mandibular drift or malocclusion. To avoid anorexia after surgery due to pain, supplemental tube feeding is recommended following mandibulectomy in cats. However, almost one in eight of the cats in the study never returned to voluntary eating after mandibulectomy [35]. Liptak et al. [34] pointed out that, in cats and dogs, dental roots and neurovascular structures comprise most of the bone volume in the mandibles and maxilla, and avoiding these structures is important during mandibular and maxillary repair. This is a further challenge in the rostral portion of the mandible because the canine dental root fills most of the mandible. The rationale for avoiding the dental roots is the high likelihood of tooth death and consequent periapical periodontitis, resulting in infection and potential implant failure. In general, this case is a good example of the future trends in cat mandibular osteosynthesis. However, we consider that surgeons would have been more concerned about the following issues: (1) neurovascular bundle and (2) the screw size used in this cat patient. Regarding the neurovascular supply, the authors did not report the fate of the neurovascular bundle in the mandibular canal nor what they did with these structures before/during/after cutting the mandibular body [34]. Also, in this case report, a bone fragment was kept in the rostral part of the mandible to fix the prosthesis (after a mandibulectomy of 40 mm in the mandibular body). It could be expected that lacking ipsilateral blood supply and innervation, this fragment would not survive. On the contrary, it showed no evidence of failure 14 months postoperatively. It is reasonable to think that this bone fragment might have achieved collateral blood supply from the mandibular symphysis vessels due to tissue hypoxia that may have promoted angiogenesis. However, in our opinion, surgeons must be more attentive to the importance of avoiding damage to the neurovascular bundle because new blood vessels may grow from pre-existing ones (angiogenesis), but the sensitive (and motor) neural components do not recuperate and could cause additional neural damage. This is important to the well-being of the cat. Regarding point (2), the 2.0 mm-diameter screws used to fix the custom-made prosthesis seemed too large and long for the cat mandible, and after traversing both bone cortexes, their tips would become embedded into other structures, such as muscles or damage any intermandibular structure. Hence, using a screw size suitable to the cat patient is highly recommended.

On one hand, the mandibular prosthesis size should be adapted to the small size of feline patients. It seems that most pet prostheses are designed for bigger specimens (except those designed for a specific cat patient, printed in titanium), they must have no sharp edges, and finally, they should have plenty of holes to place the screws and for muscle or tissue attachment. However, the screws used are usually too long, far surpassing the contralateral cortical bone.

On the other hand, materials used in osteosynthesis must be more biologically acceptable because soft tissue is not going to attach to metal [36], so a deeper understanding is needed, and further research will be required to produce more biocompatible prostheses.

To close this section, we cannot agree more with Vaughan [36], who stated "custom-shaped implants will likely be part of the future, but the process needs to be refined from a biomechanical and biological perspective".

6. Conclusions

In the body of the cat mandible, dental roots and the mandibular canal (with the vascular supply and the inferior alveolar nerve) occupy most of the volume. Therefore, in mandibular fractures (due to a variety of causes, such as periodontitis, tooth resorption, trauma, or secondary to tooth extraction), it makes it challenging to apply a plate with fixed screw positions without invading dental roots or neurovascular structures. Otherwise, it would be a very painful process, including the failure of fixation due to chronic infection and inflammation at the fixation site. In the face of all these difficulties, we proposed a suitable prosthesis design, produced by additive manufacturing, that would provide acceptably rigid biomechanical stabilisation and avoid damage to any of those structures when fixing a mandibular body fracture. The future depends on the improvement of diagnostic images and Computer-Aided Design/Computer-Aided Manufacturing (CAD/CAM) technology to manufacture custom-designed prostheses made of highly biocompatible material.

Author Contributions: Conceptualisation, M.L. and M.d.M.Y.; Writing—Original Draft Preparation, M.L., D.A.-P., M.L.-L. and M.d.M.Y.; Prosthesis conceptualisation M.L.-L. and M.L.; Writing—Review and Editing, M.L., D.A.-P., M.L.-L. and M.d.M.Y. All authors have read and agreed to the published version of the manuscript.

Funding: This research received no external funding.

Institutional Review Board Statement: The Institutional Review Board statement was not required as no animals were used to carry out this study.

Data Availability Statement: Data sharing is not applicable to this article as all data associated is available in the text.

Conflicts of Interest: The authors declare no conflict of interest.

References

1. Lombardero, M.; Alonso-Peñarando, D.; Yllera, M.M. The cat mandible (I): Anatomical basis to avoid iatrogenic damage in veterinary clinical practice. *Animals* **2021**, *11*, 405. [CrossRef] [PubMed]
2. Hsuan, L.; Biller, D.S.; Tucker-Mohl, K. Open mouth jaw locking in a cat and a literature review. *Isr. J. Vet. Med.* **2017**, *72*, 54–59.
3. Reiter, A.; Lewis, J.R. Trauma-associated musculoskeletal injury to the head. In *Manual of Trauma Management in the Dog and Cat*; Drobatz, K.J., Beal, M.W., Syring, R.S., Eds.; Willey-Blackwell: Delhi, India, 2011; pp. 255–292. ISBN 978-0-470-95831-5.

4. Glyde, M.; Lidbetter, D. Management of fractures of the mandible in small animals. *Practice* **2003**, *25*, 570–585. [CrossRef]
5. Tundo, I.; Southerden, P.; Perry, A.; Haydock, R.M. Location and distribution of craniomaxillofacial fractures in 45 cats presented for the treatment of head trauma. *J. Feline Med. Surg.* **2019**, *21*, 322–328. [CrossRef] [PubMed]
6. Knight, R.; Meeson, R.L. Feline head trauma: A CT analysis of skull fractures and their management in 75 cats. *J. Feline Med. Surg.* **2018**, *21*, 1120–1126. [CrossRef]
7. Kleftouri, S.; Panagopoulou, E.; Kouki, M.T.; Papadimitriou, S.A. Fractures of the mandible in cats. Retrospective study of 23 cases. *Hell. J. Companion Anim. Med.* **2017**, *6*, 20–28.
8. Higgins, B. Fractures of the mandible. In Proceedings of the World Small Animal Veterinary Association 38th World Congress, Auckland, New Zealand, 21–24 March 2013.
9. German, A. Upper and Lower Jaw Fracture in Cats. PetMD. Available online: https://www.petmd.com/cat/conditions/mouth/c_ct_upper_lower_jaw_fracture (accessed on 9 September 2020).
10. Umphlet, R.C.; Johnson, A.L. Mandibular Fractures in the Cat. A Retrospective Study. *Veter. Surg.* **1988**, *17*, 333–337. [CrossRef] [PubMed]
11. Umphlet, R.C.; Johnson, A.L. Mandibular Fractures in the Dog. A Retrospective Study of 157 Cases. *Veter. Surg.* **1990**, *19*, 272–275. [CrossRef] [PubMed]
12. Little, S. Dental and Oral Diseases. In *The Cat: Medicine and Management*; Elsevier-Saunders: Amsterdam, The Netherlands, 2012; pp. 329–370. ISBN 978-1-4377-0660-4.
13. Colmery, B.H.; Hale, F.A.; Hoffman, S.L.; Johnston, N.W.; Kuntsi-Vaattovaara, H.; Reiter, A.M. Concern about dental trauma after invasive jaw fracture repair. *J. Small Anim. Pract.* **2011**, *52*, 448–449. [CrossRef]
14. Steenkamp, G. Non-Invasive fracture repair techniques in dogs and cats. In Proceedings of the World Small Animal Veterinary Association, Cape Town, South Africa, 16–19 September 2014.
15. Lobprise, H.; Dodd, J.R. *Wiggs's Veterinary Dentistry*, 2nd ed.; Willey-Blackwell: Hoboken, NJ, USA, 2019; ISBN 9781118816165.
16. Southerden, P.; Haydock, R.M.; Barnes, D.M. Three Dimensional Osteometric Analysis of Mandibular Symmetry and Morphological Consistency in Cats. *Front. Veter. Sci.* **2018**, *5*, 1–7. [CrossRef] [PubMed]
17. Spodnick, G.J.; Boudrieau, R.J. Mandible: Anatomy, Biomechanical Principles, Surgical Approaches. In Proceedings of the Operative treatment of Veterinary CMF Trauma & Reconstruction, Las Vegas, NV, USA, 22–24 January 2018; pp. 47–52.
18. Lemmons, M. Clinical Feline Dental Radiography. *Veter. Clin. N. Am. Small Anim. Prac.* **2013**, *43*, 533–554. [CrossRef] [PubMed]
19. Reiter, A.M.; Soltero-Rivera, M.M. Applied Feline Oral Anatomy and Tooth Extraction Techniques. *J. Feline Med. Surg.* **2014**, *16*, 900–913. [CrossRef] [PubMed]
20. Hoffman, S.L. Increased Number of Iatrogenic Jaw Fractures. *AVMA PLIT* **2015**, *34*, 3–4. Available online: https://www.im3vet.com/iM3US/media/Documents/professional_liability_Dental_X-ray.pdf (accessed on 2 March 2021).
21. Bellows, J. The ABCs of Veterinary Dentistry: "J" Is for Jaw Fractures. 2017. Available online: https://www.dvm360.com/view/abcs-veterinary-dentistry-j-jaw-fractures (accessed on 9 September 2020).
22. Milella, L.; Smithson, A. Radiology: Dental. Felis. Available online: www.vetstream.com/treat/felis/freeform/radiology-dental (accessed on 13 November 2020).
23. Niemiec, B.A. Feline dental radiography and radiology. *J. Feline Med. Surg.* **2014**, *16*, 887–899. [CrossRef] [PubMed]
24. Hoffman, S.L.; Kressin, D.J.; Verstraete, F.J.M. Myths and misconceptions in veterinary dentistry. *J. Am. Veter. Med. Assoc.* **2007**, *231*, 1818–1824. [CrossRef]
25. Lantz, G.C. Interarcade wiring as a method of fixation for selected mandibular injuries. *J. Am. Anim. Hosp. Assoc.* **1981**, *17*, 599–603.
26. Woodbridge, N.; Owen, M. Feline Mandibular Fractures. *J. Feline Med. Surg.* **2013**, *15*, 211–218. [CrossRef]
27. Clinician's Brief. Considerations for Feline Mandibular Fractures. Available online: https://www.cliniciansbrief.com/article/considerations-feline-mandibular-fractures (accessed on 31 November 2020).
28. Greiner, C.L.; Verstraete, F.J.M.; Stover, S.M.; Garcia, T.C.; Leale, D.; Arzi, B. Biomechanical evaluation of two plating configurations for fixation of a simple transverse caudal mandibular fracture model in cats. *Am. J. Veter. Res.* **2017**, *78*, 702–711. [CrossRef]
29. Misch, C.E.; Resnik, R. Mandibular Nerve Neurosensory Impairment after Dental Implant Surgery: Management and Protocol. *Implant. Dent.* **2010**, *19*, 378–386. [CrossRef]
30. Gobel, S.; Binck, J.M. Degenerative changes in primary trigeminal axons and in neurons in nucleus caudalis following tooth pulp extirpations in the cat. *Brain Res.* **1977**, *132*, 347–354. [CrossRef]
31. Niemann, G. Cálculo de la resistencia. In *Tratado Teórico-Práctico de Elementos de Máquinas*; Editorial Labor, S.A.: Barcelona, Spain, 1973; Volume 1, pp. 46–79. ISBN 84-335-6287-8.
32. Ow, A.; Tan, W.; Pienkowski, L. Mandibular Reconstruction Using a Custom-Made Titanium Prosthesis: A Case Report on the Use of Virtual Surgical Planning and Computer-Aided Design/Computer-Aided Manufacturing. *Craniomaxillofacial Trauma Reconstr.* **2016**, *9*, 246–250. [CrossRef]
33. Heney, C.M.; Arzi, B.; Kass, P.H.; Hatcher, D.C.; Verstraete, F.J.M. The Diagnostic Yield of Dental Radiography and Cone-Beam Computed Tomography for the Identification of Dentoalveolar Lesions in Cats. *Front. Vet. Sci.* **2019**, *6*, 1–15. [CrossRef]
34. Liptak, J.M.; Thatcher, G.P.; Bray, J.P. Reconstruction of a mandibular segmental defect with a customized 3-dimensional–printed titanium prosthesis in a cat with a mandibular osteosarcoma. *J. Am. Veter. Med. Assoc.* **2017**, *250*, 900–908. [CrossRef]

35. Northrup, N.C.; Selting, K.A.; Rassnick, K.M.; Kristal, O.; O'Brien, M.G.; Dank, G.; Dhaliwal, R.S.; Jagannatha, S.; Cornell, K.K.; Gieger, T.L. Outcomes of Cats With Oral Tumors Treated With Mandibulectomy: 42 Cases. *J. Am. Anim. Hosp. Assoc.* **2006**, *42*, 350–360. [CrossRef]
36. Vaughan, D. 3-Dimensional printing in veterinary medicine. *Am. Vet. Crit. News Expert. Insights Anim. Health Med.* **2018**, 3, 19–23.

Article

Anatomic Study of the Elbow Joint in a Bengal Tiger (*Panthera tigris tigris*) Using Magnetic Resonance Imaging and Gross Dissections

Mario Encinoso [1], Jorge Orós [2], Gregorio Ramírez [3], José Raduan Jaber [2], Alejandro Artiles [1] and Alberto Arencibia [2,*]

1 Hospital Veterinario Los Tarahales, Recta de Los Tarahales 15, 35013 Las Palmas de Gran Canaria, Spain; mencinoso@gmail.com (M.E.); movilidad@hvtarahales.es (A.A.)
2 Departamento de Morfología, Facultad de Veterinaria, Universidad de Las Palmas de Gran Canaria, Trasmontaña, Arucas, 35416 Las Palmas, Spain; jorge.oros@ulpgc.es (J.O.); joseraduan.jaber@ulpgc.es (J.R.J.)
3 Departamento de Anatomía y Anatomía Patológica, Universidad de Murcia, 30100 Murcia, Spain; grzar@um.es
* Correspondence: alberto.arencibia@ulpgc.es

Received: 4 November 2019; Accepted: 28 November 2019; Published: 1 December 2019

Simple Summary: We present a report describing the normal elbow joint anatomy in a Bengal tiger (*Panthera tigris tigris*) using magnetic resonance imaging (MRI) and gross dissections of this region. Anatomical findings detected using MRI of the different bony and soft tissues were evaluated according to the characteristics of signal intensity and compared with the corresponding gross anatomical dissections. The main anatomical structures were labelled and identified. This study provides a valuable resource for veterinarians, biologists, and researchers involved in Bengal tiger (*Panthera tigris tigris*) conservation.

Abstract: The objective of our research was to describe the normal appearance of the bony and soft tissue structures of the elbow joint in a cadaver of a male mature Bengal tiger (*Panthera tigris tigris*) scanned via MRI. Using a 0.2 Tesla magnet, Spin-echo (SE) T1-weighting, and Gradient-echo short tau inversion recovery (GE-STIR), T2-weighting pulse sequences were selected to generate sagittal, transverse, and dorsal planes. In addition, gross dissections of the forelimb and its elbow joint were made. On anatomic dissections, all bony, articular, and muscular structures could be identified. The MRI images allowed us to observe the bony and many soft tissues of the tiger elbow joint. The SE T1-weighted MR images provided good anatomic detail of this joint, whereas the GE-STIR T2-weighted MR pulse sequence was best for synovial cavities. Detailed information is provided that may be used as initial anatomic reference for interpretation of MR images of the Bengal tiger (*Panthera tigris tigris*) elbow joint and in the diagnosis of disorders of this region.

Keywords: magnetic resonance imaging; anatomy; elbow joint; Bengal tiger (*Panthera tigris tigris*)

1. Introduction

Magnetic resonance imaging (MRI) is commonly used, especially to assess the musculoskeletal system in humans and animals, due to its good image resolution, superior soft tissue contrast among different anatomical structures, and use of a magnetic field rather than ionizing radiation, compared with other imaging diagnostic technique [1–3]. In veterinary medicine, MRI anatomical studies have been performed in the elbow region of equines [4] and canines [5–7]. Several reports have provided descriptions of clinical findings regarding the elbow joint using MRI [8–15].

Since 2010, the Bengal tiger has been included on the Red List of Endangered Species [16]. In zoos and wildlife rehabilitation centers, veterinarians, researchers and specialized technicians are involved in large nondomestic cat medicine, as well as the welfare and conservation of this species. In tigers, anatomical and clinical studies via MRI are limited. To our knowledge, MRI has been only used to describe the normal anatomic features of the stifle and tarsus joints [17,18], as well as brain characteristics [19]. Also, previous reports describing several neurological disorders have been published on this species [20,21]. However, no published studies were found describing the Bengal tiger elbow joint using MRI and gross anatomical dissections.

The elbow joint is a synovial joint and is important due to its location of anatomical structures such as the articular surfaces of the bones (humerus, radius, and ulna), as well as the articular cavities, ligaments, muscles, and tendons that contribute to the stability of this region. The elbow is susceptible to congenital, developmental, and traumatic musculoskeletal disorders. Therefore, an accurate interpretation and thorough understanding of regional anatomy of the MRI elbow images could be useful in the evaluation and diagnosis of different elbow joint disorders in felines [22–28].

The aim of this study was to describe the elbow joint findings in a Bengal tiger using MRI and gross anatomical dissections.

2. Materials and Methods

2.1. Animals

In this research, the cadaver of a captive female six-year-old Bengal tiger (105 kg) was referred from the Cocodrilos Park Zoo (Canary Islands, Spain) to the Veterinary Faculty of Las Palmas de Gran Canaria University. The tiger, which had no history of any disease related to articulation, was subjected to observation, and no external elbow lesions were observed. After, the animal was immediately frozen to mitigate post-mortem changes. Two days later, the tiger was defrosted to perform the MRI study. This research was authorized by the Conservation Nature Service (Seprona) of Gran Canaria at the Spanish Ministry of Interior (Protocol 2012).

2.2. Magnetic Resonance Imaging

With the objective of carrying out the MRI, a 0.2-Tesla magnet (Vet-MR Esaote, Genoa, Italy) was employed with the tiger placed in right lateral recumbence. A standard MRI protocol was used to generate spin-echo (SE) T1-weighted and Gradient-Echo short tau inversion recovery (GE-STIR) T2-weighted images in the sagittal, transverse, and dorsal anatomical planes. The sagittal plane was aligned parallel to the olecranon tuberosity and planned using the transverse and dorsal plane. The transverse plane was perpendicular to the body of humerus and ulna and planned on the sagittal and dorsal planes. The dorsal plane was aligned parallel to the head of radius and planned using the sagittal and transverse planes.

SE T1-weighted sagittal images were acquired with the following settings: Echo time (TE) = 26 ms, repetition time (TR) = 700 ms, an acquisition matrix of 320 × 187, and 4-mm slice thickness with 4.4-mm spacing between slices. For GE-STIR T2-weighted sagittal images, the TE was 25 ms, TR was 1680 ms, acquisition matrix was 256 × 172, and 4-mm slice thickness with 4.4-mm interslice spacing was used. For SE T1-weighted dorsal images, the TE was 80 ms, TR was 3000 ms, acquisition matrix was 256 × 172, and 4.5-mm slice thickness with 4.9-mm interslice spacing was used. For GE-STIR T2-weighted dorsal images, the TE was 25 ms, TR was 1680 ms, acquisition matrix was 256 × 172, and 4.5-mm slice thickness with 4.9-mm interslice spacing was used. For SE T1-weighted transverse images, the TE was 26 ms, TR was 500 ms, acquisition matrix was 256 × 172, and 5-mm slice thickness with 5.5-mm interslice spacing. For GE-STIR T2-weighted transverse images, the TE was 25 ms, TR was 1680 ms, acquisition matrix was 256 × 172, and 5-mm slice thickness with 5.5-mm interslice spacing. In our study, a medical imaging software tool (OsiriX MD, http://www.osirix-viewer.com, Geneva, Switzerland) was used to evaluate the Dicom images obtained.

2.3. Anatomical Evaluation

After the performance of the imaging procedure, lateral and medial gross anatomical dissections of the right forelimb were performed in order to facilitate the identification and comparison of brachial and antebrachial muscles and tendons. Also, dissections of the articulation were made to facilitate the identification of the ligaments and bones. These images were compared to the corresponding MRI images presented in this study. Also, we also resorted to the Bengal elbow bones and to veterinary anatomy literature [29,30]. We labelled the identified elbow joint structures according to the anatomical nomenclature [31,32].

3. Results

3.1. Gross Anatomical Dissections

Anatomical dissections images of the right forelimb and its corresponding elbow joint from different aspect are presented (Figures 1 and 2). In Figure 1, the main muscles and tendons that stabilize this joint could be identified. Thus, the triceps brachii muscle and the muscular complex composed of the longum, lateral, medial, and accessory head were visible. The longum head originates from the caudal border of scapula, the lateral head originate lateral to the neck of humerus, the medial head originate from the humerus caudal to tuberositas teres major, and the accessory head originates to the neck of humerus. The longum, lateral, medial, and accessory head were identified with their tendons inserted into the olecranon tuberosity (Figure 1a,b). The anconeus muscle, extending from the humeral epicondyles and lateral to the olecranon fossa and terminating on the olecranon, was visible in Figure 1a. The extensor carpi radialis muscle was identified to extend from the lateral supracondylar crest of the humerus to the base of the II and III metacarpal bones (Figure 1a). The extensor digitorum communis muscle arose from the lateral epicondyle of humerus and the tendon split to terminate on the distal phalanx of second to fifth digits was also visible (Figure 1a). The extensor digitorum lateralis muscle was seen in Figure 1a from the lateral collateral ligament and radius head, and its tendons terminated on the III, IV, and V digits. The extensor carpi ulnaris muscle was visible from the lateral epicondyle of humerus, and it terminated on basis of the fifth metacarpal bone, and in the accessory and carpocubital bones (Figure 1a). The abductor digiti I longus muscle arose from the radius and ulna obliquely, which terminated on the basis of 1th metacarpal bone, was also identified in the Figure 1a.

In Figure 1b, the tensor fasciae antebrachii muscle was visible aponeurotic from the tendon of insertion of the latisimus dorsi muscle and terminated on the medial collateral ligament. The biceps brachii muscle, extending from the supraglenoidal tubercle and terminating on the radial tuberosity and medial collateral ligament, was seen in Figure 1a,b. The brachialis muscle was identified to extend from the caudal surface of the humerus to the radial tuberosity (Figure 1a). The flexor carpi radialis muscle was observed from the medial condyle of humerus to the second and third metacarpal bones (Figure 1b). The flexor digitorum superficialis muscle was visible from the medial epicondyle of humerus and terminated divided in four tendons on the middle phalanx of the second to fifth digits (Figure 1b). The flexor digitorum profundus muscle is composed of three heads with combined tendon, which divides to terminate on flexor surface of each distal phalanx (Figure 1b). The flexor carpi ulnaris muscle, composed by the humeral and ulnar heads, was visible in the Figure 1b. The humeral head was identified to extend from medial epicondyle of humerus to the carpal accessory bone, whereas the ulnar head was seen from medial side of olecranon to the carpal accessory bone (Figure 1b). The pronator teres muscle, extending from the medial epicondyle of humerus and terminating on the medial surface of radius, was also identified in the Figure 1b. Also, the brachial vein and artery, the median artery and nerve, the radial artery and nerve, and the cubital nerve were well-defined in the Figure 1b.

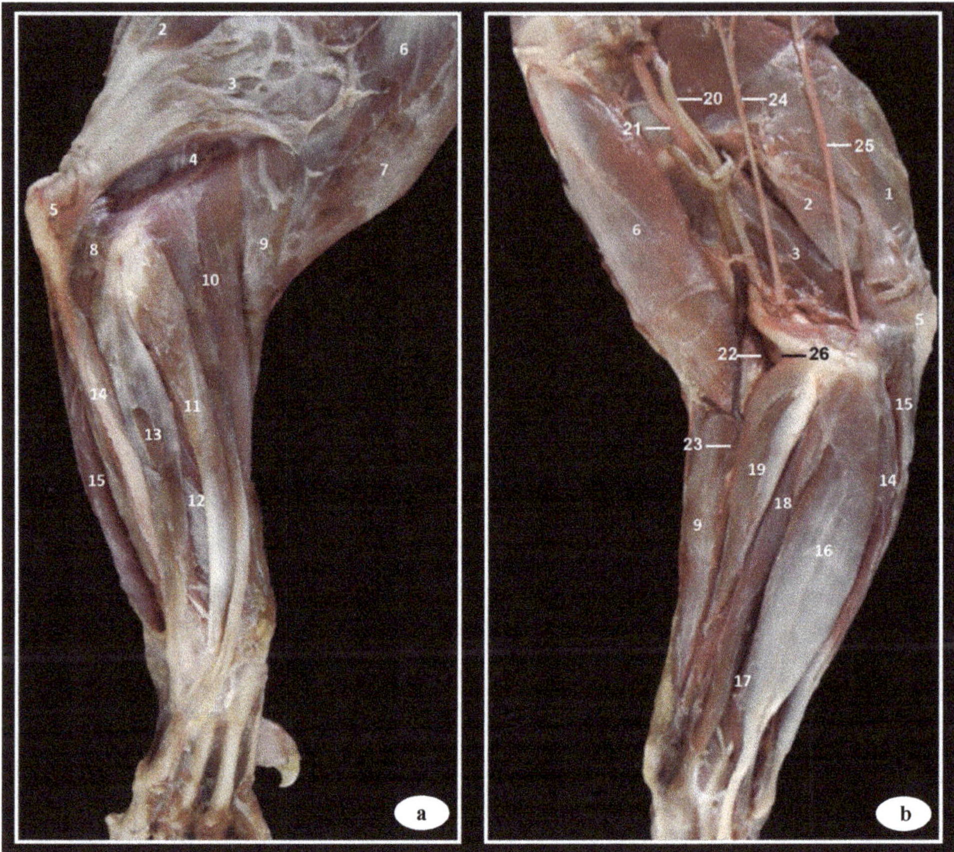

Figure 1. Dissection images of the Bengal tiger forelimb. (**a**) Lateral aspect and (**b**) medial aspect. 1. Tensor fasciae antebrachii muscle; 2. Triceps brachii muscle (long head); 3. Triceps brachii muscle (lateral head); 4. Triceps brachii muscle (accessory head); 5. Olecranon tuberosity; 6. Brachialis muscle; 7. Biceps brachii muscle; 8. Anconeus muscle; 9. Extensor carpi radialis muscle; 10. Extensor digitorum communis muscle; 11. Extensor digitorum lateralis muscle; 12. Abductor digiti I longus muscle; 13. Extensor carpi ulnaris muscle; 14. Flexor carpi ulnaris muscle (humeral head); 15. Flexor carpi ulnaris muscle (cubital head); 16. Flexor digitorum superficialis muscle; 17. Flexor digitorum profundus muscle; 18. Flexor carpi radialis muscle; 19. Pronator teres muscle; 20. Brachial vein; 21. Brachial artery; 22. Median artery; 23. Radial artery; 24. Radial nerve; 25. Cubital nerve; 26. Median nerve.

In Figure 2, corresponding to the gross dissections of the right elbow joint, several bones, ligaments, and membrane could be identified. Thus, the humerus bone (including the body, metaphysis, condyles, epicondyles, and the olecranon fossa), the radius bone (with their head, metaphysis, radial tuberosity, and articular circumference), and the ulna bone (olecranon tuberosity, anconeus process, lateral, and medial coronoid processes and body) were observed. Also, the supracondylar foramen is clearly visible in Figure 2c. Articular structures of elbow joint between humerus, radius, and ulna bones, such as the articular cavities, the lateral collateral cubital ligament that connects lateral epycondyle of humerus to radius, and the medial collateral cubital ligament that connects medial epycondyle of humerus to radius, were identified in Figure 2. Also, several structures of the proximal radiocubital joint were observed. Thus, the radial annular ligament that encircles the head of radius is attached to the edges of the radial incisure of the ulna was visible in Figure 2a–c, whereas the olecranon ligament from the medial border of olecranon fossa to olecranon was seen only in the proximal aspect (Figure 2d). In addition, the membrane interossea antebrachii that connects radius and ulna bones, and the interosseum antebrachii ligament that connects radius and ulna in the proximal half of spatium interosseum, were also visible in Figure 2a,b.

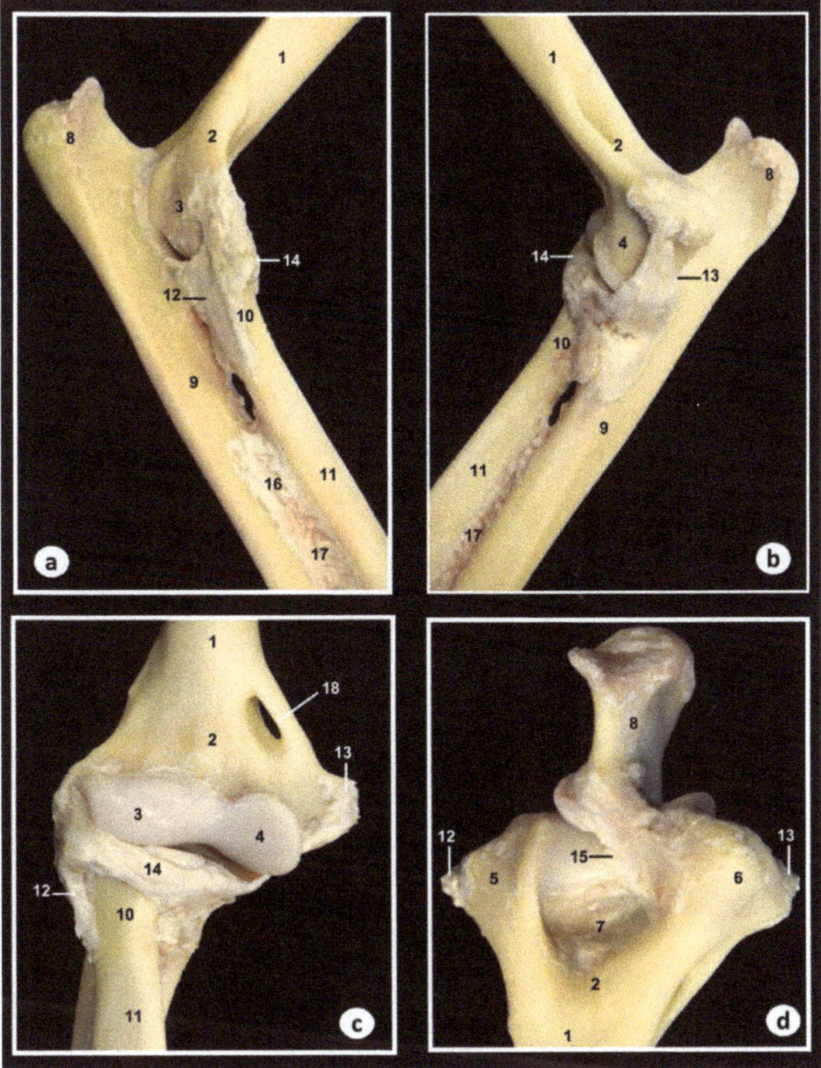

Figure 2. Dissection images of the Bengal tiger right elbow joint. (**a**) Lateral aspect, (**b**) medial aspect, (**c**) cranial aspect, and (**d**) proximal aspect. 1. Humerus (body); 2. Humerus (metaphysis); 3. Humerus (lateral condyle); 4. Humerus (medial condyle); 5. Humerus (lateral epicondyle); 6. Humerus (medial epicondyle); 7. Olecranon fossa; 8. Olecranon tuberosity; 9. Ulna (body); 10. Radius (head); 11. Radius (body); 12. Lateral collateral ligament; 13. Medial collateral ligament; 14. Radial annular ligament; 15. Olecranon ligament; 16. Interosseoum antebrachii ligament; 17. Membrana interossea antebrachia; 18. Supracondylar foramen.

3.2. *Magnetic Resonance Imaging*

Selected MR images are presented in Figures 3–5. In Figure 3, four sagittal MR images are shown in a lateromedial direction from the lateral epicondyle of humerus (level I) to medial epicondyle of humerus (level IV).

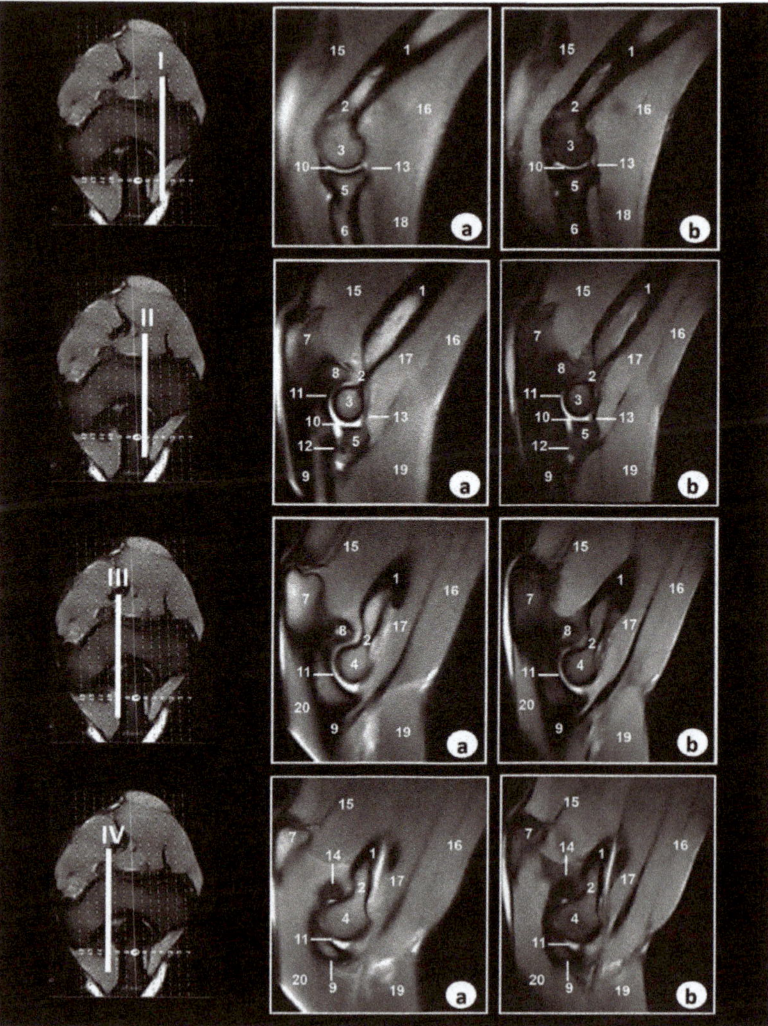

Figure 3. Sagittal MR images of the right elbow joint at the level of the lateral epicondyle of humerus (level I), lateral part of olecranon fossa (level II), medial part of olecranon fossa (level III), and medial epicondyle of humerus (level IV). The lines depict the level of section. The images are oriented so that the left is caudal and the right is cranial. (**a**) SE T1-weighted MR images and (**b**) GE-STIR T2-weighted images. 1. Humerus (body); 2. Humerus (metaphysis); 3. Humerus (lateral epicondyle); 4. Humerus (medial epicondyle); 5. Radius (head); 6. radius (body); 7. Ulna (olecranon tuberosity); 8. Ulna (anconeus process); 9. Ulna (body); 10. Humeroradial joint (articular cavity); 11. Humeroulnar joint (articular cavity); 12. Proximal radioulnar joint (articular cavity); 13. Radial annular ligament; 14. Olecranon ligament; 15. Triceps brachii muscle; 16. Biceps brachii muscle; 17. Brachialis muscle; 18. Extensor digitorum communis; 19. Extensor carpi radialis muscle; 20. Flexor carpi ulnaris muscle.

Figure 4 shows three transverse MR images presented in a proximodistal direction from the olecranon tuberosity (level I) to the proximal radioulnar joint (level III).

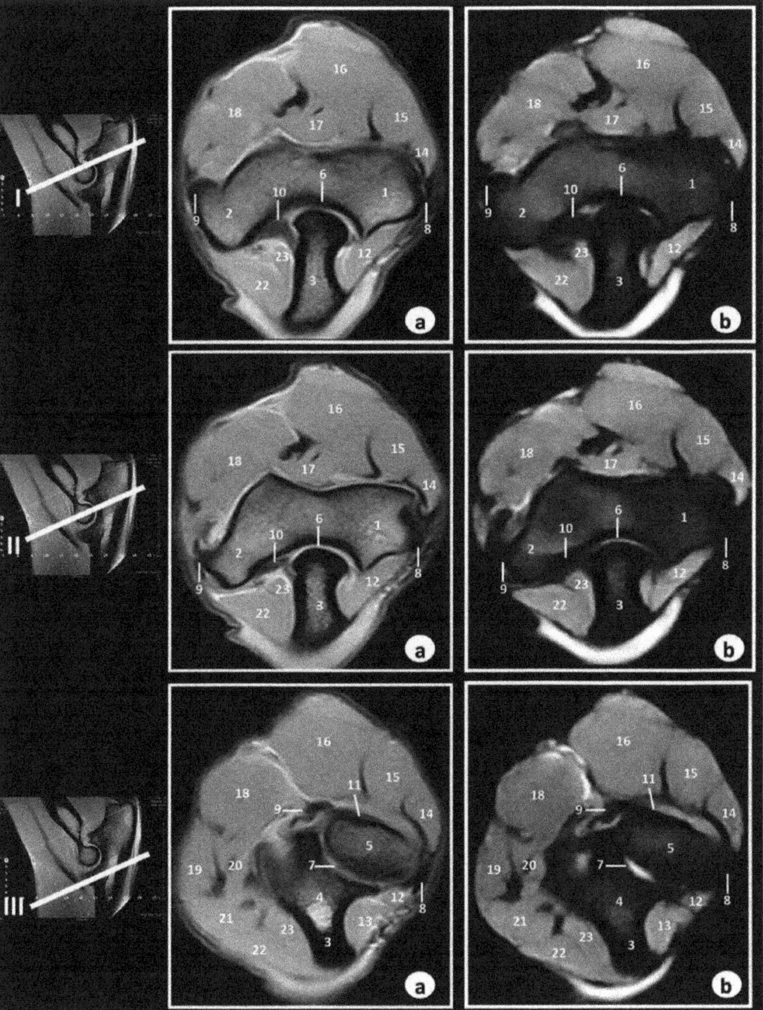

Figure 4. Transverse MR images of the right elbow joint at the level of olecranon tuberosity (level I), epicondyles of humerus (level II), and proximal radioulnar joint (level III). The lines depict the level of section. The images are oriented so that the left is medial, and the right is lateral. (**a**) SE T1-weighted MR images and (**b**) GE-STIR T2-weighted images. 1. Humerus (lateral epicondyle); 2. Humerus (medial epicondyle); 3. Ulna (olecranon tuberosity); 4. Ulna (body); 5. Radius head; 6. Humeroulnar joint (articular cavity); 7. Proximal radioulnar joint (articular cavity); 8. Lateral collateral ligament; 9. Medial collateral ligament; 10. Olecranon ligament; 11. Radial annular ligament; 12. Anconeus muscle; 13. Extensor carpi ulnaris muscle; 14. Extensor digitorum lateralis muscle; 15. Extensor digitorum communis muscle; 16. Extensor carpi radialis muscle; 17. Brachialis muscle; 18. Biceps brachii muscle; 19. Flexor digitorum superficialis muscle; 20. Flexor carpi radialis muscle; 21. Flexor digitorum profundus muscle; 22. fFexor carpi ulnaris muscle (humeral head); 23. Flexor carpi ulnaris muscle (ulnar head).

In Figure 5, three dorsal MR images are presented in a caudocranial direction from the anconeus process of humerus (level I) to the head of radius (level III).

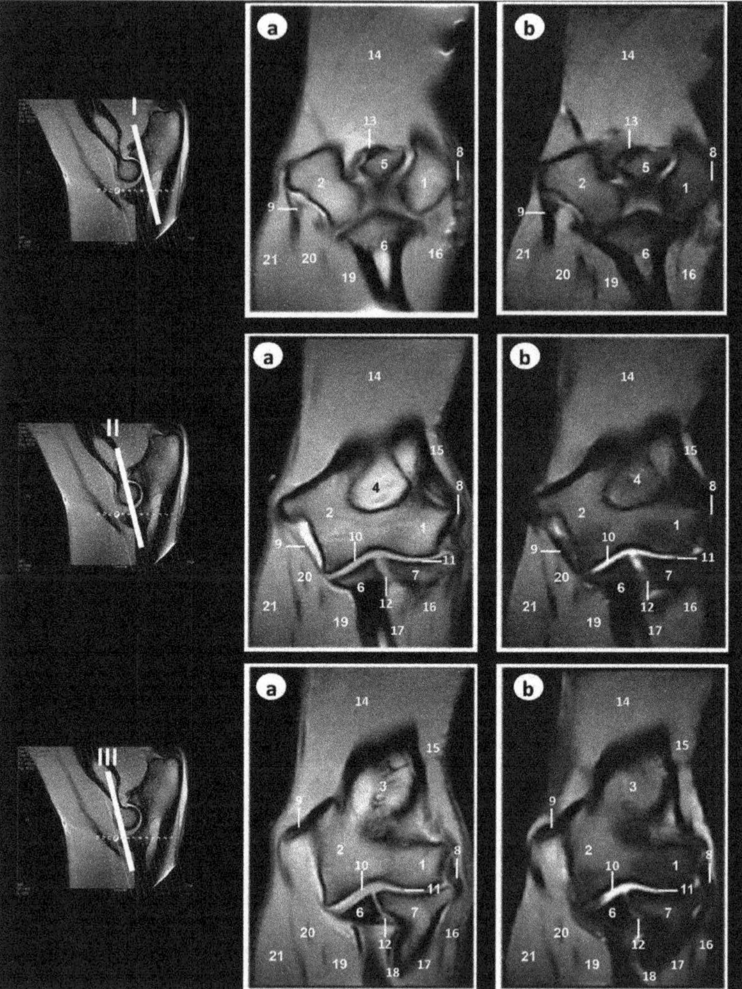

Figure 5. Dorsal MR images of the right elbow joint at the level of anconeus processus of ulna (level I), olecranon fossa (level II), and head of radius (level III). The lines depict the level of section. The images are oriented so that the right side of the image is lateral and the top is proximal. (**a**) SE T1-weighted MR images and (**b**) GE-STIR T2-weighted images. 1. Humerus (lateral epicondyle); 2. Humerus (medial epicondyle); 3. Humerus (metaphysis); 4. Olecranon fossa; 5. Anconeus process; 6. Ulna (body); 7. Radius (head); 8. Lateral collateral ligament; 9. Medial collateral ligament; 10. Humeroulnar joint (articular cavity); 11. Humerorradial joint (articular cavity); 12. Proximal radioulnar joint (articular cavity); 13. Olecranon ligament; 14. Triceps brachii muscle; 15. Anconeus muscle; 16. Extensor digitorum lateralis muscle; 17. Extensor digitorum communis muscle; 18. Extensor carpi ulnaris muscle; 19. Flexor digitorum profundus muscle; 20. Flexor carpi ulnaris muscle; 21. Flexor digitorum superficialis muscle.

On the MR images, anatomic details of the Bengal tiger elbow joint were evaluated according to the characteristics of signal intensity of the different bony and soft tissues (Table 1).

Table 1. Tissue signal intensity characteristics for MRI of the Bengal tiger (*Panthera tigris tigris*) elbow joint.

TISSUE	SE-T1 Weighted	GE-STIR-T2 Weighted
Cortical and subchondral bone	Very low	Very low
Bone marrow	High	Intermediate
Fat	High	Intermediate
Synovial fluid	Low	High
Articular cartilage	Intermediate	Intermediate
Ligament	Low	Very low
Muscle	Intermediate	Intermediate
Tendon	Low	Very low

In the SE T1-weighted sequence, the cortical and subchondral bone of the humerus, radius, and ulna appeared with very low signal intensity compared with the high signal of the bone marrow. The articular cartilage was visualized with intermediate signal intensity. By contrast, in the GE-STIR T2-weighted MR images, the bone marrow was seen with intermediate signal intensity and could be observed in the area of the negligible signal corresponding to the cortical and subchondral bones. The articular cartilage was identified by the intermediate signal characteristics on both the SE T1-weighted and GE-STIR T2-weighted images (Figures 3–5). The lateral and medial collateral ligaments of the elbow joint were readily seen on the transverse (Figure 4) and dorsal (Figure 5) planes. These ligaments were visible with low signal intensity in SE-T1-weighted images and with very low signal in the GE-STIR T2-weighted pulse sequence as linear bands similar in intensity to the cortical bone. On the other hand, articular structures of radiocubital proximal joint were also visible in the images. Thus, the radial annular ligament was observed in Figures 4 and 5, whereas the olecranon ligament was more clearly identified in the sagittal (Figure 3) and transverse (Figure 4) images. Besides, the membrana interossea antebrachii and interosseoum antebrachii ligament were seen especially in sagittal MR images (Figure 3). This ligament and membrane had low signal intensity in both sequences. Finally, synovia could be seen in the articular cavities with intermediate signal intensity on the T1-weighted MR images. By contrast, in the GE-STIR T2-weighted images, synovial fluid appeared with high signal intensity. Several main muscles, such as the brachialis, the antebrachial fascia tensor, biceps brachii, the triceps brachii, anconeus, extensor carpi radialis, extensor digitorum communis, extensor digitorum lateralis, abductor digiti I longus, extensor carpi ulnaris, flexor carpi ulnaris, flexor digitorum superficialis, flexor digitorum profundus, flexor carpi radialis, and pronator teres, were well-identified in Figures 3–5. These muscles were defined with intermediate signal intensity in both sequences. By contrast, its tendons appeared with dark grey to black signal intensities in the SE T1-weighted images and with dark grey signal intensity in the GE STIR T2-weighted images. The muscles and tendons were easily seen, especially in the sagittal (Figure 3) and transverse (Figure 4) images, compared with the dorsal plane (Figure 5).

4. Discussion

In humans, MRI has become the modality of choice for morphological assessment and the diagnosis and management of diseases of the elbow joint [33]. Advances in equipment and MRI pulse sequences have allowed for superior visualization of this joint and its adjacent structures. Numerous MRI pulse sequences have been applied to musculoskeletal system. Traditionally, spin-echo sequences have been used for elbow diagnostic imaging. However, these techniques have been supplanted by the newer fast spin-echo and gradient-echo pulse sequences, which provide anatomical and functional information with greater ability to distinguish between bony and soft tissue structures [33,34].

In veterinary practice, the exploration of anatomic structures within the elbow joint and clinical assessment of soft tissues is laborious because of the anatomic complexity. Traditionally, radiography [35] and ultrasonography [36] have been used to obtain images of the bony and the main soft-tissue structures of this region. Nevertheless, computed tomography has become the preferred imaging technique for the evaluation of the osseous structures [37] and MRI has progressively gained credit for their ability to assess the ligaments, musculotendinous, cartilaginous surfaces, and osseous structures of this anatomical region [4–7].

In recent years, the contributions of the veterinarians working with captive and free-ranging animals to prevent and/or treat diseases that threaten species' survival in wildlife conservation have increased [38]. In the Bengal tiger (*Panthera tigris tigris*), as well as in other mammals, the elbow conforms an anatomical region composed by bones, ligaments, muscles, and tendons that support this joint. Thus, the physical examinations and clinical assessments of these structures are very difficult due to its complexity. Our MRI study facilitated the identification of the main elbow anatomical structures. Also, an accurate anatomical interpretation of MRI images would be useful for assessment of elbow joint tissues such as bones, fluids, ligaments, and muscular structures, and could be used in the diagnosis of disorders of this joint.

The interpretation of elbow joint MRI studies requires an understanding of the MRI unit, basic pulse sequences, and standard imaging planes [33]. With regard to the MRI protocol used in the present study, an MRI can be used as an initial valid reference for assessment of the Bengal tiger (*Panthera tigris tigris*) elbow. However, more clinical studies, which include more specimens to assess the adequate MRI protocols for elbow disorders, are necessary. Our research was obtained via low-field MRI magnet (0.2 T), which provided a correct visualization of the main anatomical structures of this joint. Anatomical characteristics of the elbow joint using low-field MRI have been reported in dogs [6], and previous studies have evaluated this anatomical region using high-field magnet in horses [4] and dogs [5,7]. Low-field (0.2–0.4 T) MRI equipment predominates in veterinary practice due to its reduced costs, better patient access, and greater safety compared to high-field MRI units [38]. However, in veterinary musculoskeletal imaging, using high-field equipment is recommended. High-field equipment improves the signal-to-noise ratio, resulting in increased image resolution and decreased exam time of bones, ligaments, muscles, and tendons [39,40].

Manipulation of the technical parameters permits a variety of MRI pulse sequences. The SE T1-weighted MR images shows fat as a bright signal intensity. By contrast, the GE-STIR-T2-weighted sequence is a fat-suppression technique and shows water as a bright signal intensity [34]. In our study, these same MRI pulse sequences were selected. SE-T1-weighted MR images were best to distinguish the best anatomic detail, whereas GE-STIR-T2-weighted MR images showed anatomical information for synovial cavities of the humeroradial, humeroulnar, and proximal radioulnar joints. Nevertheless, in both MRI sequences, the major difficulty was the definition of articular cartilage due to the presence of synovial fluid with similar MRI signal intensity. The elbow joints of dogs [5–7] and horses [4] have been studied using similar MRI pulse sequences.

In our research, the Bengal tiger (*Panthera tigris tigris*) elbow joint was imaged in three anatomical planes: Sagittal, transverse, and dorsal. In this regard, Baeumlin et al. [6] reported these same planes in dogs [6], whereas Snaps et al. [5] and Wucherer et al. [7] showed images of the elbow joint in dogs only in the sagittal and dorsal planes, respectively. On the other hand, Tnibar et al. [4] showed images of this region in the horses only in the transverse plane. In our study, the images allowed us to see the relationship between the cortical, subchondral bone, and bone marrow. Several ligaments were identified in the MRI planes. Thus, the lateral and medial collateral ligaments of the elbow joint showed the best views in the dorsal plane. These observations have also been reported in horses [4] and dogs [5–7]. On the other hand, the transverse and dorsal planes provided better definition of the olecranon ligament, whereas the radial annular ligament was observed especially in the sagittal and transverse planes. Also, in our research, the sagittal planes provided optimal views of the proximal radioulnar joint structures, such as the membrana interossea antebrachii and interosseoum antebrachii ligament. These articular structures have not described previously. Moreover, the brachii, extensor, and flexor muscles were best seen in the different planes. Thus, the biceps brachii, brachialis, and triceps brachii were best seen on the sagittal plane. The extensor muscles of the forelimb were seen in the transverse and dorsal planes, whereas the flexor muscles were visible especially in the sagittal and dorsal planes. However, the tendons of these muscles were not visible in all images because of their very short size. These observations were previously described in dogs [6].

The efficiency and effectiveness of conservation efforts are significantly enhanced by incorporating animal health considerations into the planning, implementation, and evaluation phases of all programs (captive and wild) involved in conserving wildlife [38]. Anatomical and clinical studies on captive felines are essential activities for tiger conservation around the world. The identification of the bony, articular cavities, ligaments, muscles, and tendons presented in this study was facilitated by the use of gross anatomical dissections of the right forelimb and elbow joint. The gross anatomical dissections provided a good location of the main anatomical structures, and they are a helpful tool for the identification of MRI features. Obtaining MRIs of the Bengal tiger is severely hindered by high cost and limited availability. Nevertheless, the small risk degree which its application entails might allow us to justify its use in these endangered species. With developing technology in zoos and wildlife rehabilitation centres, MRI is a promising, noninvasive, and accurate method for tiger imaging [17–21]. Our research provides the first anatomical description of the elbow in a Bengal tiger (*Panthera tigris tigris*) via low-field MRI. In the future, it is important to carry out new studies and to establish new MRI protocols to ensure a better assessment and diagnosis of disorders of this joint using low-field and high-field MRI units.

5. Conclusions

In conclusion, imaging findings of this research indicate that MRI provides an accurate depiction of the anatomical structures of the elbow joint. These images should be useful for veterinarians, biologists, researchers, and technicians involved in Bengal tiger (*Panthera tigris tigris*) medicine, welfare, and conservation.

Author Contributions: Conceptualization, M.E. and A.A. (Alberto Arencibia); methodology, M.E., J.O., A.A. (Alejandro Artiles) and A.A. (Alberto Arencibia); formal analysis, M.E., J.O. and A.A. (Alberto Arencibia); investigation, M.E., J.R.J., G.R. and A.A. (Alberto Arencibia); writing—original draft, M.E. and A.A. (Alberto Arencibia); supervision, M.E. and A.A. (Alberto Arencibia).

Funding: This research received no external funding and the University of Las Palmas de Gran Canaria funded the study.

Acknowledgments: The authors thank Cocodrilos Park zoo (Canary Islands, Spain), for the cession of the specimen for our study.

Conflicts of Interest: The authors declare no conflict of interest.

References

1. Bredella, M.A.; Tirman, P.F.; Fritz, R.C.; Feller, J.F.; Wischer, T.K.; Genant, H.K. MR imaging findings of lateral ulnar collateral ligament abnormalities in patients with lateral epicondylitis. *Am. J. Roentgenol.* **1999**, *173*, 1379–1382. [CrossRef] [PubMed]
2. Kijowski, R.; Tuite, M.; Sanford, M. Magnetic resonance imaging of the elbow. Part I: Normal anatomy, imaging technique, and osseous abnormalities. *Skelet. Radiol.* **2004**, *33*, 685–697. [CrossRef] [PubMed]
3. Sampath, S.C.; Sampath, S.C.; Bredella, M.A. Magnetic resonance imaging of the elbow: A structured approach. *Sports Health* **2013**, *5*, 34–39. [CrossRef] [PubMed]
4. Tnibar, M.A.; Auer, J.A.; Bakkali, S. Ultrasonography of the equine elbow technique and normal appearance. *J. Equine Vet. Sci.* **2001**, *21*, 177–187. [CrossRef]
5. Snaps, F.R.; Saunders, J.H.; Park, R.D.; Daenen, B.; Balligand, M.H.; Dondelinger, R.F. Comparison of spin echo, gradient echo and fat saturation magnetic resonance imaging sequences for imaging the canine elbow. *Vet. Radiol. Ultrasound* **1998**, *39*, 518–523. [CrossRef]

6. Baeumlin, Y.; De Rycke, L.; Van Caelenberg, A.; Van Bree, H.; Gielen, I. Magnetic resonance imaging of the canine elbow: An anatomic study. *Vet. Surg.* **2010**, *39*, 566–573. [CrossRef]
7. Wucherer, K.L.; Ober, C.P.; Conzemius, M.G. The use of delayed gadolinium enhanced magnetic resonance imaging of cartilage and T2 mapping to evaluate articular cartilage in the normal canine elbow. *Vet. Radiol. Ultrasound* **2012**, *53*, 57–63. [CrossRef]
8. Cook, C.R.; Cook, J.L. Diagnostic imaging of canine elbow dysplasia: A review. *Vet. Surg.* **2009**, *38*, 144–153. [CrossRef]
9. Adamiak, Z.; Jaskólska, M.; Matyjasik, H.; Pomianowski, A.; Kwiatkowska, M. Magnetic resonance imaging of selected limb joints in dogs. *Pol. J. Vet. Sci.* **2011**, *14*, 501–505. [CrossRef]
10. Piola, V.; Posch, B.; Radke, H.; Telintelo, G.; Herrtage, M.E. Magnetic resonance imaging features of canine incomplete humeral condyle ossification. *Vet. Radiol. Ultrasound* **2012**, *53*, 560–565. [CrossRef]
11. De Bakker, E.; Gielen, I.; Kromhout, K.; van Bree, H.; Van Ryssen, B. Magnetic resonance imaging of primary and concomitant flexor enthesopathy in the canine elbow. *Vet. Radiol. Ultrasound* **2014**, *55*, 56–62. [CrossRef] [PubMed]
12. Ambrosius, L.; Arnoldy, C.; Waller, K.R., 3rd; Little, J.P.; Bleedorn, J.A. Reconstruction of chronic triceps tendon avulsion using synthetic mesh graft in a dog. *Vet. Comp. Orthop. Traumatol.* **2015**, *28*, 220–224. [CrossRef] [PubMed]
13. Coppieters, E.; Gielen, I.; Verhoeven, G.; Van Vynckt, D.; Van Ryssen, B. Erosion of the medial compartment of the canine elbow: Occurrence, diagnosis and currently available treatment options. *Vet. Comp. Orthop. Traumatol.* **2015**, *28*, 9–18. [PubMed]
14. Wavreille, V.; Fitzpatrick, N.; Drost, W.T.; Russell, D.; Allen, M.J. Correlation between histopathologic, arthroscopic, and magnetic resonance imaging findings in dogs with medial coronoid disease. *Vet. Surg.* **2015**, *44*, 501–510. [CrossRef]
15. Franklin, S.P.; Burke, E.E.; Holmes, S.P. Utility of MRI for characterizing articular cartilage pathology in dogs with medial coronoid process disease. *Front. Vet. Sci.* **2017**, *4*, 25. [CrossRef]
16. International Union for Conservation of Nature and Natural Resources Species Survival Commission (IUCN), The IUCN Red List of Threatened Species. 2011. Available online: https://www.iucnredlist.org (accessed on 1 November 2019).
17. Arencibia, A.; Encinoso, M.; Jaber, J.R.; Morales, D.; Blanco, D.; Artiles, A.; Vázquez, J.M. Magnetic resonance imaging study in a normal Bengal tiger (*Panthera tigris*) stifle joint. *BMC Vet. Res.* **2015**, *11*, 192. [CrossRef]
18. Arencibia, A.; Matos, J.; Encinoso, M.; Gil, F.; Artiles, A.; Martínez-Gomariz, F.; Vázquez, J.M. Computed tomography and magnetic resonance imaging study of a normal tarsal joint in a Bengal tiger (*Panthera tigris*). *BMC Vet. Res.* **2019**, *15*, 126. [CrossRef]
19. Jaber, J.R.; Encinoso, M.; Morales, D.; Artiles, A.; Santana, M.; Blanco, D.; Arencibia, A. Anatomic study of the normal Bengal tiger (*Panthera tigris tigris*) brain and associated structures using low field magnetic resonance imaging. *Eur. J. Anat.* **2016**, *20*, 195–203.
20. Snow, T.M.; Litster, A.L.; Gregory, R.J. Big cat scan: Magnetic resonance imaging of the tiger. *Australas. Radiol.* **2004**, *48*, 93–95. [CrossRef]
21. Zeira, O.; Briola, C.; Konar, M.; Dumas, M.P.; Wrzosek, M.A.; Papa, V. Suspected neurotoxicity due to Clostridium perfringens type B in a tiger (*Panthera tigris*). *J. Zoo Wildl. Med.* **2012**, *43*, 666–669. [CrossRef]
22. Williams, M.C.; Lambrechts, N.E. Synovial osteochondromatosis in the elbow joint of a Bengal tiger (*Panthera tigris tigris*). *Vet. Comp. Orthop. Traumatol.* **2001**, *14*, 56–59. [CrossRef]
23. Rossi, F.; Vignoli, M.; Terragni, R.; Pozzi, L.; Impallomeni, C.; Magnani, M. Bilateral elbow malformation in a cat caused by radio-ulnar synostosis. *Vet. Radiol. Ultrasound* **2003**, *44*, 283–286. [CrossRef] [PubMed]
24. Lascelles, B.D.X. Feline Degenerative Joint Disease. *Vet. Surg.* **2010**, *39*, 2–13. [CrossRef] [PubMed]
25. Medl, N.; Hurter, K. Fracture of the anconeal process in two cats. *Vet. Comp. Orthop. Traumatol.* **2010**, *23*, 124–127.
26. Tan, C.; Allan, G.S.; Barfield, D.; Krockenberger, M.B.; Howlett, R.; Malik, R. Synovial osteochondroma involving the elbow of a cat. *J. Feline Med. Surg.* **2010**, *12*, 412–417. [CrossRef]
27. Streubel, R.; Geyer, H.; Montavon, P.M. Medial humeral epicondylitis in cats. *Vet. Surg.* **2012**, *41*, 795–802. [CrossRef]
28. Monsalve Buriticá, S.; Rojano, C.; Martínez, L.M. Tumor de la vaina nerviosa periférica en un tigre de bengala (*Panthera tigris*). *J. Agric. Animal Sci.* **2012**, *1*, 70–74.

29. Latorre, R.; Gil, F.; Climent, S.; López, O.; Henry, R.; Ayala, M.; Ramírez, G.; Martínez, F.; Vázquez, J.M. Artrologia. In *Atlas en Color sobre Abordajes Quirúrgicos a Huesos y Articulaciones en el Perro y el Gato. Miembros Toracico y Pelviano*, 1th ed.; Inter-Médica: Buenos Aires, República Argentina, 2008; pp. 48–89.
30. Done, S.H.; Goody, P.C.; Evans, S.A.; Stickland, N.C. El miembro torácico. In *Atlas en Color de Anatomía Veterinaria. El Perro y el Gato*, 2nd ed.; Elsevier Mosby: Barcelona, Spain, 2010; pp. 139–184.
31. World Association of Veterinary Anatomists. Arthrologia. In *Nomina Anatomica Veterinaria*, 6th ed.; International Committee on Veterinary Gross Anatomical Nomenclature: Hanover, NH, USA, 2005; p. 50.
32. Schaller, O. Arthrologia. In *Illustrated Veterinary Anatomical Nomenclature*, 2nd ed.; Enke: Sttutgart, Germany; Ghent, Belgium; Columbia, MO, USA; Rio de Janeiro, Brazil, 2007; pp. 84–85.
33. Hauptfleisch, J.; English, C.; Murphy, D. Elbow magnetic resonance imaging: Imaging anatomy and evaluation. *Top. Magn. Reson. Imaging* **2015**, *24*, 93–107. [CrossRef]
34. Brunton, L.M.; Anderson, M.W.; Pannunzio, M.E.; Khanna, A.J.; Chhabra, A.B. Magnetic resonance imaging of the elbow: Update on current techniques and indications. *J. Hand. Surg. Am.* **2006**, *31*, 1001–1011. [CrossRef]
35. Freire, M.; Robertson, I.; Bondell, H.D.; Brown, J.; Hash, J.; Pease, A.P.; Lascelles, B.D.X. Radiographic evaluation of feline appendicular degenerative joint disease vs. macroscopic appearance of articular cartilage. *Vet. Radiol. Ultrasound* **2011**, *52*, 239–247. [CrossRef]
36. Kramer, M.; Gerwing, M.; Hach, V.; Schimke, E. Sonography of the musculoskeletal system in dogs and cats. *Vet. Radiol. Ultrasound* **1997**, *38*, 139–149. [CrossRef] [PubMed]
37. De Rycke, L.M.; Gielen, I.M.; Henri van Bree, H.; Simoens, P.J. Computed tomography of the elbow joint in clinically normal dogs. *Am. J. Vet. Res.* **2002**, *63*, 1400–1407. [CrossRef] [PubMed]
38. Deem, S.L. Role of the zoo veterinarian in the conservation of captive and free ranging wildlife. *Int. Zoo. Yearb.* **2007**, *41*, 3–11. [CrossRef]
39. Konar, M.; Lang, J. Pros and cons of low-field magnetic resonance imaging in veterinary practice. *Vet. Radiol. Ultrasound* **2011**, *52*, S5–S14. [CrossRef]
40. Basa, R.M.; Podadera, J.M.; Burland, G.; Johnson, K.A. High field magnetic resonance imaging anatomy of feline carpal ligaments is comparable to plastinated specimen anatomy. *Vet. Radiol. Ultrasound* **2018**, *59*, 597–606. [CrossRef]

Article

A Morphological and Morphometric Dental Analysis as a Forensic Tool to Identify the Iberian Wolf (*Canis Lupus Signatus*)

Víctor Toledo González [1,2,3,*], Fernando Ortega Ojeda [1,2], Gabriel M. Fonseca [4], Carmen García-Ruiz [1,2], Pablo Navarro Cáceres [5,6], Pilar Pérez-Lloret [3] and María del Pilar Marín García [3]

1 Department of Analytical Chemistry, Physical Chemistry and Chemical Engineering, University of Alcalá, 28871 Alcalá de Henares (Madrid), Spain; fernando.ortega@uah.es (F.O.O.); carmen.gruiz@uah.es (C.G.-R.)
2 University Institute of Research in Police Sciences (IUICP), University of Alcalá, 28801 Alcalá de Henares (Madrid), Spain
3 Department of Anatomy and Embryology, Faculty of Veterinary, Universidad Complutense de Madrid (UCM), 28040 Madrid, Spain; pilper01@ucm.es (P.P.-L.); pilmarin@vet.ucm.es (M.d.P.M.G.)
4 Centro de Investigación en Odontología Legal y Forense (CIO), Facultad de Odontología, Universidad de La Frontera, Temuco 4780000, Chile; gabriel.fonseca@ufrontera.cl
5 Centro de Investigación en Ciencias Odontológicas, Facultad de Odontología, Universidad de La Frontera, Temuco 4780000, Chile; pablo.navarro@ufrontera.cl
6 Universidad Autónoma de Chile, Temuco 4780000, Chile
* Correspondence: victor.toledo@edu.uah.es

Received: 22 April 2020; Accepted: 30 May 2020; Published: 3 June 2020

Simple Summary: Attacks by Iberian wolves on farm animals routinely cause conflicts with humans and threaten their economic interests related to livestock. However, wolf predation can sometimes be confused with that caused by other carnivores like dogs. Some studies have tried to identify or differentiate canids as the predators responsible for such attacks by analysing their tooth/bite marks on bone remains. Nevertheless, most of those studies have only considered a few dental measurements, and they were carried out in a palaeoecological and zooarchaeological context. As there is still limited information on Iberian wolf's dental anatomy that can be used in forensic cases, this study aimed to describe the morphology of the Iberian wolf's teeth and to provide new morphometric characteristics, as complete as possible, to collaborate in the correct interpretation of a wolf's bite marks at crime scenes. Based on the morphometric dental analysis, it was possible to differentiate female and male wolves. Moreover, the dental morphometric characteristics described can be used, at least as a reference, to identify the Iberian wolf's tooth/bite marks or to rule out other potential aggressors.

Abstract: Depredation by the Iberian wolf (*Canis lupus signatus*) is currently thought to be a problem in some areas of Spain. However, there are few technically validated forensic tools available to determine the veracity of claims with a high degree of scientific confidence, which is important given that such attacks may lead to compensation. The analysis of bite marks on attacked animals could provide scientific evidence to help identify the offender. Thus, the aim of this study was to assess the morphological and morphometric characteristics of Iberian wolf dentition. This data collection would serve as a base-point for a more accurate identification of the wolves thorough their bite marks. For the first time, 36 dental variables have been studied in wolves' skulls, employing univariate and multivariate analyses. The general morphological dental characteristics of wolves are very similar in terms of their dental formula and tooth structure to other canids, like domestic dogs. Sex differentiation was evident, principally in terms of the maxillary distance between the palatal surfaces of the canine teeth (UbC) and the width of the left mandibular canine teeth (LlCWd). New morphometric reference information was obtained that can aid the forensic identification of bite marks caused by the Iberian wolf with greater confidence.

Keywords: forensics; analysis; Iberian wolf; bite marks; veterinary; dentistry

1. Introduction

Animals are capable of inducing severe injuries through bites, which can even result in death [1]. Indeed, some families of the order Carnivora include species that are capable of attacking and killing human beings. Wild animals only rarely kill in urban areas [2], with predation on livestock representing the main source of conflict between large carnivores and humans [3]. In such cases, it is important to be able to determine if the animal bites were the true cause of death, and when this is the case, to identify the perpetrator [1]. Among mammals, carnivores are the most common scavengers of human remains [4] and their post mortem outdoor scavenging activity has traditionally been the best documented from a forensic perspective [4–6]. The analysis of bite marks can be used by forensic experts to aid in the identification of the taphonomic agent, and to interpret the behaviour, scavenging patterns, and predatory conduct of different animal species [7].

Livestock attacks attributed to wolves are becoming more and more frequent in the Iberian Peninsula [8], producing important societal problems [9]. Wolf predation can be confused with that of other carnivores, mainly free-ranging domestic dogs, a phenomenon that is more widespread in southern European countries [10]. Confusing field conditions or insufficient technical skills often make it difficult to correctly identify the perpetrators of such attacks [11], highlighting the need to develop tools to identify predators [12]. Different scavenger species within the same family can have different tooth dimensions, bite forces, jaw muscle strengths, and scavenging behaviours and patterns, which may be affected by various other factors. These features affect the type of bite marks produced on bone surfaces [13,14], making it necessary to define the morphological and morphometric features of the dentition of animals like wolves in order to be able to resolve conflicts with farmers. Indeed, up until 2015 there were no specific data available on tooth morphometry in the wolf [15]. In addition to collecting information that may help identify an animal assailant, these data may aid in cases where animal scavenging has occurred. Considering the increase in the alleged attacks on livestock attributed to the Iberian wolf and the lack of a detailed dental information available for this species, this study sets out to characterize the complete dentition of the Iberian wolf. Accordingly, the morphological features of different dental series were characterised, performing a morphometric analysis of 21 maxillary and 15 mandibular bones from wolves of either sex.

2. Materials and Methods

The dentition in the skull of 45 Iberian wolves (*Canis lupus signatus*) of both sexes (26 females and 19 males) was examined and measured. The skulls analysed in this study were obtained from the Mammalian Collection of the Spanish Natural Science Museum (Madrid, Spain). Only wolves skulls identified anatomically with mature dentition (permanent dentition), and with all teeth at eruption stage 3, were included in the study. This means that all incisors, premolars, and molar teeth had fully erupted into the occlusal plane, and that the visible cementoenamel junction was above the alveolus, in accordance with the code used by Geiger et al. [16]. No congenital skull anomalies or external dental damage were evident in the skulls sampled. All the measurements were made with a traceable digital calliper (Control Company, ISO 17025, Blacksburg, VA, USA) and registered in millimetres (mm).

2.1. Morphometric Study

Although the variables considered in this study were those based on Lemmons and Beebe's dental nomenclature, the arbitrary acronyms are used to simplify the visualization and data analysis (Table 1) [17].

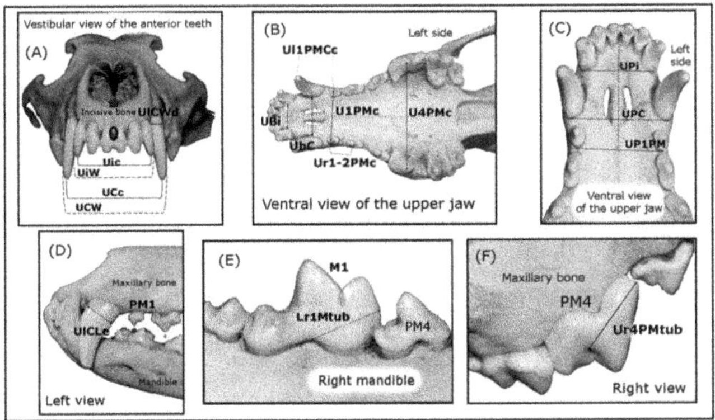

Figure 1. The studied measurements. (**A**) The canine and incisor measurements on the maxillary and incisive bone. The same canine measurements were taken from mandibular canines; (**B**) Measurements were taken on the maxillary dental arch between one tooth and its counterpart on the opposite side, and some were also taken on the mandible; (**C**) Distance from a line drawn behind the three teeth (third maxillary incisors, canines and first premolars) and a point located between both the central incisor teeth on the rostral face of the incisive bone (black circle); (**D**) ULCLe, also registering the same measurement on the right side and for the mandibular canines, (PM1: first premolar); (**E**) Lr1Mtub and the same measurement taken on the left side, (PM4: fourth premolar, M1: first molar); (**F**) Ur4PMtub and the same measurement taken on the left side.

Table 1. Acronyms of the studied variables.

Variable	Acronym	Variable	Acronym
Maximum maxillary (upper) intercanine width	UCW *	Maximum mandibular (lower) intercanines width	LCW
Maxillary distance between the cusp tip of the canine teeth	UCc *	Mandibular distance between the cusp tip of the canine teeth	LCc
Maxillary distance between the palatal surfaces of the canine teeth	UbC *	Mandibular distance between the lingual surfaces of the canine teeth	LbC
Maximum maxillary inter-third incisor teeth width	UiW *	Maxillary distance between the cusp tip of the third incisor teeth	Uic *
Maxillary distance between the palatal surfaces of the third incisor teeth	Ubi *	Maxillary distance between the cusp tip of the 1st premolar teeth	U1PMc
Mandibular distance between the cusp tip of the 1st premolar teeth	L1PMc	Maxillary distance between the cusp tip of the 4th premolar teeth	U4PMc *
Mandibular distance between the cusp tip of the 1st molar teeth	L1Mc	Distance between a line drawn behind the third maxillary incisors and a point located between both central incisor teeth on the rostral face of incisive bone	UPi *
Distance between a line drawn behind of third maxillary canines and a point located between both central incisor teeth on the rostral face of incisive bone	UPC *	Distance between a line drawn behind maxillary 1st premolars and a point located between both central incisor teeth on the rostral face of the incisive bone	UP1PM *
Right maxillary distance between the cusp tip of the 1st premolar–canine teeth	Ur1PMCc	Left maxillary distance between the cusp tip of the 1st premolar–canine teeth	Ul1PMCc *
Right mandibular distance between the cusp tip of the 1st premolar–canine teeth	Lr1PMCc	Left mandibular distance between the cusp tip of the 1st premolar–canine teeth	Ll1PMCc
Right maxillary distance between the cusp tip of the 1st–2nd premolar teeth	Ur1-2PMc *	Left maxillary distance between the cusp tip of the 1st–2nd premolar teeth	Ul1-2PMc
Right mandibular distance between the cusp tip of the 1st–2nd premolar teeth	Lr1-2PMc	Left mandibular distance between the cusp tip of the 1st–2nd premolar teeth	Ll1-2PMc
Width of the right maxillary canine teeth	UrCWd	Length of the right maxillary canine teeth	UrCLe
Width of the left maxillary canine teeth	UlCWd *	Length of the left maxillary canine teeth	UlCLe *
Width of the right mandibular canine teeth	LrCWd	Length of the right mandibular canine teeth	LrCLe
Width of the left mandibular canine teeth	LlCWd	Length of the left mandibular canine teeth	LlCLe
Length of the right maxillary 4th premolar mesial tubercle	Ur4PMtub *	Length of the left maxillary 4th premolar mesial tubercle	Ul4PMtub
Right mandibular 1st molar tubercle length(includes intermediate and mesial tubercles)	Lr1Mtub *	Length of the left mandibular 1st molar tubercle(including the intermediate and mesial tubercles)	Ll1Mtub

* Measurements as illustrated in Figure 1.

The data presented here were obtained on three different days by two independent observers. For the LCW, LCc, LbC, L1PMc, and L1Mc, 15 non-articulated mandibles were positioned and attached to the cranium in order to make accurate measurements. The damaged or worn mandibular incisor teeth were not considered in this study for the morphometric analysis. Likewise, the maxillary series of molar teeth, and the second and third mandibular molar teeth were also not considered for the morphometric analysis. The UbC, LbC (not shown), and Ubi, and the canine width and length measurements (UrCWd, UrCLe, UlCWd, UlCLe (Figure 1A–D) and, LrCWd, LrCLe, LlCWd, LlCLe, Ur4PMtub, Ul4PMtub, Lr1Mtub, and Ll1Mtub), were taken from the furcation (the point at which the roots diverge) at the level of the cementoenamel junction (the boundary between dental enamel and cement: Figure 1D–F).

The UPi, UPC, and UP1PM measurements were taken from the mandibular symphysis (between both the central incisor teeth: Figure 1A, black circle), and the distal surface of the third incisors, canines and first premolars (PM1s) respectively, to the height of the alveolar bone (Figure 1C). Finally, only teeth with pointed cusps on the largest dental tubercles were considered for the UPM4c and L1Mc measurements.

2.2. Morphological Description

The external (crown) and internal (root) morphology of the wolf's teeth was described based on the anatomic nomenclature established in Lemmons and Beebe [17] and in Schaller [18]. Dental root X-rays were obtained at a veterinary clinic (Trasportix LW AL, portatile domiciliare radiografico digitale, TXLW-4kW model, Milan, Italy) using the following parameters: KVp 54–58; mA 50 and exposure 0.08 s. To illustrate the Iberian wolf bite mark, an impression of the maxillary and mandibular dental arch was registered on dental wax and photographed with a Nikon D60 Digital camera (Nikon, Tokyo, Japan), mounted on a Hama tripod (Hama Technics, S.L., Barcelona, Spain). The patterns were registered by pressing the maxillary bone into the dental wax up to the maxillary PM1 mark and the images were acquired at a natural 90° angle (perpendicular).

2.3. Statistical Analysis

A Kolgomorov–Smirnov test was initially performed to assess normality, followed by a parametric Student t-test test for independent samples and a non-parametric U Mann-Whitney test for independent samples. Statistical analyses were performed with the IBM SPSS software v23.0 (IBM-Corp, Armonk, NY, USA) using ANOVA with a factorial design and multiple comparisons. A 95% confidence interval was considered ($p < 0.05$) and the results are presented as intervals in millimetres (mm), with their mean and standard deviation (SD).

For the multivariate analysis, a principal components analysis (PCA) was used to explore the extent to which variables studied were related, and a partial least squares discriminant analysis (PLS-DA) was used to classify the data by sex (males and females) and to identify the critical variables that allow such differentiation. The multivariate analysis was performed with Soft Independent Modelling of Class Analogies (SIMCA) (Sartorius Stedim Biotech, Göttingen, Germany) [19] and seven cross-validation groups were used to consider similar observations in the same group, validating the calculated latent variables. The data were centred and scaled (Unit Variance, UV), and the software was set to calculate the boundaries with 95 % probability [19]. Only variables that presented statistically significant differences between the sexes ($p < 0.05$) were considered for PCA and PLS-DA analysis.

3. Results

3.1. Dental Morphology

The morphological patterns of the wolves' teeth were described based on the observations presented in Figures 2–4, in addition to the relevant bibliographic information [17,18]. The dentition of canids involves two different dental arches, each of which has its own characteristics and number of teeth defined as incisors, canines, premolars, and molars: the maxillary arch (Figure 1B,C) and the mandibular arch. Molar teeth were only detected in permanent dentition (dental maturity). Each tooth described had a visible part that projected into the mouth, called a crown, and a deeper part that extended into the mandibular, incisive and maxillary bones, the root. Enamel and cement usually cover the crown and root, respectively. Between the crown and the root, there is a mild constriction that constitute the neck or cementoenamel junction (the transition between the enamel of the crown and the cement of the root: Figure 1D–F). In general, the incisor, canine, premolar, and molar teeth have four exposed surfaces, and with a cusp or ridge constituting a fifth surface. The cusp is the point or the tip of the crown of a tooth. The surfaces of the teeth that face into the vestibule (the space between the lips and incisors/canine teeth, or between the cheek and the premolar/molar teeth) are the vestibular surfaces. Moreover, the space between the lips and incisors, and between the lips and the canine teeth is commonly described as the labial surface, while the space between the cheek and the premolar/molar teeth is known as the buccal surface. The dental surfaces that are orientated towards the tongue are described as the lingual surface, although for the maxillary teeth, this surface is often referred to as the palatal surface.

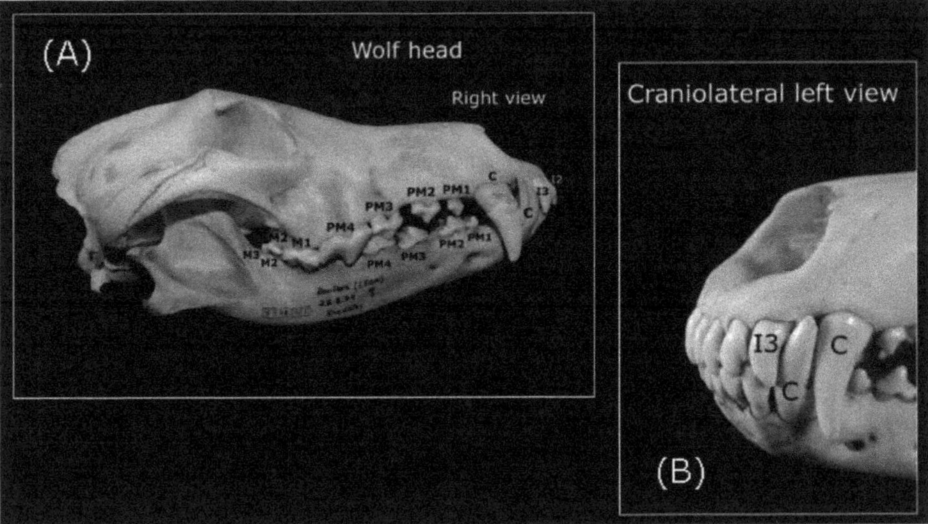

Figure 2. General vision of the Iberian wolf's teeth. C: Canine, PM2 and PM3: second and third premolar teeth, respectively; M2 and M3: second and third molar teeth, respectively. (**A**) Wolf skull with the right dental line indicated on the maxillary and mandibular bones; (**B**) Detail of the left-hand canine and I3 on the maxillary and mandibular bones.

Different permutations of the anatomical nomenclature can be used to indicate the precise position of each structure. The occlusal surface is that of the premolar and molar teeth that faces their antagonist, and that contacts the teeth in the opposite jaw on closure, while the ridges or cusps of premolars that do not make contact are known as occlusal ridges. For the incisors, the ridge along the "occlusal surface" is referred to as the incisal ridge, a chisel-like shape with a sharp edge, and the cusp of the canine teeth is generally called the cusp surface. However, premolars and molars may have multiple cusps and tubercles. The surface that faces the adjacent teeth within the same dental arch is the contact surface, a "mesial" surface if it is directed forward towards the median line and a "distal" surface if it faces away from the median line of the face. In the case of the canine and some premolar teeth, the term "surface" is replaced by the term "edges" when these are sharp. The space between two teeth is referred to as the diastema.

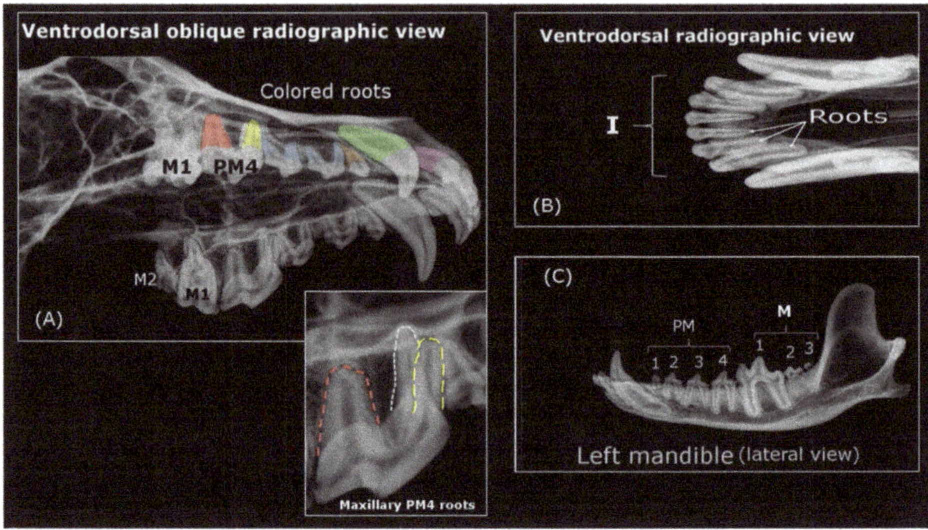

Figure 3. Radiographic views of the wolf's teeth and skull. (**A**) (Upper image) Right I3 root (purple), right canine root (green), PM1 root (orange), PM2 and PM3 roots (blue) and PM4 roots (white, red and yellow); (Lower image). Detail of the PM4 roots-mesiobuccal (yellow dotted line), mesiopalatal (white dotted line) and distal root (red dotted line); (**B**) Roots of three right maxillary incisors; (**C**) Molar and premolar disposition teeth in the mandible.

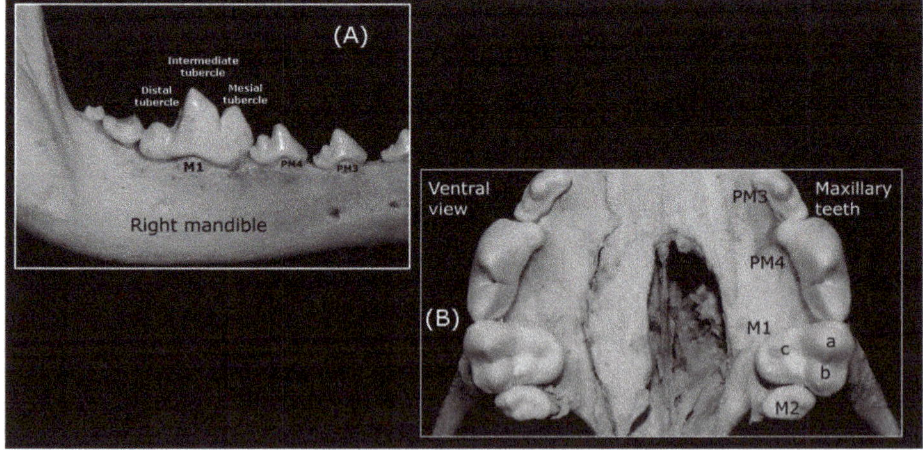

Figure 4. (**A**) Wolf's right mandible (lateral view) up to the PM4 tubercles and cusps; (**B**) The occlusive face of left maxillary M1 with its (a) mesiobuccal, (b) distobuccal and (c) distolingual tubercles and cusps.

3.1.1. The Teeth of the Iberian Wolf

The full dentition of the adult Iberian wolf is comprised of 42 teeth of four different types: 12 incisor (I), 4 canine (C), 16 premolar (PM), and 10 molar (M) teeth (Figure 2A). Thus, the permanent dental formula of the Iberian wolf: 2(I3/3, C1/1, PM 4/4 and M2/3), where the first numbers above are for the maxillary teeth and the second are for the teeth of the mandibular bone. Given that the dentition is the same on both sides, the formula lists only one side enclosed in parentheses and it is to be multiplied by 2 to give the total number of teeth.

Incisors

An analysis of the wolf's incisors identified six maxillary and six mandibular incisors (three right and three left incisors) numbered as the first (I1), second (I2) and third (I3) incisor from the midline (Figure 1A–D, Figure 2A,B, and Figure 3B). Each incisor has one root (Figure 3B), with the three maxillary teeth implanted into the incisive bone (Figure 3A). Both the crowns and roots of the maxillary and mandibular incisors increase in size (length) from I1 to I3, as is also seen for the width of the crowns (the distance between the distal and mesial surface: Figure 3B). The maxillary I1 and I2 have three protrusions along the incisal edge called mamelons, with two lateral mamelons and one central mamelon (Figure 1A), while I3 is conical with no mamelons and a sharp cusp (Figure 2B). The central mamelon is the longest one. The labial vestibule surface of the maxillary incisors is longitudinally convex (Figure 2B) and lightly concave along most of their lingual surface. From a rostral view, the incisal ridge of I1 and I2 of the maxillary bone is slightly rounded while I3 presents an incisal ridge that is more pointed. The crown of the incisor teeth has a light downward angle of inflexion up to a few millimetres ventral to the cementoenamel junction. Moreover, the base of the crown of I3 is broad compared to I1 and I2. The mandibular incisors are smaller in size (length and width) than their maxillary counterparts. They have two mamelons with a similar curvature to that of the maxillary incisors, and they are angled obliquely forward and upwards. The incisal ridge is present in I1, I2, and I3 forming a sharp edge, and I1 and I2 are laterally compressed. The roots of the maxillary and mandibular I1 and I2 are straight, twice as long as the crown is high (Figure 3B, the mandibular incisors are not shown). The root and crown of I3 are arched, displaying a similar crown and root length (Figure 3B). The distal surface of I3 is sharp and it is possible to describe a soft sideways curvature of the root. The maxillary and mandibular I3s are laterally compressed.

Canines

An analysis of the wolf's canines identified four canine teeth, each positioned behind I3, two maxillary and two mandibular canine teeth in each dental arch (one-right and one-left canine on the maxillary and mandibular bone: Figure 1A–D). These are the largest teeth in the Iberian wolf's mouth, all with a light convex mesial edge and a concave distal edge from the cementoenamel junction. The maxillary canines are wider and larger than the mandibular ones (Figure 2A,B). From the cementoenamel junction, the crowns in the maxillary canines project in a forward, downward, and slightly lateral direction (Figure 3A,C). By contrast, the mandibular canines project in a forward direction and upwards, with a clearer and larger lateral deviation than their maxillary counterparts (Figures 1D and 2B). All canines present a single tip cusp, are conical with an elliptical to triangular cross-section, and the joint to their unique root is arched. The crown on the maxillary canine teeth is less curved than that on the mandibular canine teeth. Maxillary canines are larger than the mandibular ones. Like the incisors, the canine teeth have a longer root than their crown height and they are also laterally compressed.

Premolars

In this analysis, 16 premolar teeth were identified, eight in the maxillary dental arch (Figure 3A) and eight in mandibular dental arch (four-right and four-left premolars in the maxillary and mandibular bones: (Figure 1B–F and Figure 2A). The maxillary premolar teeth are positioned more laterally than the mandibular ones. The PM1 is the smallest of the series (Figure 2A) and the crown of the maxillary PM1 is slightly lower than its counterpart in the mandibular bone. The mandibular PM1 is the most rostral premolar tooth in the series and it has only one tubercle, with one tip cusp and one root (Figure 3C), a straight yet longer root relative to the crown: the root is twice the crown height. Both, mandibular and maxillary PM2s and PM3s are similar. These premolar teeth have two tubercles (mesial and distal) and two diverging roots (mesiobuccal and mesiopalatal: Figure 3A,C). The mesial edge of the maxillary crowns faces down and backwards, while in the mandibular ones the direction of the edge is upwards and backwards. By contrast, the caudal edge of PM2 and PM3 is directed downward and onward, and up and onwards, respectively. Both the mesial and caudal edge meet to form the tip cusp. The maxillary PM4 is a carnassial tooth with two tubercles. A transverse groove on the contact surface separates the bigger mesial tubercle from the smaller distal tubercle (Figures 1F and 4A,B). The direction of the mesial edge is similar to that in PM2 and PM3. PM4 has three roots, two diverging roots belonging to the mesial tubercle (mesiobuccal and mesiopalatal) and one distal root belonging to the distal tubercle (Figure 3A).

Molar Teeth

Regarding the molar teeth, there were two right and two left molars in the maxillary bone (Figures 2A and 3A) and three right and three left molars in the mandibular bone (Figure 3C), a total of five left and five right molar teeth. In both the maxillary and mandibular bones, the M1 is the largest of the series and it is a carnassial tooth (Figure 1E, Figure 3A,C). The maxillary M1 and M2 have three tubercles and three roots, and these three tubercles are distributed as two vestibular tubercles (mesiobuccal and distobuccal) and one smaller lingual tubercle (Figure 4B). The mesiobuccal tubercle is bigger than the distobuccal tubercle, and a deep groove separates these two vestibular tubercles. By contrast, an occlusion surface separates the distolingual tubercle from others. In terms of the roots, those from the vestibular tubercles diverge while the root of the distolingual tubercle travels to the palatine bone (hard palate). The mandibular M1 has three tubercles (mesial, intermediate, and distal: Figure 4A), of which the intermediate is the biggest. A transverse groove separates the first two and a small occlusion surface separates the smaller distal tubercle from the intermediate tubercle. Like the mandibular M2, this molar has two diverging roots (mesial and distal) that pertain to their respective tubercles. The mandibular M2 has two tubercles (mesial and distal) separated by a small occlusion surface. The mandibular M3 is the smallest of the series, and it has one tubercle and one root. Diastemas in the incisor, premolar, and molar teeth are usually of varying length or they may be absent.

3.1.2. Normal Occlusion (the Closing of the Jaw)

In normal occlusion, the edges of the maxillary incisors cover those of the mandibular incisor apices, which means that the mandibular incisors come into contact with the lingual surface of the maxillary incisors, producing a scissor effect (Figure 2A,B). The maxillary I3s tooth interdigitates between the mandibular I3s and mandibular canines. The sharp mesial edges of the maxillary canines come into direct contact with the sharp distal edges of the mandibular canines (Figure 2A,B). The mandibular canines adopt a position caudal to the maxillary I3s and they come into direct contact with their sharp distal surface. The maxillary canine teeth adopt a position more lateral to the mandibular ones during occlusion (Figure 2A,B). The first two maxillary and last three mandibular premolars are interspersed. The more lateral position of the maxillary premolar teeth relative to the mandibular ones is more evident in occlusion. Finally, the molar teeth make contact along their occlusal surface. The occlusal surface of the maxillary M1 tooth comes into direct contact with the distal cusp that pertains to the distal tubercle of the mandibular M1 tooth. The distal cusp of the mandibular M2 tooth comes into direct contact with the occlusal surface of the maxillary M2, while the mandibular M1 tooth comes into direct contact with the lingual surface of the maxillary PM4 and M1. Finally, the mandibular M2 tooth contacts the lingual surface of the maxillaryM1 and M2.

When a wolves' bite marks on dental wax were evaluated, the incisor and canine teeth form a curved arch on maxillary and mandibular jaw. The maxillary incisor and canine teeth are registered in the superficial (Figure 5A) and deep bite (Figure 5B). Figure 5C shows the maxillary incisor and canine teeth registered in the deep bite. In a deeper bite, the diameter of the PM4 and the canine bite marks appear to be slightly larger (Figure 5A vs. Figure 5B). Thus, the distance between the palatal (UbC) or lingual (LbC) surface of both canine teeth decreases with depth. A degree of tooth crowding is visible among the Iberian wolf's incisor teeth on both bones (maxillary and mandible bones), both in superficial and deep bite marks. Mandibular incisors bite marks are smaller than maxillary ones in deep marks.

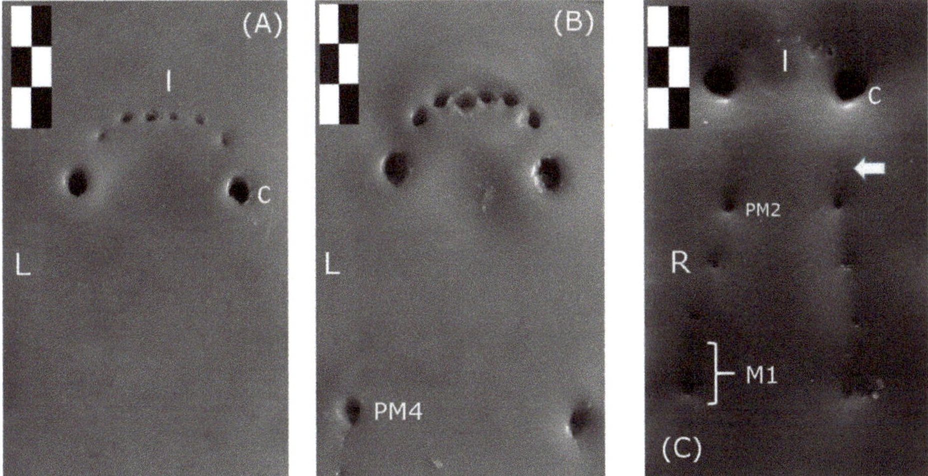

Figure 5. (**A**) Superficial bite mark caused by the maxillary dental arch of the Iberian wolf; (**B**) Deeper bite mark caused by the maxillary dental arch; (**C**) Deep bite mark caused by the mandibular dental arch (mandible). The length of each black/white rectangle at the top left corresponds to one centimetre.

The bite marks of the Iberian wolf's mandibular premolars become progressively bigger towards the distal and maxillary PM4 (carnassial tooth), the latter the biggest of the series while the PM1 is the smallest. The maxillary bite mark is a little larger than the mandibular one, which explains the deeper bite. Premolar teeth have a straighter disposition in both the maxillary (not visualized) and mandibular arch (Figure 4A).

3.2. Morphometric Data

Although significant differences were detected among the morphometric variables, sex differentiation was only possible for UCW, LCW, UCc, LCc, UbC, LbC, UiW, Uic, Ubi, U1PMc, U4PMc, L1Mc, UP1PM, Ur1PMCc, Ul1PMCc, UrCWd, UrCLe, UlCWd, UlCLe, LrCLe, LlCLe, Ur4PMtub, Ul4PMtub, Lr1Mtub, and Ll1Mtub ($p < 0.05$). Of all the variables for which there were significant differences between the sexes ($p < 0.05$), the averages for males was larger than the mean measurements registered in females (Figure 6A–D). In addition, males reached larger minimal and maximum absolute values than females, although the variables in the females were more variable (Figure 6). However, the maximum absolute LrCle and LlCle values were larger in females and thus, they were considered as outliers (Figure 6C: variables that did not present a significant difference are not shown: $p > 0.05$).

We consider some of the observations as outliers (outside the 95 % confidence level), especially in females (Figure 6), which may explain why the maximum intervals for the LrCle and LlCle were larger in females than for males (Figure 6C). While the exclusion of these outliers from the statistical analysis affected some intervals, there were no significant changes in the average values of males and females, i.e., the relationship between sexes remained stable (Figure S1 and Table S1). In addition, when comparing the two sexes there were variables with no area of overlap and others with a differing degree of overlap (Figure 6). Accordingly, certain limits can be set for sex differentiation (Table 2).

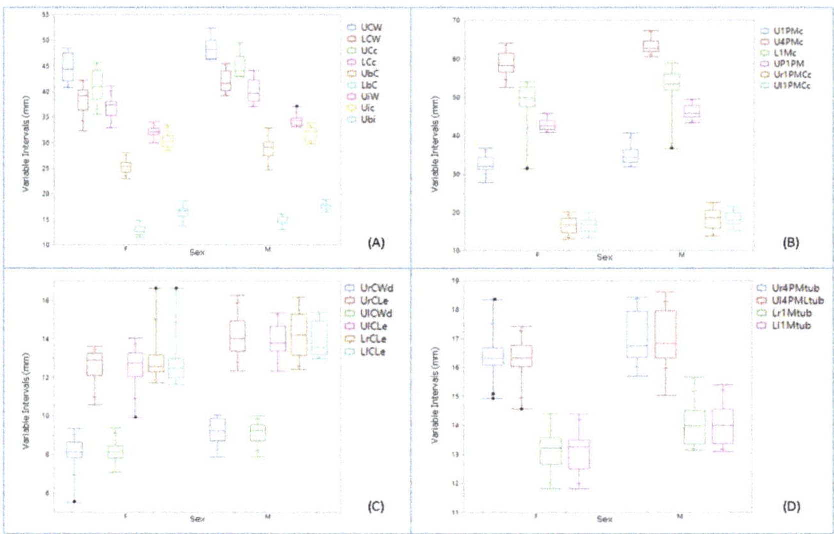

Figure 6. Intervals of the dental measurements considering sex as a source of variation. The minimum, maximum, mean, and quartiles values for each variable according to sex. The black points are the outlier values. (**A**) Measurements taken between right and left canine and between right and left incisors on maxillary and mandibular bones; (**B**) Measurements taken between right and left premolars and between molars; (**C**) Measurements taken on canines, (**D**). Measurements taken on premolars and molars teeth.

Table 2. Absolute limits of the variables for the wolf sex differentiation (95% confidence), considering 26 female and 19 male skulls. Va: variable; Lv: Limit value for the sex differentiation; F: females; M: males; mm: millimetres; F/M: sex differentiation was not possible.

Va	Lv (mm)	Mean	Va	Lv (mm)	Mean
UCW	F < 46.25 M > 48.55	44.7 48.4	Ur1PMCc	F < 13.76 M > 20.13	16.48 18.4
LCW	F < 39.15 M > 24.23	38.66 41.91	Ul1PMCc	F < 15.27 M > 20.13	16.6 18.39
UCc	F < 42.84 M > 45.66	41.27 45.01	UrCWd	F < 7.87 M > 9.36	8.28 9.23
LCc	F < 37.06 M > 41.07	36.91 40.09	UrCLe	F < 12.34 M > 13.62	12.62 14.2
UbC	F < 24.65 M > 28.04	25.39 29.14	UlCWd	F < 7.88 M > 9.37	8.13 9.11
LbC	F < 13.06 M > 14.87	12.91 14.71	UlCLe	F < 12.36 M > 14.05	12.82 13.99
UiW	F < 33.10 M > 34.16	32.04 34.18	LrCLe	F < 12.44 M > 14.36	12.7 14.27
Uic	F < 29.81 M > 33.39	30.46 31.84	LlCLe	F < 13 M > 13.69	12.48 M: 14.02
Ubi	F < 16.47 M > 18.61	16.38 17.68	Ur4PMtub	F/M < 17.43 M > 17.43	F: 16.41 17.02
U1PMc	F < 31.83 M > 36.71	32.64 34.94	Ul4PMtub	F/M < 7.43 M > 7.43	F: 16.32 17
U4PMc	F < 60.55 M > 64.08	58.59 63.57	Lr1Mtub	F < 3.15 M > 4.41	13.06 14.04
L1Mc	F/M < 54.11 M > 54.11	F: 50.54 54.28	Ll1Mtub	F < 13.10 M > 14.4	13.09 14.02
UP1PM	F < 43.35 M > 45.84	42.92 46.18			

In percentage terms, the mean UCW value was approximately 7.65% lower in females than in males, while the mean UCc value was 8.31% higher in males than females. The same variables registered on the mandible (LCW and LCc) were approximately 11% and 13% lower in females and males than the maxillary values. The mean values for UbC, LbC, Ur1PMCc, Ul1PMCc, UrCLe, UlCLe, LrCLe and LlCLe were 9–13% lower in females, and the mean values for the remaining variables that were significantly different ($p < 0.05$) were also lower in females (4–8%). There were no significant differences ($p > 0.05$) between the same variables measured on either side of the maxilla and mandible. However, they were different in absolute terms.

A PCA exploration allowed two relatively well separated groups with some overlap between them to be visualized (data not shown), and while the PLS-DA also showed a certain degree of overlap between the two groups, it allowed two well-defined groups to be discriminated. The relationship and degree of dispersion among the scores (specimens) within and between both groups was classified by sex.

The exception was one specimen, a female skull, which seemed to share characteristics of both sexes. This unusual specimen was excluded from the multivariate analysis as it was considered an outlier (outside the 95 % confidence level) (Figure S2). This specimen exaggerates the high variability in the female group, yet its exclusion did not cause any variation in the distribution of the variables within and between the two sexes. Consequently, the three latent variables in Figure 7, t(1) t(2), and t(3), explained 51.1%, 9.13%, and 4.4 % of the model's variability, respectively. Therefore, the model explained about 64.6% of the variance between the sexes.

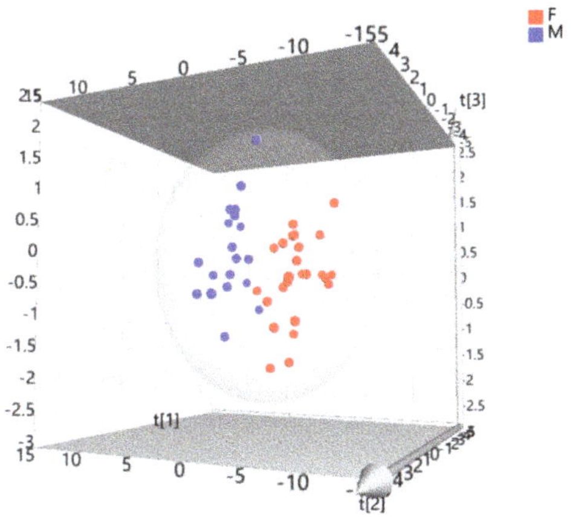

Figure 7. The 3D scatter PLS-DA plot showing the sex differentiation. The 3D scatter PLS-DA plot of the scores showing the distribution and relationship of the variables within and between sexes. The initial latent variables t(1), t(2) and t(3), explained 51.1, 9.13 and 4.4% of the variability, respectively. The whole model explained 64.6 % of the variance.

The contribution plot visualizes the variables and their contribution (weight) to the differentiation between sexes (Figure 8). The analysis showed that the most important variable for this differentiation was UrCWd (orange column), which presented the strongest variation between sexes. Sequentially, the next most important variables were UrCWd, UbC, UiW, LbC, U4PMc, UP1PM, and Ubi.

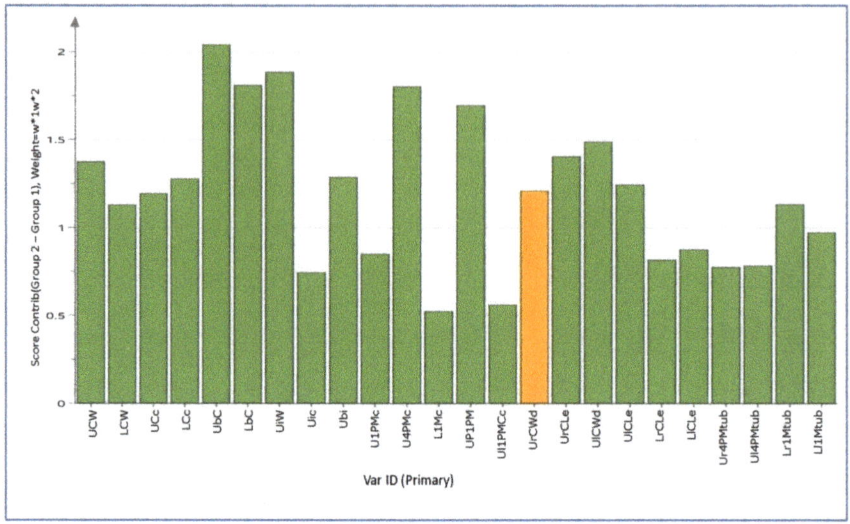

Figure 8. Contribution variables for the sex differentiation. Contribution plot indicating the most significant variables in differentiating males from females with the strongest mean differences ($p < 0.05$). Since the data was a scaled unit vector, the vertical scale units were presented in terms of standard deviations (SDs). The dominant bars show which variables deviate most from the reference point (in this case, the average). The orange colour shows which variables lie outside the 3 SD range, i.e., that which is most influential.

4. Discussion

In this study, we have performed a detailed analysis of the dentition of the Iberian Wolf, defining the different types of teeth, their morphology and morphometric features, as well as any sexual divergence in these parameters. An earlier study of 53 skulls from nine different Gray wolf sub-species (17 females, 21 males, and 14 indeterminate sex: Murmann et al. 2006 [2] took into account three dental measurements on the maxilla and two on the mandible, studying the MCW, maxillary and mandible tip (corresponding to the UCW, UCc, and LCc in our study, respectively). In this earlier study [2], the term mesial bone height (MBH) was employed for the measurement taken next to the most mesial portion of the canine between the canine teeth on the maxillary and mandibular bones. Here, we referred to this variable as the maxillary palatal intercanine surface neck height (UbC) and the mandibular lingual canine surface neck height (LbC), since the measurements were taken between palatal/lingual surfaces of the canine teeth, respectively, and not between their mesial surfaces. Nevertheless, that earlier study identified larger intervals for the UCW, UCc, LCc, UbC, and LbC parameters (adapted to our nomenclature). In the Iberian wolf, the limits of the LCc interval were lower than those recorded for Gray wolves and indeed, the minimum and maximum values of UCW, UCc, and LCc for female Iberian wolves were lower than those registered in Gray wolves. Finally, our UbC and LbC values were lower in both sexes. When these dental measurements were compared with those in different domestic dog breeds (*Canis familiaris*), they were generally smaller in the wolves [2]. However, a reliable comparison with our outcomes is not possible as the outcomes from the dog breeds were not classified by sex, breed or size, nor by sub-species, and no statistical analysis was performed. In light of the above, for reference purposes, we can only indicate that our results for UCW and UCc in male Iberian wolves largely overlapped with those of the nine sub-species of Gray wolves.

In addition to classifying our specimens, we used millimetres (mm) and not centimetres (cm) as the unit of measurement allowing us to obtain more precise results. Although the dog (*Canis lupus familiaris*) is a direct descendant of the Gray wolf [20,21], it displays greater morphological diversity than any other species, which makes it extremely difficult to compare between the two animals, especially when the data are not classified. This must be taken into consideration when performing more complete comparative studies to correctly differentiate between Iberian wolves and domestic dogs through dental analyses. Although the outcomes from this earlier study were broad and not sufficiently specific to the wolves' dental morphometry, it did focus on canids and it established benchmarks for later efforts to identify offending animals thorough bite marks. In recent years, indirect dental morphometric studies have compared bite marks caused by wolves (American and Iberian wolves) and other species registered on various types of bone, and using different approaches [22–24]. It should be noted that the average size of the Iberian wolf lies somewhere between that of the American and European wolves [25].

The general dimensions (breadth and length), distribution and proportions of the classic tooth marks (pits, punctures, scores and furrows) have been described for wolves [22,24,26–31]. Similar studies have been performed on dogs and a few others have compared wolf's and dog's bite marks [2,32–34], yet none have assessed the distances between their teeth considering a large number of dental variables. Characteristic features of the animals and samples that could have influenced the measurements have been considered (e.g., overlapping), such as environmental factors. Canid dental morphology allows this species to be differentiated from other carnivores and characteristic post-mortem lesions caused by domestic dogs can now be recognized, reflecting the dog's dental anatomy [10]. Most of these past studies on bite marks, focused principally on archaeological and zooarchaeological samples, and contexts, with ambiguous outcomes. This represented a huge effort undertaken by those authors whom worked with non-standardized samples, exposed to unknown external conditions, which may have caused the results variability they obtained. Indeed, they focused on bite mark characteristics caused by only a few teeth. To complement and extend these studies, here, we collected as many dental variables as possible in order to make it possible to identify the bite mark patterns caused by Iberian wolves in a forensic context. This inspired us to define dental reference values for the Iberian wolf, a possible aggressor or scavenger, which should help to identify its activity through the analysis of its bite marks. The skulls of domestic dogs are more varied than that of its ancestor, the Gray wolf, approximating to the variation found in the wild [35]. Some arch sizes can produce intercanine distances (ICDs), referred to as the U/LCc here, that vary as much as 70 mm in large breeds, like the Presa Canarios. Some dog breeds have been considered to be "non-modified breeds", meaning that they retain characteristics more similar to ancestral wolves [36], and they constitute the wolf-like group. These include the German Shepherd based on its skull metrics [37] and in fact, wolves have been said to have a semblance to German Shepherds but smaller [38]. The bite mark patterns registered on dental wax of the adult German Shepherd dog (weight > 25 kg) were analysed recently [39]. These were more than 95% coincident with measurements taken on plaster casts of dog's dentition, with all measurements made within five hours to avoid structural changes to the wax. Comparing these results with our outcomes, all the mean values, intervals and maximum values registered for the German Shepherd were greater than those recorded in Iberian wolves for both sexes. However, different overlaps were evident for UCW and LCW intervals of between males of both species, yet only for the UCW in females. Although a direct descendant of Gray wolves, dogs show some morphological differences, such as smaller dentition [20]. Despite rapidly recording the data after registering the bite, we must consider some environmental factors that could possibly influence the results from the wax imprints, like temperature and humidity, as well as those related to the characteristics of the substrate (dental wax in our case), such as the elasticity, hardness, etc.

The dental variables analysed here displayed sexual dimorphism and over 95% of the measurements were larger in males, although some such differences may not have reached statistical significance. This sexual dimorphism was also noted when evaluating dental measures in three different dog breeds [39] and in an earlier study on Iberian wolves [25]. Sexual dimorphism has been mentioned previously when comparing the "anteroposterior diameter of mandibular canines", corresponding to the Lr/Ll-CLe here [40]. Our data are to some extent consistent with those results, although we found thorough a PLS-DA analysis that these variables are certainly not the only or the most important variables to differentiate the two sexes. Besides, these measurements can be influenced by other factors as discussed below. Sexual dimorphism in Iberian wolves has been studied via dental and cranial dimensions [40] and although the sample size was small, sex differentiation was possible by combining dental and cranial measurements. We found measurements like UP1PM present significance differences between the sexes, with males displaying a larger snout than females from the canine teeth forwards. Other dental measurements also show sexual dimorphism [41], although the utility of these measurements to differentiate sexes has been questioned [42]. In accordance with our nomenclature, they registered very similar mean values for males and females as those presented here for the UrCLe, UlCLe, LlCLe, and LrCLe, even though their measurements were taken at different points of the canines. Other measurements they analysed also showed similar sex differences as those seen here (e.g., the UiW and UCW), such as when comparing the distance separating both the maxillary PM4 (carnassial teeth [40]) with the maxillary distance between the PM4s cusp tip (U4PMc) measured here, producing similar results. Nevertheless, there is some overlap for certain variables between sexes, as seen here, and some ambiguity may exist when morphological characteristics are not due to a given sex or age but rather, reflect individual asymmetry for example (see below).

There is considerable and extensive information about the ecology, conservation and anatomical aspects of Iberian wolves [25]. Anatomically, seven variables on Iberian wolf skulls have been measured, corresponding to UCW, UCc, UbC, LCc, LCW, LCc, and LbC in our study. For the UCW, intervals of 4.5–5 cm were found in males and 3.5–4 cm in females, while the measurements for the UCc were 3–5 mm less than the UCW, and the UBC and LbC were 1.5–2 cm less than the UCW. This study also indicated the measurements for the mandibular canine teeth were 15–20% lower than those from the maxillary arch. Here, we also found lower mandible values than maxillary measurements but we registered a slightly wider variation for these variables (13–20% less in mandible). Greater differences (46–50%) were found between other variables analysed in our study than in this earlier study [25]. The quantification of more dental morphometric variables and their PLS-DA analysis made it possible to graphically display the behaviour of the variables analysed. Despite the mild overlap between sexes and the homogeneous dispersion of variables in each group, the PLS-DA analysis identified two well-defined groups (sexes), as also evident by comparing defined limits to differentiate sexes for each variable. In the light of the results presented here, considering only a few dental dimensions to differentiate sexes may be inaccurate. Thus, it is necessary to process as much data as possible to reduce uncertainty when differentiating sexes. Complementary analysis should determine the most important variables to differentiate sexes and how they influence each other. Wolves and domestic animals are known to exhibit sexual dimorphism [20,43], although to our knowledge, the present study is the first to analyse such extensive morphometric dental data through multivariate methods in order to differentiate the sexes of Iberian wolves.

A degree of asymmetry was evident between values registered on the right and left maxillary and mandibular bones. On the one hand, this could explain why some variables for particular teeth may not be useful to differentiate sexes, generating overlap between the sexes, while on the other hand, these asymmetries may also be reflect sex differentiation. Either way, this asymmetry provides a unique dental morphology for each animal that would permit, if necessary, the identification of any specific animal through its unique dental morphology, as also seen previously [44]. In many cases, a simple tooth can be used for identification if it contains enough unique characteristics [45]. Therefore, morphometric and morphological studies could be combined in a complementary manner for better identification. It has been pointed out that the wear of teeth progresses with age, affecting all teeth [15,46] and especially incisor teeth, the teeth most likely to suffer breaks. This effect was seen in our morphometric study but most specifically, in the mandibular incisor teeth. We noted that all the mandibular incisor teeth in our skulls had a high degree of wear, and if they were absent or fractured, they were not included in our study. The degree of wear of teeth is indicative of the intensity and character of tooth use [15,46], and both these studies indicated that deterioration of a wolf's teeth may change their bite pattern and their eating habits. Briefly, I1 and I2 have sharp edges that are used like chisels to take small bites when gnawing at bones, cutting away pieces of food, and in grooming. The canines and I3s serve for deep penetration into the prey's tissues and for cutting through the tissues, while the premolars strip away food with a cutting action, except for the PM1s (not functional). The small crowns and weakly developed roots of these teeth probably render them useless to hold and carry parts of prey. The molars are designed to crush and pulverize bones to expose the marrow [46], except for the mandibular M3 that is not functional. Therefore, we assume that the roles played by these teeth may be the source of the different degrees of asymmetry found in our study, although they may simply be due to individual characteristics, for example the specimens with outlier values. Despite the asymmetries, two different groups could be defined, namely males and females.

The data suggest that the crowns and the base (cementoenamel junction) of the I3s, canines and PM1s are points of minimal wear. Therefore, variables measured at these points should be more stable over time and more useful to differentiate between sexes. In fact, all the variables measured between both sides of the maxillary and mandible bones, and the variables measured at the cementoenamel junction (UCW, UCc, UbC, LCW, LCc, LbC, UiW, Uic, Ubi, U1PMc, Ur4PMtub, Ul4PMtub, Lr1Mtub, and Ll1Mtub, UrCwd, UrCLe, UlCWd, UlCLe, LrCLe, LlCLe) appeared to be extremely well-suited to differentiate between the sexes. By contrast, the width of the right and left mandibular canine teeth (LrCWd and LlCWd, respectively) was not useful for sexual differentiation, and the L1PMc did not differ significantly between the sexes. Both the maxillary and mandibular PM1s are considered rudimentary teeth, not functional teeth, and they commonly break [15]. In fact, the teeth that most often absent were the maxillary and mandibular PM1s [47]. Here, over 50% of the skulls lacked the L1PM and this reduced number of observations for L1PM may explain why this variable did not differentiate between sexes. Alternatively, we considered that the LrCwd and LlCWd were not suitable to differentiate between sexes due to individual characteristics and not due to the wear of these teeth. In fact, LrCWd and LlCWd showed no significant differences on either side. Contrary to expectations, U4PMc and L1Mc were useful for the sex differentiation, even though the point or tip of their crown is constantly used to crush and pulverize bones. However, we chose accurate cusps and not flat ones, with as little wear as possible. Moreover, according to an earlier study [48], into the wear of teeth the maxillary premolars and M1s in the present study would belong to wolves under six years of age. We must also consider that during occlusion, the cusps of the maxillary premolars and molars do not contact their counterpart on the mandible, producing less wear.

The dispersion of the dental variables analysed here in females was a little higher than in the male group. Similar outcomes were obtained elsewhere when eight mandibular measurements were compared in different kinds of African canids [49]. Here, we used one measurement that was also used in this earlier study, L1Mc, which also proved to be important to differentiate between sexes.

We must remember that some specimens presented values designated as "outliers" in our work, demonstrating high variability for some determined variables, especially in females. The exclusion of these outliers from the morphometric analysis did not cause significant changes in the average values for both sexes, nor did it cause changes in the distribution of the variables within and between the two sexes. However, in other cases these "unusual data" could to alter significantly the distributions. We must also consider the possibility of a degree of reverse dimorphism as an explanation for these outliers, whereby females present higher variability than males. Although Ralls [50] and Reiss [51] listed animal species in which reverse dimorphism could be observed, the only Canidae Family member included was the small-eared zorro (*Atelocynus microtis*). Diverse nutrition, environmental, evaluative, or genetic factors could explain the diversity in sexual dimorphism, as mentioned previously [52], or reverse dimorphism [50,51]. However, we found values considered as outliers for different variables and in different specimens. Moreover, outliers were often asymmetric, with most of them only present on one side (left or right) or in a different jaw (maxillary or mandibular). It would be expected that under normal conditions, genetic, nutritional, evolutive or environmental factors would influence such variables (e.g., teeth) in a more homogeneous way and less asymmetrically. For this reason, we consider that the outlier values obtained here are characteristics of each specimen and they are unlikely to reflect reverse dimorphism. Indeed, the absence of sufficient data regarding the general nature of the sample make it difficult to associate any specific factor to these values.

In Iberian wolves the lateral disposition of the canine teeth's cusp is important in forensic terms. The depth of the bite will vary in morphological and morphometric terms, with UCW/LCW and UCc/LCc the features most often analysed in bite marks. We see these in superficial bite marks, whereas in a deeper bite, the UbC or LbC may be the most accurate features to measure. Therefore, using intervals could reduce the error due to the depth of the bite. Greater tooth crowding (or smaller diastema sizes) has been seen in Iberian wolves than in the domestic dog, a differentiating characteristic between both canid sub-species [23]. By contrast, some studies have demonstrated that they cannot be confidently separated based on tooth crowding alone [36]. A variation in the length or the absence of diastemas in skull of Iberian wolves was not a consistent characteristic in our study.

A reduction in tooth size has been accepted as the principal signature with which to track domestication [36]. Dog's teeth are smaller than those of the Iberian wolf [25]. Indeed, the range of sizes among dogs extends much more than that of wolves, distinguishing dogs as the most morphologically diverse terrestrial mammalian species known [53]. For this reason, we must more precisely define the comparisons made in future studies (e.g., wolf-like dogs, wild dogs, domestic, breeds, age, size, etc.).

This study allowed the Iberian wolf's dentition to be characterized scientifically in morphological and morphometric terms. In a forensic context, the morphological characterization of teeth, including those which had not been morphometrically analysed, will permit their recognition should some of the teeth fall out during the attack or biting process. The intervals defined may be useful to characterize bite marks left on inanimate and/or biological substrates, which have little deformation, such as skeletal or bone remains. The bite marks found may be generated during an attack or in defence, or when scavenging. Their characteristics would also depend on other factors, such as the muscle mass of the victim, the number of animals eating or attacking, the time available for scavenging, etc. [23]. The high variability among dog breeds can lead to a high degree of overlap, such that the use of intervals would be a better option than the average values to compare bite mark patterns in order to determine the guilty party. We have already mentioned some articles in which attempts were made to differentiate and identify diverse families or the genus of carnivorous mammals based on an analysis of the bite marks found on bone remains. The results were mixed, with a wide degree of overlap among them and no studies reporting clear measurements among those patterns. Such analyses become more difficult when bite mark patterns are registered on other biological substrates like skin, which allow significant deformation. To differentiate the same gender in two sub-species is still more difficult, as seen for the Iberian wolf and domestic dog. The differentiation of sexes may be important from an archaeological, population and conservationist point of view, among others. Our study suggest that it should be possible to differentiate the bite marks generated by a male or female Iberian wolf, especially on hard surfaces (as indicated above). In forensic terms, this differentiation can give us important information about the possible ethological nature of an attack (reproductive season?, hormonal factors?, competition for food sources?, etc ...). Our data can help to incriminate or exculpate to the Iberian wolf as the possible aggressor against false accusations. Forensic odontology is an important discipline, although caution must be exerted when interpreting the results. To establish a clear separation between sub-species thorough dental morphometry can be risky. Although only some teeth act during biting, a dental analysis must incorporate as many of the variables as possible in order to differentiate both sub-species. As indicated above, ethological, environmental, and population issues must also be considered in forensic investigations. For this and other reasons, the analysis of animal's bite marks must involve multidisciplinary teams from different forensic areas, such as biologists, veterinarians, dentists, pathologists, and police officers.

The data obtained in this study marks another step in the differentiation between bite mark patterns caused by female and male Iberian wolves. These results would be useful as a base complementary forensic tool for future studies that allow discriminating the Iberian wolf from other sub-species of the

Canidae family, such as domestic dogs. New and more standardized and comparison studies must be carried out among Iberian wolves' dentition vs. wolf-like dogs and large mesocephalic dog breeds, two subspecies that have received little blame without any scientific evidence.

5. Conclusions

This study defines the general morphological characteristics of the Iberian wolf dentition in terms of the dental formula and tooth structure. The Iberian wolf presents evident sexual dimorphism in its morphometric dental features. Variables measured on less worn points of teeth are useful in sex differentiation, despite some asymmetry. Therefore, it should be possible to differentiate the bite marks from a male and female Iberian wolf registered on hard/inanimate surfaces with low capacity of deformation (e.g., bones).

Both dental morphometric and morphological analysis are complementary forensic tools, and these features may help identify the Iberian wolf as a suspected aggressor in a more effective and reliable way.

The use of intervals instead of average values seems to be more useful to diferentiate Iberian wolves' tooth marks between sexes.

Dental characteristics can be used along with other evidence obtained at a site of attack to better interpret the events that occurred.

Given the large number of conflicts caused by the alleged attack of Iberian wolves on domestic cattle, this study's results may serve as a base complementary forensic tool for future studies. It would ideally allow discriminating the Iberian wolf from other carnivores with similar phenotypic characteristics by means of further comparative studies using bite mark patterns.

Supplementary Materials: The following are available online at http://www.mdpi.com/2076-2615/10/6/975/s1, Figure S1: Boxplots for the dental measurements without the outlier values considering sex as a source of variation, Figure S2: 3D scatter PLS-DA plot. Sex differentiation before the outlier female sample (green arrow) was excluded from the multivariate analysis, Table S1: Comparison of the mean values for the wolf sex differentiation considering the existence or exclusion of the outlier values present only in seven specific variables.

Author Contributions: Conceptualization, project administration, writing-original draft, and visualization, V.T.G.; Formal analysis, V.T.G., F.O.O. and P.N.C.; Investigation, V.T.G. and P.P.-L.; Methodology, V.T.G., G.M.F. and F.O.O.; Resources, F.O.O., C.G.-R., P.N.C., P.P.-L. and M.d.P.M.G.; Supervision, V.T.G., G.M.F., F.O.O. and C.G.-R.; Validation, M.d.P.-L. and M.P.M.; Writing-review & editing, V.T.G., F.O.O., G.M.F. and C.G.-R. All authors have read and agreed to the published version of the manuscript.

Funding: This research received no external funding.

Acknowledgments: The authors would like to thank Mr Ángel Garvía Rodríguez (director) and Luis Castelo (technician) at the Mammalian Collection of the National Natural Science Museum (Madrid, Spain), and José España and Ángel Iglesias for their cooperation in this study.

Conflicts of Interest: The authors declare no conflict of interest.

References

1. Souviron, R. Forensic odontology. In *Forensic Pathology: Principles and Practice*, 1st ed.; Dolinak, D., Matshes, E., Lew, E., Eds.; Elsevier Academic Press: Burlington, MA, USA, 2005; pp. 605–629.
2. Murmann, D.C.; Brumit, P.C.; Schrader, B.A.; Senn, D.R. A comparison of animal jaws and bite mark patterns. *J. Forensic Sci.* **2006**, *51*, 846–860. [CrossRef] [PubMed]
3. MacDonald, D.W.; Loveridge, A.J. *Biology and Conservation of Wild Felids*; Oxford University Press Inc.: New York, NY, USA, 2010.
4. Moraitis, K.; Spiliopoulou, C. Forensic implications of carnivore scavenging on human remains recovered from outdoor locations in Greece. *J. Forensic Leg. Med.* **2010**, *17*, 298–303. [CrossRef] [PubMed]
5. Haglund, W.D. Dogs and coyotes: Postmortem involvement with human remains. In *Forensic Taphonomy: The Postmortem Fate of Human Remains*; Haglund, W.D., Sorg, M.H., Eds.; CRC Press LLC: Boca Raton, FL, USA, 1997; pp. 386–400.
6. Morton, R.; Lord, W. Taphonomy of child-sized remains: A study of scattering and scavenging in Virginia, USA. *J. Forensic Sci.* **2006**, *51*, 475–479. [CrossRef] [PubMed]

7. Santoro, V.; Smaldone, G.; Lozito, P.; Smaldone, M.; Introna, F. A forensic approach to fatal dog attacks. A case study and review of the literature. *Forensic Sci. Int.* **2011**, *206*, e37–e42. [CrossRef]
8. Pimenta, V.; Barroso, I.; Boitani, L.; Beja, P. Wolf predation on cattle in Portugal: Assessing the effects of husbandry systems. *Biol. Conserv.* **2017**, *207*, 17–26. [CrossRef]
9. González, J.A.; Carvalho, A.M.; Vallejo, J.R.; Amich, F. Plant-based remedies for wolf bites and rituals against wolves in the Iberian Peninsula: Therapeutic opportunities and cultural values for the conservation of biocultural diversity. *J. Ethnopharmacol.* **2017**, *209*, 124–139. [CrossRef]
10. Garff, E.L.; Mesli, V.; Delannoy, Y.; Pollard, J.; Becart, A.; Hedouin, V. Domestic predation of an elder: A fatal dog attack case. *J. Forensic Sci.* **2017**, *62*, 1379–1382. [CrossRef]
11. Cozza, K.; Fico, R.; Battistini, M.L.; Rogers, E. The damage-conservation interface illustrated by predation on domestic livestock in central Italy. *Biol. Conserv.* **1996**, *78*, 329–336. [CrossRef]
12. Steffens, K.; Sanders, M.; Gleeson, D.; Pullen, K.; Stowe, C. Identification of predators at black-fronted tern *Chlidonias albostriatus* nests, using mtDNA analysis and digital video recorders. *N. Z. J. Ecol.* **2012**, *36*, 48–55.
13. Gidna, A.; Yravedra, J.; Domínguez-Rodrigo, M. A cautionary note on the use of captive carnivores to model wild predator behavior: A comparison of bone modification patterns on long bones by captive and wild lions. *J. Archaeol. Sci.* **2013**, *40*, 1903–1910. [CrossRef]
14. Young, A.; Stillman, R.; Smith, M.J.; Korstjens, A.H. An experimental study of vertebrate scavenging behavior in a northwest European woodland context. *J. Forensic Sci.* **2014**, *59*, 1333–1342. [CrossRef] [PubMed]
15. Severtsov, A.S.; Kormylitsin, A.A.; Severtsova, E.A.; Yatsuk, I.A. Functional differentiation of teeth in the wolf (*Canis lupus*, Canidae, Carnivora). *Biol. Bull. Russ. Acad. Sci.* **2016**, *43*, 1271–1280. [CrossRef]
16. Geiger, M.; Gendron, K.; Willmitzer, F.; Sánchez-Villagra, M. Unaltered sequence of dental, skeletal, and sexual maturity in domestic dogs compared to the wolf. *Zool. Lett.* **2016**, *2*, 16. [CrossRef] [PubMed]
17. Lemmons, M.; Beebe, D. Oral anatomy and physiology. In *Wigg's Veterinary Dentistry. Principles and Practice*, 2nd ed.; Lobprise, H., Dodd, J., Eds.; Wiley Blackwell: Hoboken, NJ, USA, 2019; pp. 1–15.
18. Schaller, O. *Nomenclatura Anatómica Veterinaria Ilustrada*, 1st ed.; Acribia, S.A.: Zaragoza, Spain, 1996.
19. Morales, A.D.; Osorio, A.C.; Fernández, R.F.; Dennes, E.L. Desarrollo de un modelo SIMCA para la clasificación de kerosinas mediante el empleo de la espectroscopía infrarroja. *Química Nova.* **2008**, *31*, 1573–1576. [CrossRef]
20. Carrasco, J.J.; Georgevsky, D.; Valenzuela, M.; McGreevy, P.D. A pilot study of sexual dimorphism in the head morphology of domestic dogs. *J. Vet. Behav.* **2014**, *9*, 43–46. [CrossRef]
21. Vila, C.; Savolainen, P.; Maldonado, J.E.; Amorim, I.R.; Rice, J.E.; Honeycutt, R.L.; Crandall, K.A.; Lundeberg, J.; Wayne, R.K. Multiple and ancient origins of the domestic dog. *Science* **1997**, *276*, 1687–1689. [CrossRef]
22. Yravedra, J.; García-Vargas, E.; Maté-González, M.A.; Aramendi, J.; Palomeque-González, J.F.; Vallés-Iriso, J.; Matesanz-Vicente, J.; González-Aguilera, D.; Domínguez-Rodrigo, M. The use of micro-photogrammetry and geometric morphometrics for identifying carnivore agency in bone assemblages. *J. Archaeol. Sci. Rep.* **2017**, *14*, 106–115. [CrossRef]
23. Yravedra, J.; Lagos, L.; Bárcena, F. A taphonomic study of wild wolf (*Canis lupus*). Modification of horse bones in northwestern Spain. *J. Taphon.* **2011**, *9*, 37–65.
24. Aramendi, J.; Maté-González, M.A.; Yravedra, J.; Ortega, M.C.; Arriaza, M.C.; González-Aguilera, D.; Baquedano, E.; Domínguez-Rodrigo, M. Discerning carnivore agency through the three-dimensional study of tooth pits: Revisiting crocodile feeding behaviour at FLK- Zinj and FLK NN3 (Olduvai Gorge, Tanzania). *Paleoecol. Geogr. Palaeoclimatol. Palaeoecol.* **2017**, *488*, 93–102. [CrossRef]
25. Iglesias, A.; España, A.J.; España, J. *Lobos Ibéricos. Anatomía, Ecología Y Conservación*, 1st ed.; Náyade Nature: Valladolid, Spain, 2017; p. 532.
26. Haynes, G. Evidence of carnivore gnawing on Pleistocene and recent mammalian bones. *Paleobiology* **1980**, *6*, 341–351. [CrossRef]
27. Binford, L.R. *Bones: Ancient Men and Modern Myths (Studies in Archeology)*, 1st ed.; Academic Press: London, UK, 1981.
28. Haynes, G. A guide for differentiating mammalian carnivore taxa responsible for gnaw damage to herbivore limb bones. *Paleobiology* **1983**, *9*, 164–172. [CrossRef]
29. Fosse, P.; Wajrak, A.; Fourvel, J.B.; Madelaine, S.; Esteban-Nadal, M.; Cáceres, I.; Yravedra, J.; Brugal, J.P.; Prucca, A.; Haynes, G. Bone modification by modern wolf (*Canis lupus*): A taphonomic study from their natural feeding places. *J. Taphon.* **2012**, *10*, 197–217.

30. Parkinson, J.; Plummer, T.; Bose, R. A Gis-based approach to documenting large canid damage to bones. *Palaeogeogr. Palaeoclimatol. Palaeoecol.* **2014**, *409*, 57–71. [CrossRef]
31. Sala, N.; Arsuaga, J.L.; Haynes, G. Taphonomic comparison of bone modifications caused by wild and captive wolves (*Canis lupus*). *Quat. Int.* **2014**, *330*, 126–135. [CrossRef]
32. Andres, M.; Gidna, A.; Yravedra, J.; Dominguez-Rodrigo, M. A study of dimensional differences of tooth marks (pits and scores) on bones modified by small and large carnivores. *Archaeol. Anthropol. Sci.* **2012**, *4*, 209–219. [CrossRef]
33. Yravedra, J.; Andrés, M.; Domínguez-Rodrigo, M. A taphonomic study of the African wild dog (*Lycaon pictus*). *Archaeol. Anthropol. Sci.* **2014**, *6*, 113–124. [CrossRef]
34. Yravedra, J.; Maté-González, M.Á.; Courtenay, L.A.; González-Aguilera, D.; Fernández, M.F. The use of canid tooth marks on bone for the identification of livestock predation. *Sci. Rep.* **2019**, *9*, 16301. [CrossRef]
35. Drake, A.G.; Klingenberg, C.P. Large-scale diversification of skull shape in domestic dogs: Disparity and modularity. *Am. Nat.* **2010**, *175*, 289–301. [CrossRef]
36. Ameen, C.; Hulme-Beaman, A.; Evin, A.; Germonpré, M.; Britton, K.; Cucchi, T.; Larson, G.; Dobney, K. A landmark-based approach for assessing the reliability of mandibular tooth crowding as a marker of dog domestication. *J. Archaeol. Sci.* **2017**, *85*, 41–50. [CrossRef]
37. Germonpré, M.; Sablin, M.V.; Stevens, R.E.; Hedges, R.E.M.; Hofreiter, M.; Stiller, M.; Després, V.R. Fossil dogs and wolves from Paleolithic sites in Belgium, the Ukraine and Russia: Osteometry, ancient DNA and stable isotopes. *J. Archaeol. Sci.* **2009**, *36*, 473–490. [CrossRef]
38. Blanco, J.C. Lobo-*Canis lupus* Linnaeus, 1758. In *Enciclopedia Virtual de los Vertebrados Españoles. Sociedad de Amigos del MNCN – MNCN – CSIC*; Salvador, A., Barja, I., Eds.; Museo Nacional de Ciencias Naturales CSIC: Madrid, Spain, 2017; pp. 1–25.
39. Toledo, G.V.; Ortega, O.F.; Fonseca, G.M.; García-Ruíz, C.; Pérez-LLoret, P. Morphometric analysis of bite mark patterns caused by domestic dogs (*Canis lupus familiaris*) using dental wax registers. *Int. J. Morphol.* **2019**, *37*, 885–893. [CrossRef]
40. Guitián-Rivera, J.; Sánchez-Canals, J.L.; de Castro, A.; Bas López, S. Nota sobre dimorfismo sexual en algunos cráneos de lobo (*Canis lupus* L) de Galicia. In *Trabajos Compostelanos de Biología 8*; Universidad de Santiago de Compostela: La Coruña, Spain, 1979; pp. 87–94.
41. Valverde, J.; Hidalgo, A. Sobre el lobo (*Canis lupus*) ibérico: I. Dimorfismo sexual en cráneos. *Doñana Acta Vert* **1974**, *1*, 233–244.
42. Machado, M.G. *Cánidos Ibéricos: Evaluación de la Morfología Cefálica Con Métodos Clásicos y Actuales de Diagnóstico Por Imagen del lobo Ibérico, Canis Lupus Signatus*; Universidad de León España: León, Spain, 2004.
43. Curth, S.; Fischer, M.S.; Kupczik, K. Patterns of integration in the canine skull: An inside view into the relationship of the skull modules of domestic dogs and wolves. *Zoology (Jena, Germany)* **2017**, *125*, 1–9. [CrossRef] [PubMed]
44. Toledo, G.V.; Ibarra, M.L.; Rojas, E.V.; Ciocca, G.L.; Rocha, D.N.; Jara, V.G. Estudio preliminar de patrones de mordedura según forma del cráneo, mediante el análisis morfológico y morfométrico de semiarcadas dentarias de perro doméstico (*Canis familiaris*) con fines de identificación. *Int. J. Morphol.* **2012**, *30*, 222–229. [CrossRef]
45. Sweet, D.; Pretty, I.A. A look at forensic dentistry-part 2: Teeth as weapons of violence-identification of bitemark perpetrators. *Br. Dent. J.* **2001**, *190*, 415–418. [CrossRef]
46. Long, K. *Wolves. A Wildlife Handbook*; Johnson Books: Boulder, CO, USA, 1996.
47. Doring, S.; Arzi, B.; Winer, J.N.; Kass, P.H.; Verstraete, F.J.M. Dental and temporomandibular joint pathology of the grey wolf (*Canis lupus*). *J. Comp. Pathol.* **2018**, *160*, 56–70. [CrossRef]
48. Gipson, P.; Ballard, W.; Nowak, R.; Mech, L. Accuracy and precision of estimating age of gray wolves by tooth wear. *J. Wildl. Manag.* **2000**, *64*, 752–758. [CrossRef]
49. Kieser, J.A.; Groeneveld, H.T. Mandibulodental allometry in the African wild dog, *Lycaon pictus*. *J. Anat.* **1992**, *181 Pt 1*, 133–137.
50. Ralls, K. Mammals in which females are larger than males. *Q. Rev. Biol.* **1976**, *51*, 245–276. [CrossRef]
51. Reiss, M. Males bigger, females biggest. *New Sci.* **1982**, *96*, 226–229.

52. Frayer, D.; Wolpoff, M. Sexual dimorphism. *Annu. Rev. Anthropol.* **2003**, *14*, 429–473. [CrossRef]
53. Stockard, C.R.; Anderson, O.D.; James, W.T.; Wistar Institute of Anatomy and Biology. *The Genetic and Endocrinic Basis for Differences in form and Behavior: As Elucidated by Studies of Contrasted Pure-Line Dog Breeds and Their Hybrids*, 1st ed.; The Wistar Institute of Anatomy and Biology: Philadelphia, PA, USA, 1941.

Article

A Study of the Head during Prenatal and Perinatal Development of Two Fetuses and One Newborn Striped Dolphin (*Stenella coeruleoalba*, Meyen 1833) Using Dissections, Sectional Anatomy, CT, and MRI: Anatomical and Functional Implications in Cetaceans and Terrestrial Mammals

Álvaro García de los Ríos y Loshuertos [1], Alberto Arencibia Espinosa [2], Marta Soler Laguía [3], Francisco Gil Cano [1], Francisco Martínez Gomariz [1], Alfredo López Fernández [4] and Gregorio Ramírez Zarzosa [1,*]

1 Departamento de Anatomía y Anatomía Patológica Comparadas, Facultad de Veterinaria, Universidad de Murcia, 30100 Murcia, Spain; agrios@ceuta.es (A.G.d.l.R.y.L.); cano@um.es (F.G.C.); f.gomariz@colvet.es (F.M.G.)

2 Departamento de Morfología. Anatomía y Embriología, Facultad de Veterinaria, Universidad de Las Palmas de Gran Canaria, Trasmontaña, Arucas, 35416 Las Palmas de Gran Canaria, Spain; alberto.arencibia@ulpgc.es

3 Departamento de Medicina y Cirugía, Facultad de Veterinaria, Universidad de Murcia, 30100 Murcia, Spain; mtasoler@um.es

4 Departamento de Biología—CESAM, Universidade de Aveiro, Campus Universitario de Santiago, 3810-193 Aveiro, Portugal; a.lopez@ua.pt

* Correspondence: grzar@um.es; Tel.: +34-868887546; Fax: +34-868884147

Received: 24 October 2019; Accepted: 9 December 2019; Published: 13 December 2019

Simple Summary: The head region of the dolphin has been studied widely to identify its anatomical structures and to compare it with other marine and terrestrial mammals. In this study, specimens stranded off the Spanish coast were used. Our study analyzes four dolphin heads during fetal and perinatal development. All specimens were scanned using modern imaging techniques to study their internal organs and to preserve the specimens, which are difficult to obtain. Only one fetus was transversely cross-sectioned to help us to identify critical organs. The developmental study shows several anatomical structures that are compared with cetaceans and terrestrial mammals. During development of the oral cavity, it was observed that the rostral maxillary and mandible teeth (incisive area) had not completely erupted, in contrast with the rest of teeth, which have done so. Also, the main chewing muscle (masseter) was not observed. In addition, we describe the absence of major salivary glands during these developmental stages. Furthermore, we explain the characteristics of the orbit and its relation to the eyeball. In addition, the fetal dolphin's ear is connected with pharynx in a way similar to that in horses. We conclude that these developmental studies will help cetacean conservation.

Abstract: Our objective was to analyze the main anatomical structures of the dolphin head during its developmental stages. Most dolphin studies use only one fetal specimen due to the difficulty in obtaining these materials. Magnetic resonance imaging (MRI) and computed tomography (CT) of two fetuses (younger and older) and a perinatal specimen cadaver of striped dolphins were scanned. Only the older fetus was frozen and then was transversely cross-sectioned. In addition, gross dissections of the head were made on a perinatal and an adult specimen. In the oral cavity, only the mandible and maxilla teeth have started to erupt, while the most rostral teeth have not yet erupted. No salivary glands and masseter muscle were observed. The melon was well identified in CT/MRI images at early stages of development. CT and MRI images allowed observation of the maxillary sinus. The orbit and eyeball were analyzed and the absence of infraorbital rim together with the temporal process of the zygomatic bone holding periorbit were described. An enlarged auditory tube was identified using anatomical sections, CT, and MRI. We also compare the dolphin head anatomy with some mammals, trying to underline the anatomical and physiological changes and explain them from an ontogenic point of view.

Keywords: striped dolphin (*Stenella coeruleoalba*); fetal development; PET/SPECT/CT; MRI; sectional anatomy; head anatomy; ontogenesis

---

1. Introduction

Cetaceans are a group of mammals well adapted to their marine environment and whose evolutionary changes are especially marked in the development of the structures of the head. In both suborders of living cetaceans, the skull has been highly modified by changes in feeding apparatus and the elimination or reduction of many structures [1]. The relationship of the bones in the skull to one another is altered due to the caudal migration of the nasal opening, a process known as telescoping [2–4]. In addition, differences occur in the location of the external nasal passages and the structure of the middle and the inner ear.

The study of an extensive collection of embryos and fetuses of these species has produced valuable information about the ontogeny of most of the body systems and about musculoskeletal development. Comparisons with other mammals detected the time lag in ossification, retardation of odontogeny, and the origin and development of the fluke, dorsal fin, and flipper [5].

Nevertheless, the studies performed so far lack information on the prenatal and perinatal development due to the difficulty in establishing the differences in ontogenetic development of cetaceans [6]. The precise time intervals of such development and any distinctive growth trajectories are basically unknown [7].

Even now, studies of cranial anatomy by anatomical sections seem to be scarce, with most studies performed in odontocetes (due to their smaller size) either in adults, for instance in common and striped dolphins [8,9], or in newborn bottlenose dolphins [10,11], pacific spotted dolphins, common dolphins, and narwhals [12], and in the fetal narwhal, common dolphin, Atlantic white-sided dolphin [13–15], and Beaked whale [16]. Fetal studies are least common due to the lack of stranded pregnant females. In the case of mysticetes, the few existent studies are focused almost exclusively on external anatomy: eye, nose, hair, and throat of a neonate gray whale [2]; osteology: skull anatomy in fetal specimens of whales of the genera *Megaptera* and *Balaenoptera* [17]; musculoskeletal: musculoskeletal anatomy of the head of a neonate gray whale [18] or vascular [18]. An exception to this is the work of Schute [19] in which a monograph study of the fetal anatomy of the Sei whale (*Balaenoptera borealis*) was done.

In both orders, outlines of organs are observed during the embryonic period and the organs are almost developed or in a development phase in the fetal period, which is of key importance as this is almost the only different period compared to the adult stage, because, for survival reasons, cetaceans give birth to precocial newborn. During these stages (end of the fetal and all the perinatal period), we can obtain valuable data on the species' ontogeny, so we concur with [7] about the many applications in the fields of biology and animal medicine. One of these applications could be determining an organ's development during the fetal period, helping researchers to calculate approximately the time of fetal development in odontocetes based on anatomical changes during gestation, similar to the Carnegie system designed for the human fetus [20] or for terrestrial mammals [3]. So far, we can only estimate cetacean parameters such as the gestation time by using a mathematical formula in *Stenella longirostris* [21], calculate the time of parturition using ultrasonography in Bottlenose dolphins [22], or estimate the adult's age through dental growing lines in striped dolphins [23].

In the current study, we analyze the head anatomy of two striped dolphins' (*Stenella coreuleoalba*) fetuses and one newborn of the same species. In each case, anatomical sections were correlated with computed tomography (CT) and magnetic resonance imaging (MRI).

Our goal is to accomplish several objectives: (a) to create a cephalic anatomy atlas of images during the fetal period up to the perinatal period, which could have benefits for cetacean conservation; (b) to collaborate with other studies dealing with the chronology of fetal development of these species; (c) to clarify some functional aspects of the anatomical structures of the head during prenatal and perinatal dolphin development; and (d) to accurately describe the structures of the head following the Illustrated Veterinary Anatomical Nomenclature [24].

2. Materials and Methods

2.1. Animals

A total of four pre- and perinatal specimens and one adult striped dolphin (*Stenella coeruleoalba*, Meyen 1833) were used in this study (Table 1). The mother of the youngest fetus was stranded on the Spanish Atlantic coast. The mother of the older fetus and two newborn specimens were stranded on the Spanish African coast. The adult specimen was stranded on the Spanish Mediterranean coast. Stranded specimens were found dead and ethics committee clearance was not necessary. Both fetuses and the newborn specimen were transported to the CT and MRI units to perform CT and MRI scans.

Table 1. Fetal specimens of striped dolphin used in this study.

Stranding Reference and Study Code	Sex, Length, Weight and Estimated Gestation Time [7,25,26]	Anatomical and Imaging Diagnostic Techniques	Preservation Techniques
SCOG CEMMA sco1	Female fetus, 32.5 cm, 508 g, 4.5 months	MRI, PET/SPECT/CT,	Fixation: formaldehyde 10%
SCOCE1 CECAM sco2	Male fetus, 48 cm, 1.535 kg, 7 months	MRI, CT, anatomical head sections	Fixation: formaldehyde 10%
SCOCE2 CECAM sco3	Female newborn, 95 cm, 10.84 kg	MRI, CT	Freezing × 20 °C
SCOCE3 CECAM sco4	Male newborn, 85 cm, 9.2 kg	Head dissection	Fixation: formaldehyde 10%
SCOMU CRFS sco5	Adult female, 1.91 cm, 63.65 kg	Head dissection	Embalming: formaldehyde, glycerine, isopropyl alcohol, phenol

SCOG: *Stenella coeruleoalba* from Pontevedra, Spain; *SCOCE*: *S. coeruleoalba* from Ceuta, Spain; *SCOMU*: *S. coeruleoalba* from Murcia, Spain; MRI: Magnetic resonance imaging; CT: Computed Tomography, *CEMMA*: Coordinator Center for the study of the marine mammals, Galicia; *CECAM*: Center for the study and conservation of marine animals, Ceuta; *CRFS*: Wildlife rehabilitation Center, Murcia.

2.2. Computed Tomography

The sco1 was scanned with Positron Emission Tomography (PET), Single Photon Emission Computed Tomography (SPECT)-Computed Tomography (CT) (PET/SPECT/CT Albira^{TM} Systems, Valencia, Spain; Centro de Investigación Biomédica, Universidad de Murcia, Spain); single-slice: 1 detector arrays; type of acquisition: helical; thickness: 0.125 mm; image reconstruction interval or index: 0.0125 mm; pitch: 0; tube rotation time: 0.12; mA: 0.4; Kv: 45; FOV 68 cm, Matrix dimensions 2240 × 2360; reconstruction algorithm: FBP filtered back projection; WW: 600/WL: 300). The SPECT-CT images were transferred to a Dicom workstation, while sco2 was scanned with CT (General Electric Medical Systems, Schenectady, NY, USA; Clínica Virgen de Africa, Ceuta, Spain); multislice: 4 detector arrays; type of acquisition: helical; thickness: 5 mm; index: 3.2 mm; pitch: 0.45; tube rotation time: 0.33; mA: 30; kV: 120; FOV 35 cm; matrix dimensions: 512 × 512, reconstruction algorithm: bone; WW: 350/WL:

221; WW 650/WL −34. Finally, sco3 was scanned with a CT (General Electric Medical Systems-HiSpeed dual, Schenectady, NA, USA; Hospital Clínico Veterinario, Universidad de Murcia, Spain); multislice: 2 detector arrays; type of acquisition: helical; thickness: 5 mm; index: 2.5 mm; pitch: 0.35; tube rotation time: 1; mA: 100; Kv: 120; Image field of view FOV 40 cm; matrix dimensions: 512 × 512, reconstruction algorithm: standard; WW: 350/WL: 221; WW 650/WL −34). All dolphin specimens were positioned in ventral recumbency. All CT images were transferred to a DICOM workstation and CT images were analyzed with Radiant DICOM viewer and Osiris 4.0 for Windows. A vascular window setting (WW 600/WL 300) was applied to obtain PET/SPECT/CT images. Mediastinum-vascular window (WW 350/WL 221) and soft-tissue window settings (WW 650/WL −34) were applied to obtain Ceuta and Murcia CT images, respectively.

2.3. Magnetic Resonance Imaging

In sco1, Magnetic Resonance (MR) images were obtained with a high-field MR apparatus (General Electric Sigma Excite, Schenectady, NA, USA; Centro Veterinario de Diagnóstico por Imagen de Levante, Ciudad Quesada, Alicante, Spain), 1.5 Tesla using a human wrist coil. T1-weighted spin eco (SE) and T2-weighted fast spin scho (FSE) pulse sequences were used. T1-weighted (SE) images were obtained in transverse plane and 2D acquisition, using the following parameters: TE 13 ms, TR 640 ms, TI 0, NEX 1, slices thickness 1 mm, interslice gap 1.3, field of view 75 and matrix dimensions 0\256\192\0. T2-weighted (FSE) images were obtained in transverse plane and 2D acquisition, using the following parameters: TE 84 ms, TR 8100 ms, TI 0, NEX 1, slice thickness 1 mm, interslice gap 1.3, field of view 60, and matrix dimensions 192\0\0\192.

In sco2, MR images were obtained with a high-field apparatus (Philips Medical System Intera, Eindhoven, The Netherlands; Clínica Radiológica, Ceuta, Spain), 1.5 Tesla using a sense-body coil. T1-weighted fast field echo (FFE) and T1-weighted out-of-phase (OOP) gradient echo (GRE) pulse sequences were used. T1-weighted (FFE) images were obtained in transverse plane and 2D acquisition using the following parameters: TE 4.6 ms, TR 183 ms, TI 6, NEX 6, slice thickness 8 mm, interslice gap 9, field of view 68.6, matrix dimensions 0\204\155\0. T1-weighted (OOP) images were obtained in transverse plane and 2D acquisition using the following parameters: TE 2.3 ms, TR 130.3 ms, TI 0, NEX 5, 9 mm slice thickness, interslice gap 10, field of view 69.6, and matrix dimensions 132\0\0\103.

In sco3, MR images were obtained with a high-field apparatus (General Electric Sigma Excite, Schenectady, USA; Centro Veterinario de Diagnóstico por Imagen de Levante, Ciudad Quesada, Alicante, Spain), 1.5 Tesla using a human head coil. T1-weighted spin echo (SE) and T2-weighted fast spin echo (FSE) pulse sequences were used. T1-weighted (SE) images were obtained in transverse plane and 2D acquisition using the following parameters: TE 11 ms, TR 640 ms, TI 0, NEX 1, slice thickness 4 mm, interslice gap 4.5, field of view 75, and matrix dimensions 0\192\192\0. T2-weighted (FSE) images were obtained in transverse plane and 2D acquisition using the following parameters: TE 93.8 ms, TR 6020 ms, TI 0, NEX 1, slice thickness 4 mm, interslice gap 4.5, field of view 75, and matrix dimensions 0\192\192\0. All dolphin specimens were positioned in ventral recumbency. The MR images were transferred to a DICOM workstation. MR images were analyzed with Radiant DICOM viewer and Osiris 4.0 for Windows.

2.4. Anatomic Evaluation

Sco1, sco3, and sco4 were preserved by immersion in formaldehyde (10%). Sco5 was fixed with embalming solution (formaldehyde, glycerine, isopropyl alcohol, phenol) injecting the right and left carotid arteries and left and right external jugular veins. After 48 h, the carotid arteries and jugular veins were injected with red and blue latex, respectively. These specimens were stored in the Department of Anatomy and Embryology's freezer chamber, Facultad de Veterinaria, Murcia, Spain. Sco3 was preserved frozen (−20 °C) in the Department of Anatomy and Embryology's cooling chamber, Facultad de Veterinaria, Murcia, Spain.

Sco2 was frozen at −80 °C and then taken out to obtain cross sections cut with a band saw (Anatomical Lab, Department of Anatomy and Embryology, Universidad de Murcia, Murcia, Spain), obtaining 0.7–1 cm thick slices, which were then photographed giving us 57 transverse images used to correlate the sections with CT and MR images. Slices were immersed in acetone for plastination preservation and then stored in a freezer chamber at the Department of Anatomy and Embryology, Facultad de Veterinaria, Universidad de Murcia, Spain.

2.5. Gross Dissections

A deep head dissection of sco4 showed the melon. At its midpoint, the melon was cut in transverse and horizontal sections, which showed the nucleus and peripheral tissue ring. The nasal plug and nasal cavity were observed after removing the nasal vestibule and spiracle.

The head, face, and adjacent areas of sco5 were superficially dissected showing frontal and facial fat, the melon surface, and the mandible and superficial facial muscles. After carefully removing superficial fat and fibrous tissues, the venous drainage, several depressor mandible muscles, tongue muscles, and the rudiment of external acoustic meatus were exposed.

3. Results

3.1. Oral Cavity

The oral cavity of the three studied specimens clearly showed the tongue in anatomical sections, CT, and MRI (Figures 1–5). In sco1, the lateral sublingual recesses were observed only in MRI (Figure 2 Row (from now on R)(R1D-E)) while in sco2 it was identified in CT and anatomical sections Figure 2(R2A–C)). Under the lateral sublingual recess, it was not possible to distinguish sublingual salivary glands (neither polystomatic nor monostomatic). Histological analysis of tissue from this region showed a mixture of adipose and striated muscular tissue.

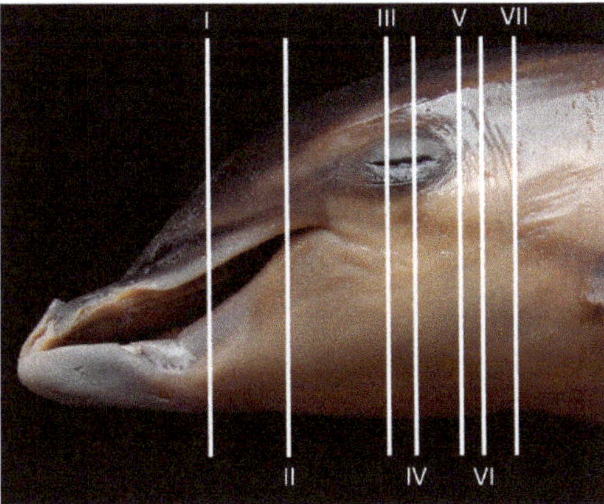

Figure 1. Approximated level sections of fetus dolphin head. Lines represent the location for each transverse anatomical section, CT, and MR images (I–VII).

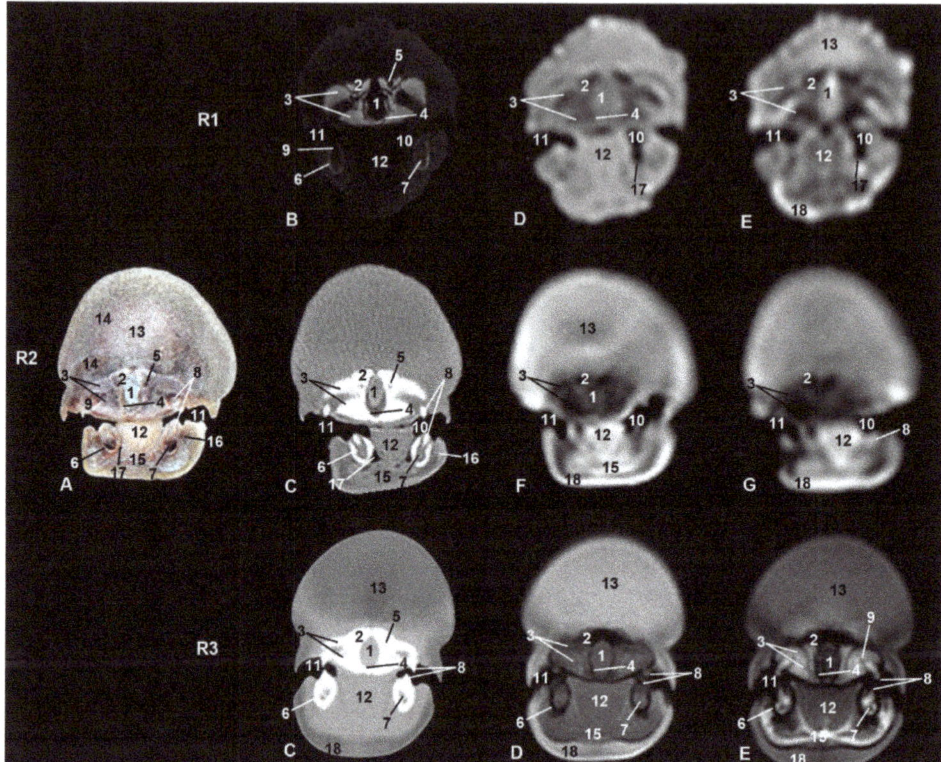

Figure 2. Representative transverse images of the snout made at the level of the rostral portion of the melon and oral cavity. Level I. Images are oriented so that the left side of the head is to the right and dorsal is at the top. Row (from now on R) 1, sco1; R2, sco2; R3, sco3. (**A**) Anatomical section. (**B**) Vascular window PET/SPECT/CT image. (**C**) Soft-tissue window CT image. (**D**) T1-weighted SE sequence. (**E**) T2-weighted fast spin echo (FSE) sequence. (**F**) T1-weighted fast field echo (FFE) sequence. (**G**) T1-weighted out of phase (OOP) gradient echo (GRE) sequence. 1, Mesethmoid cartilage; 2, incisive bone; 3, maxillary bone; 4, vomer bone; 5, supraorbital canal; 6, mandible; 7, canal and mandibular fat; 8, tooth in development; 9, socket of tooth; 10, oral cavity; 11, oral vestibule; 12, tongue; 13, melon; 14, melon rostral muscles; 15, mylohyoid muscle; 16, buccinator and depressor of the lower lip muscles; 17, lateral sublingual recess; 18, epidermis and dermis.

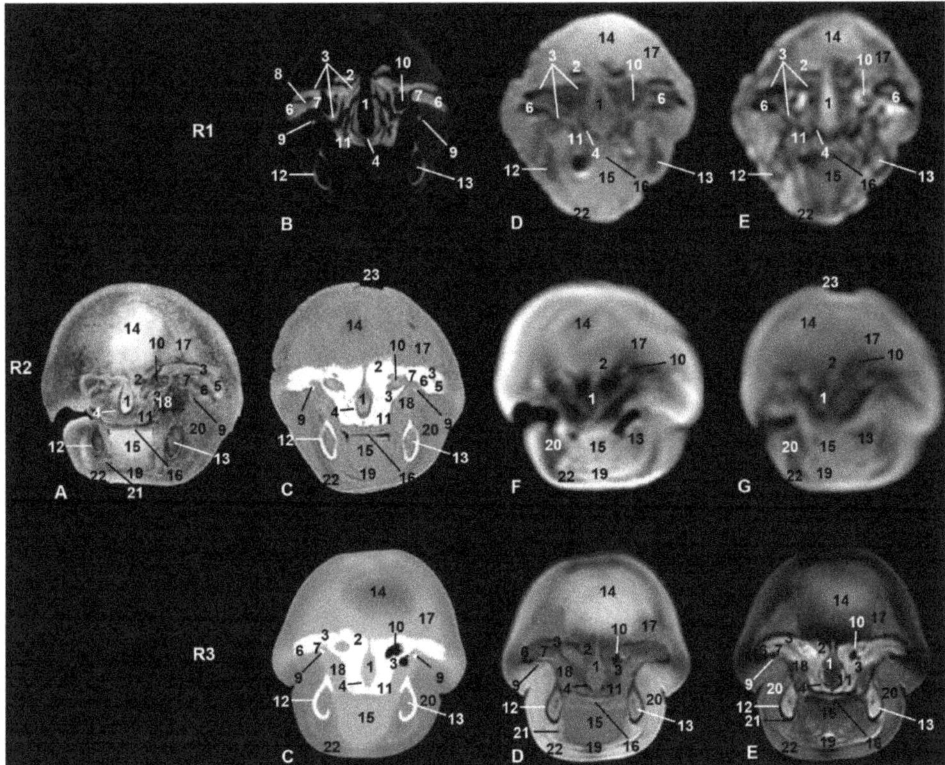

Figure 3. Representative transverse images made at the level of root of the snout, caudal portion of the melon, and oral cavity. Level II. Images are oriented so that the left side of the head is to the right and dorsal is at the top. R1, sco1; R2, sco2; R3, sco3. (**A**) Anatomical section. (**B**) Vascular window PET/SPECT/CT image. (**C**) Soft-tissue window CT image. (**D**) T1-weighted SE sequence. (**E**) T2-weighted FSE sequence. (**F**) T1-weighted FFE sequence. (**G**) T1-weighted OOP GRE sequence. 1, Mesethmoid cartilage; 2, incisive bone; 3, maxillary bone; 4, vomer bone; 5, frontal bone; 6, lacrimal bone; 7, zygomatic bone; 8, lacrimal-zygomatic synchondrosis; 9, zygomatic bone: temporal process; 10, maxillary sinus; 11, palatine bone; 12, mandible; 13, canal and mandibular fat; 14, melon; 15, tongue; 16, oral cavity; 17, melon rostral muscles; 18, pterygoid muscles; 19, mylohyoid muscle; 20, digastric muscle; 21, fat and striated muscle; 22, epidermis, dermis and subcutaneous tissue; 23, nostrils.

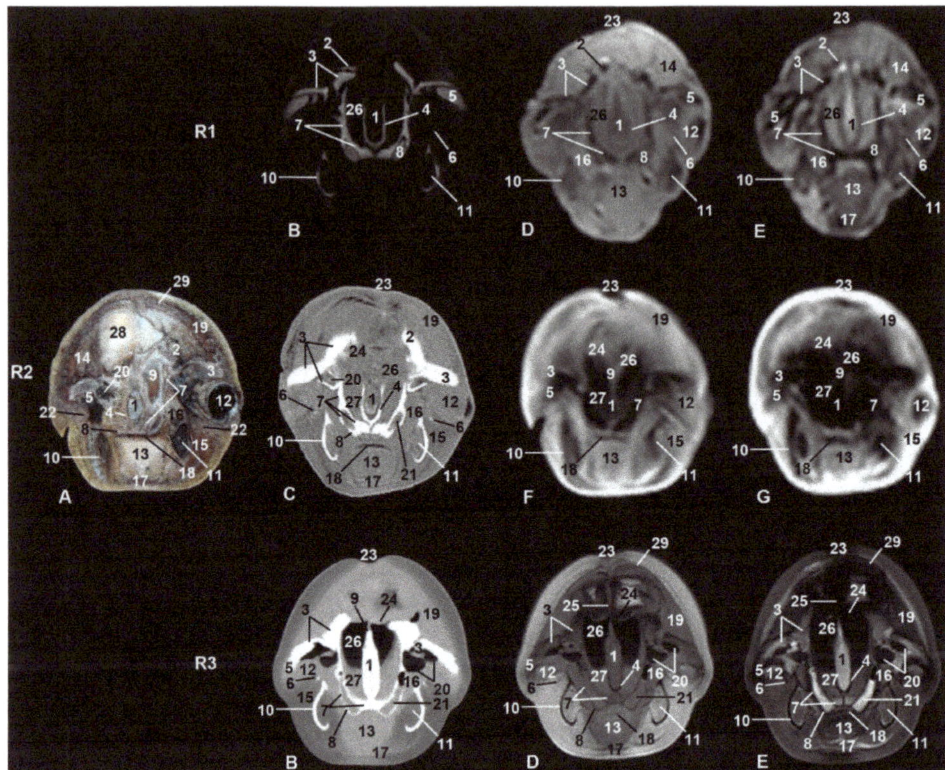

Figure 4. Representative transverse images made at the level of nasal, oral cavities, and orbital craniofacial fossa. Level III. Images are oriented so that the left side of the head is to the right and dorsal is at the top. R1, sco1; R2, sco2; R3, sco3. (**A**) Anatomical section. (**B**) Vascular window PET/SPECT/CT image. (**C**) Soft-tissue window CT image. (**D**) T1-weighted SE sequence. (**E**) T2-weighted FSE sequence. (**F**) T1-weighted FFE sequence. (**G**) T1-weighted OOP GRE sequence. 1, Mesethmoid cartilage; 2, incisive bone; 3, maxillary bone; 4, vomer bone; 5, lacrimal bone; 6, zygomatic bone: temporal process; 7, palatine bone; 8, pterygoid bone; 9, ethmoid bone; 10, mandible; 11, canal and mandibular fat; 12, periorbit and eyeball; 13, tongue; 14, melon external fiber ring; 15, digastric muscle; 16, pterygoid muscle; 17, mylohyoid muscle; 18, oral cavity; 19, melon caudal muscles; 20, frontal bone: orbital recess; 21, pterygopalatine recess; 22, fat and striated muscle; 23, nostrils; 24, nasal diverticulum and nasal plug (arrow); 25, membranous part of nasal septum; 26, nasal cavity; 27, nasal mucosa; 28, melon; 29, nasal vestibule muscles.

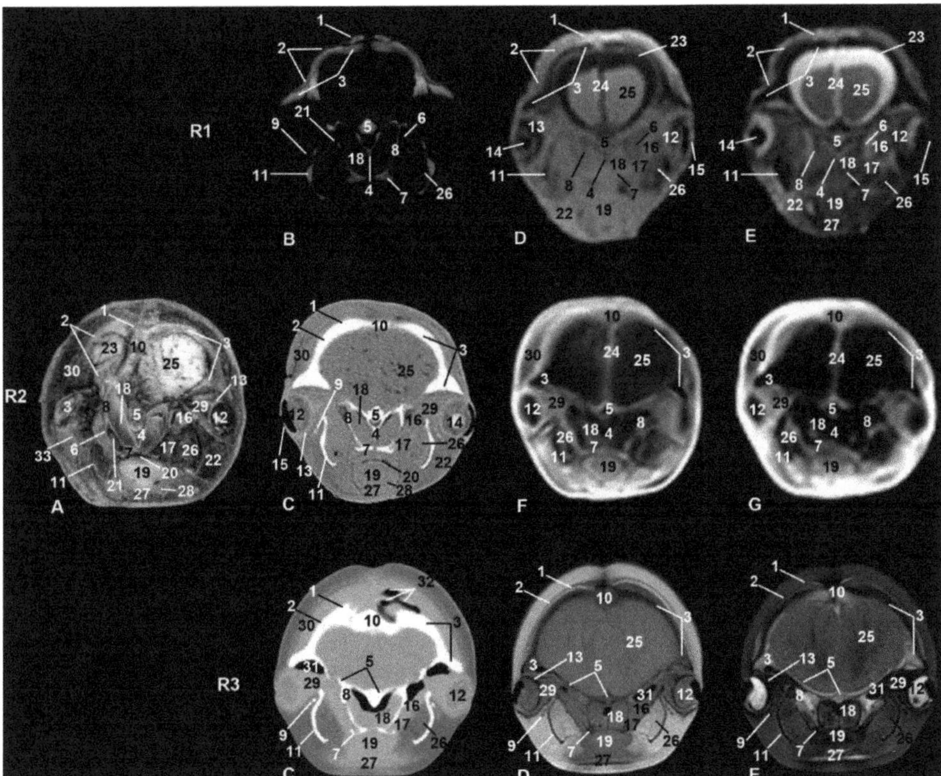

Figure 5. Representative transverse images made at the level of the rostral part of the cranial cavity, choanas, and eyeball. Level IV. Images are oriented so that the left side of the head is to the right and dorsal is at the top. R1, sco1; R2, sco2; R3, sco3; (**A**) Anatomical section. (**B**) Vascular window PET/SPECT/CT image. (**C**) Soft-tissue window CT image. (**D**) T1-weighted SE sequence. (**E**) T2-weighted FSE sequence. (**F**) T1-weighted FFE sequence. (**G**) T1-weighted OOP GRE sequence. 1, Incisive bone; 2, maxillary bone; 3, frontal bone; 4, vomer bone; 5, presphenoid bone: body and wings; 6, palatine bone; 7, pterygoid bone; 8, basisphenoid bone: pterygoid crest; 9, zygomatic bone: temporal process; 10, ethmoid bone; 11, mandible; 12, eyeball; 13, sclera; 14, lens; 15, eyelids; 16, lateral pterygoid muscle; 17, medial pterygoid muscle; 18, choanae and nasopharyngeal sphincter muscle; 19, tongue; 20, oral cavity; 21, pterygopalatine recess; 22, digastric muscle; 23, subarachnoid space; 24, longitudinal brain fissure; 25, brain: frontal lobe; 26, mandibular canal; 27, mylohyoid muscle; 28, hyoglossus muscle; 29, extraocular muscles; 30, melon caudal muscles; 31, frontal bone: orbital recess; 32, nasal diverticulum; 33, fat and striated muscle.

In sco1, CT showed clearly the dental alveolus dorsal to the mandibular canal but not in the maxillary bone (Figure 2(R1B)), while in sco2, CT and anatomical sections showed the most caudal teeth growing covered by gums in both the mandible and maxillary bones (Figure 2(R2C)). In the CT and MR images of sco3, the mid caudal teeth were forming in both dentary arches. In the three specimens, rostral maxillary and mandible teeth (incisive area) had not completely erupted, whereas the rest of mandible and maxillary teeth have done so in sco3 (Figure 2(R3C)).

Only two of the three pairs of muscles of mastication were identified: temporal and pterygoid (Figure 4). The third, the masseter muscle, originates from the facial crest or maxillary tuber and zygomatic arch which were absent in the studied specimens. Its insertion on the masseter fossa and the caudal and ventral portion of the mandible was not observed. A mixture of adipose tissue and muscle fibers on the caudolateral aspect of the body of the mandible was observed. The cheek area was vestigial and so the buccinator muscle (oral part) and depressor of the lower lip were displaced rostrally under the lower lip (Figures 4–7). The orbicularis oris muscle was absent. Medial to the temporomandibular joint, the pterygoid muscles were easily seen (Figures 3–6). The mandible depressor muscles, digastric and mylohyoid were well developed. The digastric muscle insertion enlarges until the most latero-rostral sections of the mandible body (Figures 3–8). Muscles were easy to differentiate in anatomical sections. CT showed them moderately hypoattenuated and in MR images, slightly hypointense.

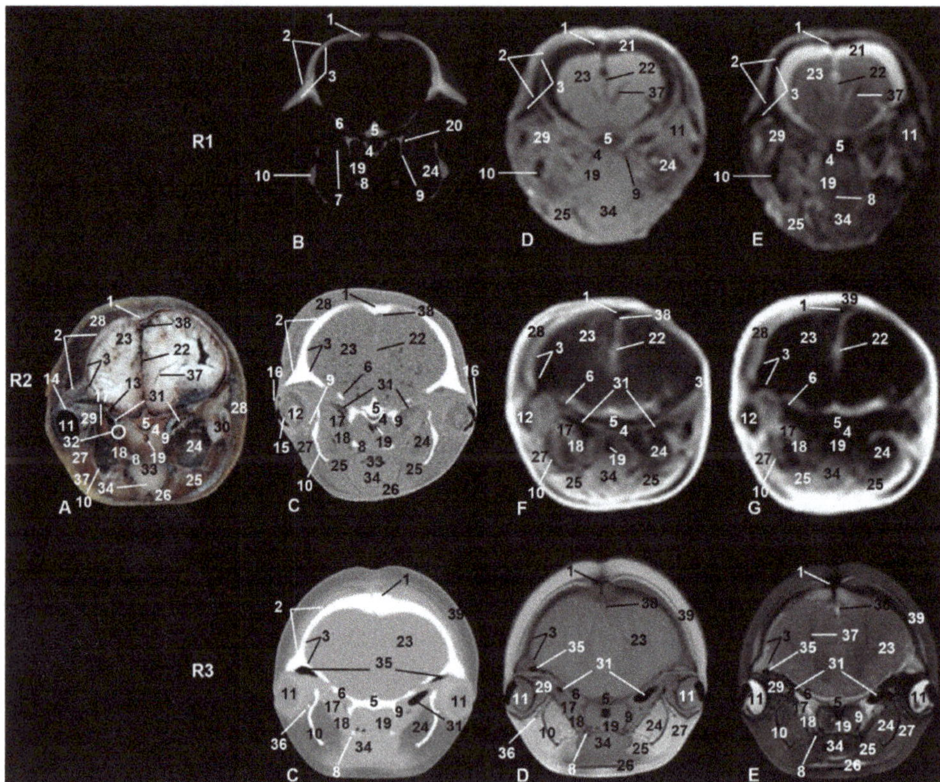

Figure 6. Representative transverse images made at the level of the pharynx and the caudal part of the orbit. Level V. Images are oriented so that the left side of the head is to the right and dorsal is at the top. R1, sco1; R2, sco2; R3, sco3. (**A**) Anatomical section. (**B**) Vascular window PET/SPECT/CT image. (**C**) Soft-tissue window CT image. (**D**) T1-weighted SE sequence. (**E**) T2-weighted FSE sequence. (**F**) T1-weighted FFE sequence. (**G**) T1-weighted OOP GRE sequence. 1, Nasal bone; 2, maxillary bone; 3, frontal bone; 4, vomer bone; 5, presphenoid bone: body; 6, presphenoid bone: wings; 7, palatine bone; 8, pterygoid bone: hook-like process; 9, pterygoid bone: pterygoid crest; 10, mandible; 11, eyeball; 12, lens; 13, optic nerve; 14, sclera; 15, cornea; 16, eyelids; 17, lateral pterygoid muscle; 18, medial pterygoid muscle; 19, nasopharynx and nasopharyngeal sphincter muscle; 20, pterygopalatine fossa; 21, subarachnoid space; 22, longitudinal brain fissure; 23, brain: temporal lobe; 24, mandibular canal; 25, digastric muscle; 26, sternohyoid muscle; 27, fat and striated muscle; 28, melon caudal muscles; 29, extraocular muscles; 30, temporomandibular joint; 31, auditory tube; 32, pharyngeal opening of the auditory tube; 33, oropharynx; 34, tongue; 35, frontal bone: orbital recess; 36, zygomatic bone: temporal process; 37, brain: lateral ventricle; 38, sagittal dorsal sinus; 39, epidermis, dermis, and subcutaneous tissue.

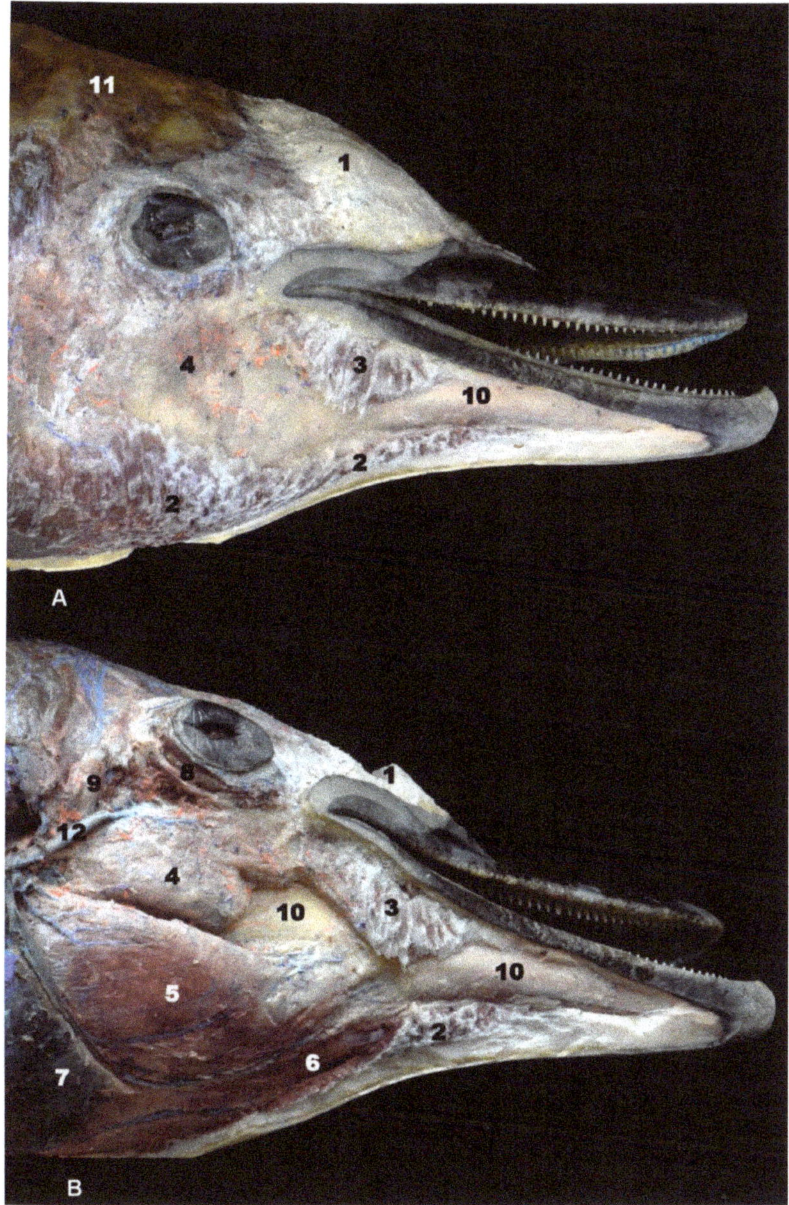

Figure 7. (**A**) Superficial and (**B**) middle head dissection made after removing melon and fat of sco5. 1, Melon; 2, mylohyoid muscle; 3, buccinator and depressor of the lower lip muscles; 4, fat and striated muscle; 5, digastric muscle; 6, geniohyoid muscle; 7, sternohyoid and sternothyroid muscles; 8, orbicularis oculi muscle; 9, external acoustic meatus (cartilaginous); 10, mandible: body; 11, subcutaneous tissue; 12, maxillary vein.

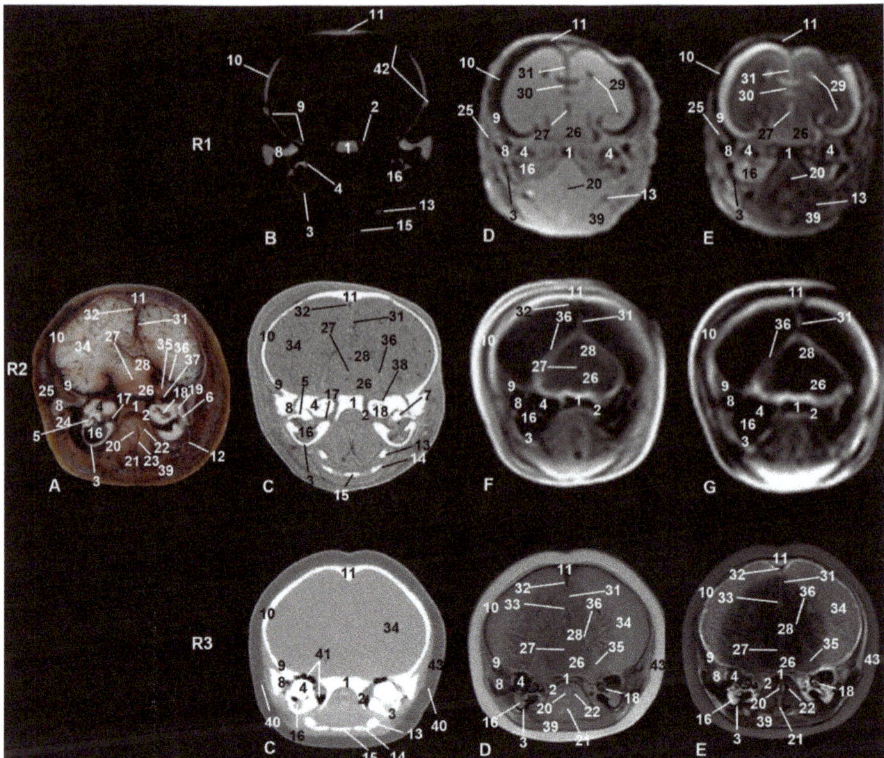

Figure 8. Representative transverse images at the level of the cranial vault of the skull involving the temporal lobe of the brain, mesencephalon, middle and inner ear, larynx and hyoid apparatus. Level VI. Images are oriented so that the left side of the head is to the right and dorsal is at the top. R1, sco1; R2, sco2; R3, sco3. (**A**) Anatomical section. (**B**) Vascular window PET/SPECT/CT image. (**C**) Soft-tissue window CT image. (**D**) T1-weighted SE sequence. (**E**) T2-weighted FSE sequence. (**F**) T1-weighted FFE sequence. (**G**) T1-weighted OOP GRE sequence. 1, Occipital bone: basilar part; 2, pterygoid crest; 3, temporal bone: tympanic part; 4, temporal bone: petrous part; 5, auditory ossicles of middle ear (malleus and incus); 6, auditory ossicles of middle ear (stapes); 7, auditory ossicles of middle ear (incus); 8, frontal process of temporal bone; 9, squamous part of temporal bone; 10, parietal bone; 11, interparietal bone; 12, tympanohyoid cartilage; 13, stylohyoid bone; 14, thyrohyoid bone; 15, basihyoid bone; 16, middle ear: tympanic cavity; 17, middle ear: musculotubarius canal; 18, inner ear: cochlea (spiral canal); 19, inner ear: vestibule; 20, arytenoid cartilage; 21, epyglotic cartilage; 22, nasopharynx: intrapharyngeal orifice; 23, laryngopharynx: piriform recess; 24, ramus of the mandible: condylar process; 25, temporal muscle; 26, mesencephalon: tegmentum; 27, mesencephalon: aqueduct; 28, mesencephalon: colliculus; 29, lateral ventricles; 30, corpus callosum; 31, falx cerebri; 32, dorsal sagittal sinus; 33, sinus transversus; 34, brain hemisphere: temporal lobe; 35, cerebellum: cerebellar hemispheres; 36, meninx: cerebellar tentorium; 37, facial and vestibulocochlear nerves and labyrinthic artery; 38, orifice and internal acoustic meatus; 39, sternohyoid muscle; 40, external acoustic meatus: cartilaginous rudiment; 41, peribullar sinus; 42, fontanelles; 43, epidermis, dermis, and subcutaneous tissue.

The mandible was hyperattenuated in CT and hypointense in MRI. The mandibular fat showed some dark color in anatomical sections; it was slightly hyperattenuated in CT sections, slightly hyperintense in T1-weighted, and slightly hypointense in T2-weighted sequences (Figures 3–6). The mandibular canal was observed patent and wide in the three specimens studied, as it is usual in odontocetes.

3.2. Rostrum (Snout)

Dorsal to the oral cavity and below melon the rostrum is observed. The mesethmoidal cartilage could be seen amongst vomer, maxillary, and incisive bones, being supported only on the groove of the vomer bone. This cartilage acts as an adhesive joining these bones to each other (Figures 2–4). CT showed one of two infraorbital canals inside the incisive bone of sco1 (Figure 2(R1B)). In the anatomical section, under the lateral sublingual recess, only fat and striated muscular tissue was seen, instead of glandular tissue, and this tissue was seen as a slightly hyper/hypointense area depending on the MRI sequence used (Figure 3).

3.3. Melon

The first two-level sections of the snout showed this particular anatomical structure of cetaceans (Figures 2 and 3). Insco1, the most rostral part of the melon could already be observed (Figure 2(R1E)). The melon was observed hyperintense in T2-weighted FSE sequence but was not detected in PET/SPECT/CT and T1-weighted SE sequence. In sco2, the melon was well appreciated in anatomical sections as well as hypointense structure in MRI sequences (Figure 2(R2A, F)). In sco3, the melon was seen as a large diffuse area in both CT and MRI (Figure 2(R3)).

The caudal part of the melon encloses the nasal cavity. In sco1, it was slightly hyperintense in MRI sequences (Figure 3(R1D, E)) and it was not observed in PET/SPECT/CT (Figure 3(R1B)). In sco2, the anatomical section showed the white nucleus of the melon surrounded by connective tissue and muscles. However, only the nucleus was identified in CT and MRI sequences (Figure 3(R2)). In sco3, CT images showed the central nucleus of the melon as moderately hypoattenuated and its external fibrous ring as slightly hyperattenuated; in a similar way, the nucleus was hyperintense and the external fibrous ring hyper/hypointense depending on the MRI sequence (Figure 3(R3)). Dissection of the melon showed the nucleus and external fibrous ring as in CT and MRI (Figure 9).

3.4. Nasal Cavity and Pasanasal Sinuses

From the beginning of fetal development until birth, an opening was observed between the melon and the frontal bone. On external examination, the nostrils, also named the spiracle or blowhole, gives the common impression of an access to odd nasal vestibule. But two nostrils closed by a musculomembranous fold were observed. Under the nostrils was the dorsal part of the nasal cavity named the vestibule of the nose, which is divided in two (left and right) by the membranous part of the nasal septum. Between the nasal vestibules and nasal plugs, different diverticula were observed. In sco3, the nasal septum, nasal diverticula (hypoattenuated areas), and plugs were clearly visualized in both CT and MR sequences (Figure 4(R3)). Nasal plugs show a cartilaginous appearance (slightly hyperattenuated in CT and hypointense in MRI), with muscular fibers and mucosa (slightly hypoattenuated in CT, slightly hypointense in T1-weighted SE sequence, and hypointense in T2-weighted FSE sequence) (Figures 4 and 5(R3)).

Under the nasal plugs (Figure 9), two nasal cavities were observed. The rostral boundary is formed by the maxillary bones, medially and caudally the vomer bone and perpendicular lamina of ethmoid bone dominate, respectively, while the lateral boundary is the pterygoid bone. The nasal septum is formed mainly by the vomer bone and by the perpendicular lamina of the ethmoid bone. In sco1, the vomer was observed hypoattenuated in CT and slightly hypointense in MR sequences and the mesethmoidal cartilage was seen hypoattenuated in CT and moderately hyperintense in MR sequences (Figure 4(R1)). In sco2, the vomer together with the ethmoid bone and mesethmoidal cartilage were identified in anatomical section. The vomer was hyperattenuated in CT and hypointense in all MR sequences. The mesethmoidal cartilage was clearly differentiated in all images except in MR sequences (Figure 4(R2)). In sco3, the vomer bone and mesethmoidal cartilage were hyperattenuated in CT, slightly hypointense in T1-weigthed MR SE sequence, and moderately hyperintense in T2-weigthed MR FSE sequence (Figure 4(R3)).

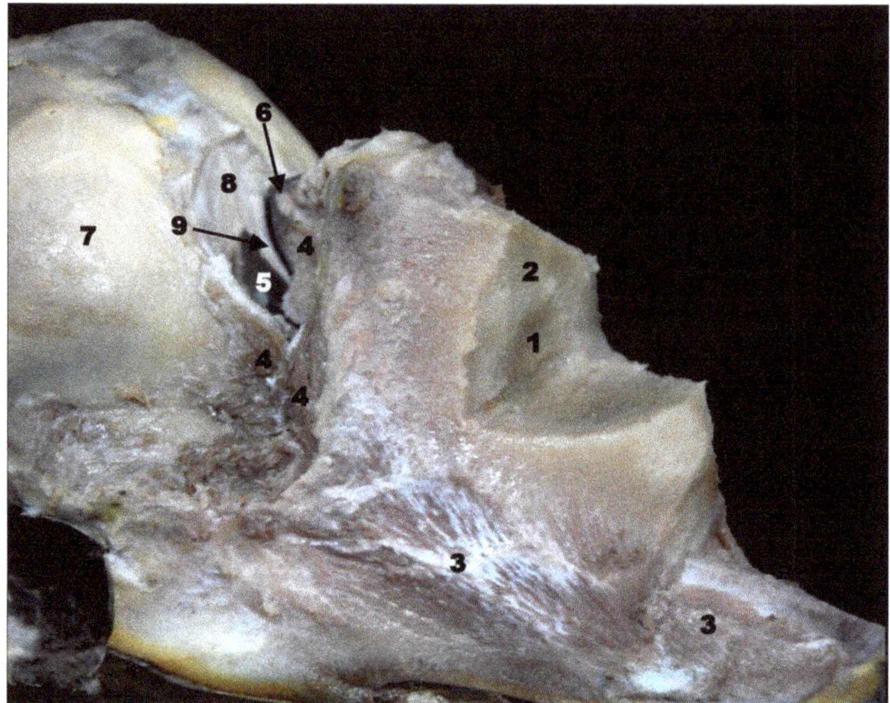

Figure 9. Deep head dissection made at the level of the nasal vestibule and melon of sco4. 1, Melon: nucleus; 2, melon: external fiber ring; 3, melon rostral muscles; 4, melon caudal muscles; 5, right nasal cavity (after removing nasal plug); 6, left nasal plug; 7, maxillary bone; 8, ethmoid bone; 9, nasal septum.

In all stages studied and with CT and MRI techniques, the maxillary sinus was observed as a small cavity within the maxillary bone. This sinus was observed in anatomical sections (sco2) filled with a heterogeneous substance observed in the fetal specimens examined in our study (Figure 3(R2A)).

Choanae are openings between the nasal cavity and the nasopharynx, and this space full of air was observed hypoattenuated in CT and hypointense in MRI sequences. In sco1, the mucosa and nasopharyngeal muscle were hypoattenuated in PET/SPECT/CT and slightly hyper/hypointense depending on the MRI sequence (Figure 5(R1)). Anatomical sections of sco2 showed mucosa and nasopharyngeal muscle with a small air space; in CT it was moderately hyperattenuated. MRI sequences showed it moderately hypointense or hypointense (Figure 5(R2)). In sco3, it could be seen as slightly hyperattenuated in CT and moderately hypointense in MRI sequences (Figure 5(R3)).

3.5. Orbit and Eyeball

CT images showed that the orbit is formed by an incomplete bony rim composed of a supraorbital part formed by the frontal bone; no infraorbital rim was observed. The rostral limit of the supraorbital rim is formed by the zygomatic and lacrimal bones. In sco1, a junction (synchondrosis) between the zygomatic and lacrimal bones was observed in PET/SPECT/CT sections (Figure 3(R1B)). Nevertheless, in sco2 (Figure 3(R2A)), anatomical sections showed both bones undergoing ossification (synostosis). Using CT, it was not possible to differentiate between these bones as the image was very hyperattenuated.

An oblique, thin temporal process of the zygomatic bone crossing under the periorbit and holding it was observed. CT showed the temporal process in sco1 with little bony density (hypoattenuated). Nevertheless, it was observed slightly hyperattenuated in sco2 and sco3 (Figures 3–5).

The eyeball was observed in all MR sequences (Figures 5 and 6) except in PET/SPECT/CT (Figures 5 and 6(R1)). CT showed the lens hyperattenuated and hypointense in MR sequences (Figure 5(R2)).

The tapetum lucidum was not appreciated in anatomical sections, but it was clearly seen in the dissection of sco4.

3.6. Central Nervous System

Brain hemispheres divided by the longitudinal brain fissure (hypointense in MR sequences) were observed in sco1 and sco2 but were not clear in sco3 (Figure 5). The sagittal dorsal sinus at the temporal level was also observed (Figures 6 and 8). In sco1, the lateral ventricle could be distinguished in MR sequences, but it was more difficult to identify in sco2 and sco3, as well as in anatomical sections, CT, and MR sequences. The meninx was observed in both anatomical sections and MR sequences (Figure 8). The mesencephalic aqueduct was clearly seen in all MR sequences. In anatomical sections, the vestibulocochlear and facial nerves as well as the labyrinthine artery passing through the inner auditory meatus were identified (Figure 8(R2)); however, these anatomical structures were not observed in CT and MR sequences.

The cerebellar tentorium appeared hyperattenuated in CT and slightly hyperintense in both MR sequences (Figure 8(R2)). The vermis and cerebellar hemispheres were observed at the level of the cerebellar fossa held by the cerebellar tentorium (Figure 10(R2A)).

3.7. Ear

The petrous part of the temporal bone was sectioned at the bony labyrinth level, which contains the bony spiral canal, bony vestibule, and spiral canal of the cochlea. Although both petrous and tympanic parts of the temporal bone were seen with both diagnostic imaging techniques (CT and MRI), components such as the auditory ossicles of the middle ear, the spiral canal of the cochlea, and the bony vestibule were visible only in CT and in anatomical sections. The auditory ossicles of the middle ear were observed hypointense in MR sequences and hyperattenuated when using CT. Nevertheless, it was possible to see the malleus, incus, and stapes using anatomical sections and CT (Figure 8). The rudimentary cartilaginous part of the external acoustic meatus was a slightly hyperattenuated small area in sco3 using CT (Figure 8(R3B)) and was also observed in dissections of sco4 and sco5 (Figure 7). CT showed a fatty content slightly hypoattenuated in the tympanic cavity (Figure 8).

Connecting the tympanic cavity with the nasopharynx was an enlargement of the auditory tube which was observed only in sco2 and sco3 (Figure 6). In anatomical sections and in MR sequences, the pharynx was appreciated surrounding aditus laryngis (epiglottic and arytenoid cartilages), and the esophageal vestibule lies dorsally (Figure 8).

3.8. Larynx

Surrounding the laryngeal cartilages, the laryngopharynx was observed, as well as the hyoid apparatus in their relationship with the ear (Figure 8). In sco1, CT images of the hyoid bones were shown slightly hyperattenuated (Figure 8(R1)). In sco2 (Figure 8(R2)), the hyoid bones were larger, and in sco3 (Figure 8(R3)) were very hyperattenuated; in anatomical sections, the tympanohyoid bone was well appreciated (Figure 8(R2A)). Also observed in this area were the temporomandibular joint and the mandibular canal fat which was very close to the middle ear, making contact with the tympanic wall. The stylohyoid, basihyoid, and thyrohyoid bones were well observed using CT and dissections because they were ossified, while the epihyoid, ceratohyoid, and tympanohyoid bones were not seen because they still remain cartilaginous. The caudal tip of the thyrohyoid bone was not ossified at birth.

3.9. Cranial Cavity

Fontanelles were wide in sco1 (Figure 8(R1)), closing in sco2 (Figure 8(R2)) and almost closed in sco3. CT in both fetuses showed clearly the fontanelles. Bones of the cranial cavity were analyzed mainly using CT and anatomical sections, since MR sequences showed bones as hypointense in all cases.

The occipital bone has three parts: basilar, lateral, and squamous. The basilar part was not observed in sco1 at this level section (Figure 10(R1)) but fontanelles were clearly seen. The basilar part was observed in sco2 (Figure 10(R2)). In sco3 (Figure 10(R3)) fontanelles are closing, though the bones surrounding the foramen magnum are not totally ossified.

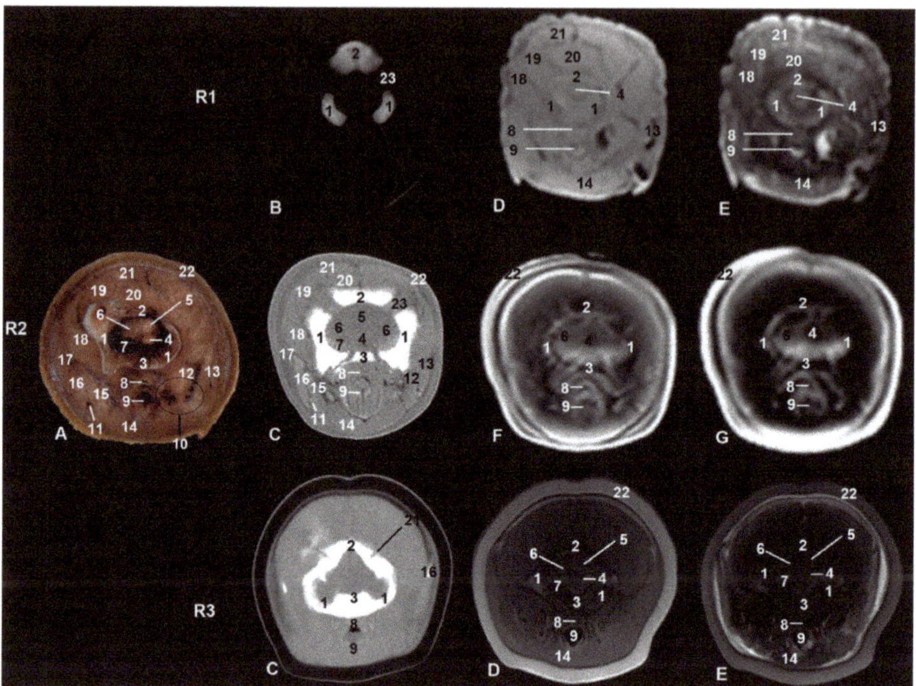

Figure 10. Representative transverse images made at the level of the occipital bone, cerebellum, and trunk of encephalon. Level VII. Images are oriented so that the left side of the head is to the right and dorsal is at the top. R1, sco1; R2, sco2; R3, sco3. (**A**) Anatomical section. (**B**) Vascular window PET/SPECT/CT image. (**C**) Soft-tissue window CT image. (**D**) T1-weighted SE sequence. (**E**) T2-weighted FSE sequence. (**F**) T1-weighted FFE sequence. (**G**) T1-weighted OOP GRE sequence. 1, Occipital bone: lateral part; 2, occipital bone: squamous part; 3, occipital bone: basilar part; 4, myelencephalon; 5, cerebellum: vermis; 6, cerebellum: cerebellar hemispheres; 7, subarachnoid space; 8 = esophagus; 9, laryngeal cavity: glottis; 10, vascular and nerve structures of the pharynx and larynx; 11, external jugular vein; 12, longus capitis muscle; 13, scapula; 14, sternohyoid and sternothyroid muscles; 15, cleidocephalic muscle: mastoid part; 16, sternocephalic muscle: mastoid part; 17, longissimus capitis muscle; 18, splenius capitis muscle; 19, semispinalis capitis muscle (digastric); 20, semispinalis capitis muscle (complex); 21, spinalis capitis muscle; 22, epidermis, dermis, and subcutaneous tissue; 23, fontanelles.

4. Discussion

4.1. Anatomical and Functional Considerations

4.1.1. Oral Cavity

Comparing our study's anatomical sections and PET/SPECT/CT images, we observed that in the fetus and newborn, the teeth of the rostral alveoli (equivalent to incisive teeth) erupt later than the more caudal alveoli (equivalent to premolars and molars). This would have a functional application in the lactation (perinatal period), where the rostral teeth erupting afterwards would to serve to help to suction milk and at the same time to hold the mother's nipple without harming it, thanks to the fact that the teeth are not yet completely formed (Figures 1–4 and 7). Odontocetes are (eu)homodont with conical teeth without a complete root, polydont, monophyodont [27] and tecodont, isognathous with centric occlusion of both dentary archs and with prognatism even during fetal stages, and are designed to catch prey [28].

We were unable to find the masseter muscle in fetal, newborn, and adult striped dolphin specimens, though it has been described by some authors in the striped dolphin [8,11], in a juvenile common dolphin [9], and in a bottlenose dolphin [29,30]; images from these studies suggest that the papers are describing the buccinator (oral part) and depressor muscles of the lower lip. (Figure 2(R2A) and Figure 9). After performing dissection of the head muscles in an adult striped dolphin, we conclude that the muscle atrophies, finding only remnants of adipose tissue and muscle fibers in its anatomical position (Figure 7). The origin of this muscle (mandible elevator) in domestic mammals extends from the maxillary tuber or facial crest to approximately the middle of the zygomatic arch, the two latter structures being absent in the pre- and perinatal studied specimens. Only a very thin temporal process of the zygomatic bone (jugal bone) is joined to the temporal bone by a symphysis, making it an unsuitable location for the attachment of a strong muscle, which acts to close the mandibles (Figures 3–5). Also, the insertion in domestic animals is the masseteric fossa and medially on the ramus of the mandible [31,32], and the masseteric fossa is absent in odontocetes but not totally in mysticetes. Functionally, the masseter muscle is an extremely powerful muscle of mastication, which varies among species in terms of its topography [32]. Odontocetes have lost this feature throughout evolution, because most of the cetaceans, with the remarkable exception of orcas [33] swallow their prey intact without chewing. Reference [30] described a residual masseter muscle in odontocetes along with a more developed temporal muscle. In addition, mysticetes catch the krill between the whalebone and raise the mandibles full of weight (mostly water) to filter it (Humpback whales [34] and rorquals [35]). The masseter muscle was also described by [19] in the boreal whale fetus and by [36] in the Minke whale.

In this study, salivary glands and lymphatic nodules were not observed in either the fetal or the newborn dolphin heads. Under the lateral sublingual recess and folds (monostomatic and polystomatic), sublingual salivary glands were not observed, only a mass of striated muscle fibers and adipose tissue (Figures 2 and 3). During dissection of the head and neck of an adult striped dolphin, only the superficial cervical lymphatic nodes were identified after histological analysis. This is logical in homodont animals, which swallow whole prey. Reference [30] showed the absence of organized salivary glands in dolphins. These same authors also describe lymphatic nodules of head to be smaller in size (about 1 cm) and therefore difficult to observe. In whales, the lack of salivary glands is described as well as the persistence of its ducts [26]. Strangely enough, an adenocarcinoma affecting the microscopic lingual salivary glands has been described in Beluga whales [37].

4.1.2. Rostrum (Snout)

Telescoping refers to the overlap of the incisive and maxilla bones and the retraction of nasal bones on top of the frontal bone, as well as to the reduction of the temporal fossa and the rostral displacement of some muscles [3,12,28]. However, in the dolphin's snout, there is a link between the vomer, maxilla, and incisive bones. The mesethmoid cartilage serves as an anti-concussive structure. The cartilage appears hypoattenuated in sco1 CT but moderately hyperattenuated in both sco1 and sco2. It does not show more signal intensity in T1-weighted SE sequence as described by [12], except in sco1. In the domestic mammal nose, the nasal septum cartilage extends rostrally to the nasal openings, but in cetaceans the nasal cavity and snout are at different levels due to telescoping. The snout tip has three functions in odontocetes: tactile (protopatic), offensive (together with tip of mandible), and, as a consequence of the mesethmoid cartilage joining these three bones, a shock absorber [8].

4.1.3. Melon

The rostral muscles are inserted in the fibrous external area of the melon in the Tursiops fetus [11] and are called maxilonasolabialis muscles in adult specimens of striped dolphin [8], in the common dolphin [12], and in bottlenose dolphin [38]. These muscles are present in the beluga whale [39]. In our study, topographically, these muscles were located ventrolateral to the melon and belong to the residual group of the facial neuromuscular system [31,32,38] (Figures 2 and 3).

4.1.4. Nasal Cavity and Paranasal Sinuses

After emerging from water, the dolphin exhales air full of CO_2 and steam through the nasal openings, nostrils. Surrounding the nostrils are striated muscle fibers opening and closing the nasal openings, acting as sphincter muscles. In our study, we could not clearly identify this group of muscles as they are mixed with melon muscles (Figure 4). In anatomical sections and MRI in a common dolphin, [9] identified the musculature of external sacs system and [40] specially studied them, which concluded that these muscles originate from the residual facial neuromuscular group, for example, orbicularis oris, levator nasolabialis, and the levator of the upper lip.

On dissecting the nasal vestibule in a fetal specimen, we observed that both nostrils have a common area under a musculomembranous sphincter, but ventrocaudally the membranous part of the nasal septum separates the left and right nasal vestibules. Reference [12] stated that the superficial blowhole emerges into an unpaired vestibulum. Ventral to the nostrils, nasal vestibules with several air sacs (diverticula) [41] were observed in the studied specimens (Figures 4 and 5). The nasal vestibule is the rostral part of the nasal cavity, related to the nasal diverticulum (horses) and lined by stratified squamous epithelium [24]. This epithelium is also referred by [30] as "similar to the epidermis of the animal's back" as in the equine vestibule. The nasal diverticulum has been described in horses as a cutaneous blind pouch inside the vestibule. The alar fold is a mucosa supported by the medial accessory cartilage covered dorsally by lateral nasal muscle [24,31,32] in the nasal vestibule in horses. In dolphins, the nasal plug, alar fold, and the nasal vestibule have been modified for sound generation and as a water reservoir. Different authors have named the nasal diverticula as sacs or air sacs or sinuses. References [8,42] described diverticula as vestibular, tubular, and premaxillary sacs. Reference [43] added the nasofrontal, accessory, and connecting air sacs and shows the premaxillary sac under the nasal plug and inside the nasal cavity [8,44]. Reference [12] mentioned three vestibular sacs. Reference [41] added the pterygoid and laryngeal air sacs. Several authors [45–48] explained that the dolphin head shows a very complex structure with unique air sacs and special sound-conducting fats. At the same time, Reference [49] claimed that no paranasal air sinuses form within the skull either prenatally or postnatally. In addition, [8] described the presence of pterygoid and maxillary sinuses with a heterogeneous substance in sections III and IV as we found in sco2 (Figure 3(R2A)). These sinuses may be non-functional. In contrast, references [26,41] describe only the pterygoid sinus (filled with a heterogeneous substance) in CT and anatomical sections in the bottlenose dolphin, whereas the maxillary sinus was the only sinus detected in our study. The paranasal sinuses heat the air and decrease the skull weight in domestic mammals. In cetaceans, moving in a less gravid environment could lead to regression of these cavities. Reference [1] said that there are no paranasal sinuses in *Tursiops truncatus*, and that in the remaining odontocetes, the maxillary sinus is described (Figure 3) (probably without functionality), though it is not well studied in whales. On the other hand, [10] described the ethmoidal sinus and [50] the pterygoid sinuses, which we have not observed in our specimens studied. We have located in striped dolphin some small orifices in the frontal wall of the nasal skull that clearly connect the nasal cavity with the maxillary sinus; however, we do not know if nasal mucosa closes these orifices completely or are vascular nutrition orifices.

The nasal cavities are symmetrical and present a similar diameter compared to those of domestic mammals [31,32]. In odontocetes, there is a skull asymmetry where structures of one side (fluctuating, that means, in one individual a structure on the right side is larger, whereas in another individual from the same species, a structure on the left side is larger) are consistently larger than those on the other [51]. This asymmetry is assumed to be based on biosonar production and reception, but it has been suggested that the asymmetry is directly proportional to the prey size, so through evolution, the larynx and the hyoid apparatus have been pushed to the left in order to extend the pharyngeal canal and allow larger prey to be ingested [52].. The ethmoid bone shows small foramina opening into the nasal mucosa equivalent to the cribriform plate observed in domestic mammals. Nasal conchae and ethmoturbinates are not observed as are found in other domestic mammals [1]. These same authors state that the ethmoid bone in cetaceans forms the dorsocaudal wall of the nasal cavity. In a striped dolphin newborn skull, we observed an ethmoid bone like the cribriform plate in domestic mammals; therefore, we believe it is possible that olfactory mucosa may be present. We agree with [30] who described the presence of "minute foramina" in very young individuals, reminiscent of the ethmoid bone cribriform plate. Reference [28] also talked about a certain olfactory capacity in cetaceans. On the other hand, [1] claims that, according to [53], in Tursiops and other odontocetes the cribriform plate appears unperforated. We did not observed the vomeronasal organ in our anatomical sections or MR sequences under the nasal mucosa and vomer bone (Figure 4).

Reference [28] described the pterygoid bone as enlarged and much pneumatized at the ventral part of nasal cavities (choanae) in beaked whales, showing medial and lateral bone laminae. According to [54], some cetaceans lost the lateral bone lamina and this lamina is replaced by the tendinous lamina, which holds the palate muscles.

4.1.5. Orbit and Eyeball

Reference [55] described an early stage of development (8–15 cm long.) in the pantropical spotted dolphin fetus, noting the lacrimal bone and later (21–22.5 cm long) the zygomatic (jugal) bone well-ossified and rostrally fused firmly with the lacrimal bone. However, in our study using PET/SPECT/CT, it is possible to observe in sco1, ossification centers which are not fused at this early stage (Figure 3). According to [1,12,30], in odontocetes (except beaked whales), the lacrimal and zygomatic (jugal) bones are fused forming the lacrimozygomatic (lacrimojugal) bone, a fact that we have confirmed in this study. The microtomographies performed on sco1 reveals that these two bones show a slightly hypoattenuated contact line indicating that it is a synchondrosis that will become a synostosis, meaning that they will be fused. As in domestic mammals, the lacrimal bone is placed lateral with respect to the zygomatic bone. The caudal projection of the zygomatic bone, the temporal process (zygomatic arch) [30], holds the periorbit and establishes a symphysis with the orbital surface of the temporal bone. The temporal process of the zygomatic bone in domestic mammals forms the infraorbital border together with the periorbit (fibrous fascia), which holds the eyeball and periorbit. In odontocetes, there is no bony infraorbital border and the only structure running under the ventral eyelid is the facial nerve. In this study, the temporal process of the zygomatic bone holding the periorbit at its middle point was observed in dissections and anatomical sections (Figures 3, 4, 5, 6 and 7B).

4.1.6. Central Nervous System

MR sequences of the brain of sco1 begin to differentiate the diencephalon and telencephalon, though the trunk of the encephalon (medulla oblongata, pons, and mesencephalon) and cerebellum are less defined. In sco2, the different parts of the trunk of the encephalon are better defined, except for the cerebellum. In addition, we have observed the cerebellum and brain hemisphere as described in perinatal dolphins [12]. The lateral ventricles and the mesencephalic aqueduct were appreciated in sco1, sco2, and sco3. We have not observed the fourth ventricle using MR sequences either in sco1 and sco2 or in sco3, even though [15] described it in a sub-adult specimen using sagittal MR sequences (Figures 5–7 and 10). Both developmental fetal stages allow us to differentiate the cerebellar tentorium as hyperintense in MR sequences. Therefore, we have observed that the tentorium cerebelli starts ossification during the fetal stage until the process is complete in adult odontocetes, unlike domestic mammals, where it remains membranous, except in cats [31,32]. We suggest the functional explanation for this relates to both species needing a well-held cerebellum due to their activity during swimming (locomotion), jumping, climbing, etc. Fetal and adult dolphin cerebellar hemispheres are situated laterally and parallel staying almost at the same level of the ventral surface of the encephalon trunk, as it can be observed in the 3D reconstruction of a common dolphin fetal brain [14].

4.1.7. Ear

The external acoustic meatus and the auricular cartilage have their origin from the projection of the first two pharyngeal arches toward the pharyngeal cleft which forms the external ear [56]. The external ear is absent in cetaceans and only remains as a small epidermal depression, barely visible in fetal specimens and newborns. At subcutaneous levels, cartilaginous nests of the external acoustic meatus appear hyperattenuated in CT in sco3. The middle ear or tympanic cavity is formed by the dilatation of the bottom of the first pharyngeal pouch and the remainder of the pouch forms the auditory tube, while the inner ear has an ectodermal origin [57]. We see both middle and inner ears in anatomical sections and tomographies as hyperattenuated (Figure 8), compact bone with thin bony walls in the fetus, with the walls thicker and hyperattenuated at the prenatal stage in agreement with [12]. In our study, MR sequences shows as hypointense the middle and inner ear walls while in our three MR sequences the hyperintense areas correspond with the ossicles and tympanic cavity described by [12] in MR T1-weighted sequence.

At the tympanic cavity, we have observed a semi-open musculotubal canal in sco2 [1], whose opening is the tympanic orifice located in the carotid or rostral wall of middle ear. This continues with the auditory tube to its opening at the pharyngeal orifice in the nasopharynx (Figure 6).

The auditory tube in sco4 could be observed in dissections as a mucosal-bony dilated space close to the perpendicular lamina of the palatine bone, extending to the pharyngeal orifice of auditory tube (eustachian notch) [1] where it opens into the nasopharynx by the pharyngeal orifice of the auditory tube. We agree with [58] about the presence of a maxillary sinus and also about the existence of a connection with the middle ear (auditory tube that we observed dilated), but not about the set of aerial sacs (premaxillary, vestibular, and nasofrontal) [30].

In our fetal study, we have just observed in CT some hypoattenuated areas (air cavities) close to the musculotubal canal of the middle ear, which could be a "pseudo-diverticulum" similar to the horse "guttural pouch" or a pharyngeal diverticulum of the auditory tube [30,31]. That would form a rudimentary osteomucomembranous space or guttural pouch, also placed in the cetaceans under the base of the cranium, connecting the tympanic cavity with the pharyngeal orifice of the auditory tube inside the nasopharynx (Figure 6).

4.1.8. Larynx

During feeding, cetaceans need the hyoid apparatus to expand, be flexible and extensible, and to project the larynx caudally (retraction) and the tongue ventrally (depression) in order to allow food to pass into the esophagus. If the hyoid apparatus was rigid, fractures could occur during feeding. In odontocetes and eschrichtiids (gray whales), increased tongue musculature and enlarged hyoids allow grasping and/or lingual depression to generate intraoral suction for prey ingestion [33]. On the other hand, balaenopterids need the mandible to be open to 90° so that the oral cavity holds up to 60 m^3 of water, so these specialized mechanisms also affect the anatomical model of the mandible and maxilla [59], as happens, for instance, in the humpback whale [60].

Nevertheless, in fetal specimens, we have observed that the entrance to the larynx (larynx peak) is very circumscribed and is formed by the rostral tips of the epiglottic and arytenoid cartilages. Both cartilages are very enlarged and oriented dorsorostrally toward the choanae. The arytenoid cartilages should not be described as cuneiform cartilages [12,29,30] and perhaps as corniculated tubercles [31]. The cuneiform cartilages are only described in the horse epiglottic cartilage and the dog's arytenoid cartilages. The corniculate tubercules are indeed mucosal eminences formed by the corniculated process of the arytenoid cartilages [24,31,32], a feature not observed in the specimens studied (Figure 8).

In the dolphin anatomical section and dissection, we have observed a cartilaginous tympanohyoid (Figure 8(R2A)) connecting the stylohyoid bone with the tympanic part of the temporal bone, placed externally on the mastoid wall of the middle ear [1,61]. Its function is to hold the tongue root and the larynx to the base of cranium as in the domestic mammals [31] while the thyrohyoid attaches to the paracondylar process of the lateral part of the occipital bone—paraoccipital process of the cranial basal bones according to [30], which is different from domestic mammals, where the thyrohyoid attaches to the rostral horn of thyroid cartilage of the larynx [31]. In addition, in the dolphins studied, the caudal tip of the thyrohyoid does not become ossified at birth, thus remaining in the adult as in gray whale [62].

4.1.9. Cranial Cavity

Three fontanelles were observed in our odontocetes fetus studied: occipital, frontal, and mastoid [1] (the last one less clear) and are confused in *Stenella attenuata* [55]. Studies in mysticetes could not add more information for comparison with dolphins [63].

5. Conclusions

Fetal anatomical sections have been very important to ensure that certain anatomical structures were correctly identified, but we needed the dissections to confirm the presence of these structures. CT was used to identify the bony and cartilaginous features of both the fetal and newborn specimens. On the other hand, different MRI sequences were used to recognize and differentiate visceral structures, which will help clinicians to diagnose different pathologies in the dolphin's head region.

We have also observed that rostral teeth erupt after lactation and the perinatal period helping suctioning milk and to protect the mother's nipples. Moreover, we have observed the absence or atrophy of the masseter muscle from fetal to adult stage in striped dolphins, mainly due to the presence of adipose tissue mixed with random muscle fibers in its anatomical position and because they swallow their prey.

No major salivary glands and lymphatic nodes were observed during developmental stages in dolphin heads, only a mixed mass of muscle fibers and fat.

A maxillary sinus has been observed filled with a heterogeneous content in our study from fetal to perinatal stage and could be non-functional.

The fusion between the lacrimal and the zygomatic bones was observed in the early fetal specimen. The temporal process of the zygomatic bone holding the periorbit in fetal dolphins has been described.

Finally, we can conclude that a "pseudo-diverticulum" similar to the "guttural pouch" connecting the tympanic cavity (middle ear) with the nasopharynx was observed in fetal anatomical sections.

Author Contributions: Conceptualization, A.G.d.l.R.y.L. and G.R.Z.; formal analysis, A.A.E. and F.G.C.; investigation A.G.d.l.R.y.L. and F.G.C.; methodology, A.A.E., M.S.L., F.M.G., and A.L.F.; resources, F.M.G. and A.L.F.; supervision, A.G.d.l.R.y.L., A.A.E., and G.R.Z.; writing—original draft, A.G.d.l.R.y.L. and G.R.Z.; writing—review and editing, A.G.d.l.R.y.L., A.A.E., M.S.L., and G.R.Z.

Funding: CT and MRI acquisitions were financed by Departamento de Anatomía y Anatomía Patológica Comparadas. Facultad de Veterinaria. Universidad de Murcia. Spain.

Acknowledgments: Many thanks to Nuria García Carrillo and Andrés Parrilla for PET/SPECT/CT image support and scan performed in CEIB, Murcia, Spain. We are grateful to José Mª Gómez-Lama López for the CT and MRI scan performed in la Policlínica Virgen de Africa, Ceuta, Spain. Also, grateful to María Leotte Sánchez, for dolphin head dissections at the Departamento de Anatomía y Anatomía Patológica Comparadas, Facultad de Veterinaria, Murcia, Spain. We are also thankful to image technician Oscar Blázquez Pérez for the MRI scan performed at Centro Veterinario de Diagnostico por Imagen del Levante, Ciudad Quesada, Rojales Alicante, Spain. We give special thanks to Mª José Gens Abujas (Oficina de impulso Socioeconómico del Medio Ambiente, Dirección General de Medio Natural, Consejería de Empleo, Universidades, Empresas y Medio Ambiente, Región de Murcia, Spain). We also thank the CRFS Veterinary Team, in a special way, Fernando Escribano Cánovas, Luisa Lara Rosales y Alicia Gómez de Ramón Ballesta, El Valle, Murcia, Spain, for allowing us to have access to the carcasses stranded in their regional area.

Conflicts of Interest: The authors of this manuscript have no conflict of interest to declare.

References

1. Mead, J.G.; Fordyce, R.E. The therian skull. A lexicon with emphasis on the odontocetes. *Smithson. Contr. Zool.* **2009**, *627*, 1–248. [CrossRef]

2. Berta, A.; Sumich, J.L.; Kovacs, K.M. Cetacean evolution and systematics. In *Marine Mammal Evolutionary Biology*, 2nd ed.; Academic Press: San Diego, CA, USA, 2005; pp. 165–209.
3. Roston, R.A.; Roth, V.L. Cetacean Skull Telescoping Brings Evolution of Cranial Sutures into Focus. *Anat. Rec.* **2019**. [CrossRef] [PubMed]
4. Armfield, B.A.; George, J.C.; Vinyard, C.J.; Thewissen, J.G.M. Allometric patterns of fetal head growth in mysticetes and odontocetes: Comparison of *Balaena mysticetus* and *Stenella attenuata*. *Marine Mammal Sci.* **2011**, *27*, 819–827. [CrossRef]
5. Sterba, O.; Klima, M.; Schildger, B. Embryology of dolphins. Staging and ageing of embryos and fetuses of some cetaceans. *Adv. Anat. Embryol. Cell. Biol.* **2000**, *157*, 1–133.
6. Thewisen, J.G.M.; Heyning, J.E. Embryogenesis and development in *Stenella atenuatta* and other cetaceans. In *Reproductive Biology and Phylogeny in Cetacea, Whales, Dolphins and Porpoises*; Miller, D.L., Ed.; Science Publishers: Enflield, NH, USA, 2007; pp. 307–330.
7. Reidenberg, J.S.; Laitman, J.T. Prenatal development in cetaceans. In *Encyclopedia of Marine Mammals*, 2nd ed.; Perrin, W.F., Würsig, B., Thewissen, J.G.M., Eds.; Academic Press: San Diego, CA, USA, 2009; pp. 220–230.
8. Hosokawa, H.; Kamiya, T. Sections of the dolphin's head (*Stenella coeruleoalba*). *Sci. Rep. Whales Res. Inst.* **1965**, *19*, 105–133.
9. Alonso-Farre, J.M.; Gonzalo-Orden, M.; Barreiro-Vázquez, J.D.; Barreiro-Lois, A.; André Morell, M.; Llarena-Reino, M.; Monreal-Pawlowsky, T.; Degollada, E. Cross-sectional anatomy, computed tomography and magnetic resonance imaging of the head of common dolphin (*Delphinus delphis*) and striped dolphin (*Stenella coeruleoalba*). *J. Vet. Med. Anat. Histol. Embryol.* **2014**, *43*, 1–9. [CrossRef]
10. Corpa, J.M.; Peris, B.; Palacio, J.; Liste, F.; Ribes, V. Hydrocephalus in a newborn bottlenosed dolphin (*Tursiops truncatus*). *Vet. Rec.* **2004**, *155*, 208–210. [CrossRef]
11. Liste, F.; Palacio, J.; Ribes, V.; Alvarez-Clau, A.; Fernández, L.; Corpa, J.M. Anatomic and Computed tomography atlas of the head of the newborn bottlenosed dolphin (*Tursiops truncatus*). *Vet. Radiol. Ultrasound.* **2006**, *47*, 453–460. [CrossRef]
12. Rauschmann, M.A.; Huggenberger, S.; Kossatz, L.S.; Oelschläger, H.H.A. Head morphology in perinatal dolphins: A window into phylogeny and ontogeny. *J. Morphol.* **2006**, *267*, 1295–1315. [CrossRef]
13. Eales, N.B. The skull of the foetal narwhal *Monodon monoceros* L. *Phil. Trans. R. Soc.* **1950**, *235*, 1–33.
14. Marino, L.; Murphy, T.L.; Lyad, G.; Johnson, J.I. Magnetic resonance imaging and three-dimensional reconstructions of the brain of a fetal common dolphin, *Delphinus delphis*. *Anat. Embryol.* **2001**, *203*, 393–402. [CrossRef] [PubMed]
15. Montie, E.W.; Schneider, G.E.; Ketten, D.R.; Marino, L.; Touhey, K.E.; Hahn, M.E. Neuroanatomy of the subadult and fetal brain of the atlantic white-sided dolphin (*Lagenorhynchus acutus*) from in situ magnetic resonance images. *Anat. Rec.* **2007**, *290*, 1459–1479. [CrossRef] [PubMed]
16. Costidis, A.M.; Rommel, S.A. The extracranial arterial system in the heads of beaked whales, with implications on diving physiology and pathogenesis. *J. Morphol.* **2016**, *277*, 5–33. [CrossRef] [PubMed]
17. Ridewood, W.G. Observations on the skull in foetal specimens of whales of the genera *Megaptera* and *Balaenoptera*. *Phil. Trans. R. Soc.* **1922**, *211*, 209–272. [CrossRef]
18. Johnston, C.; Deméré, T.A.; Berta, A.; Yonas, J.; Leger, J. Observations on the musculoskeletal anatomy of the head of a neonate gray whale (*Eschrichtius robustus*). *Mar. Mammal. Sci.* **2010**, *26*, 186–194. [CrossRef]
19. Schulte, H.W. The sei whale (*Balaenoptera borealis*, Lesson). Anatomy of a foetus of *Balaenoptera borealis*. Monographs of the Pacific *Cetacea*. *Mem. Am. Mus. Nat. Hist. New Ser.* **1916**, *1*, 389–499.
20. O'Rahilly, R.; Müller, F. *Developmental Stages in Human Embryos*; Carnegie Institute: Washington, DC, USA, 1987; pp. 1–637.
21. Perrin, W.F.; Holts, D.B.; Miller, R.B. Growth and reproduction of the eastern spinner dolphin, a geographical form of *Stenella longirostris* in the eastern tropical pacific. *Fish Bull.* **1977**, *75*, 725–750.
22. Lacave, G.; Eggermont, M.; Verslycke, T.; Kinoshita, R. Prediction from ultrasonographic measurements of the expected delivery date in two species of bottlenose. *Vet. Rec.* **2004**, *154*, 228–233. [CrossRef]
23. Hohn, A.A. Age estimation. In *Encyclopedia of Marine Mammals*, 2nd ed.; Perrin, W.F., Würsig, B., Thewissen, J.G.M., Eds.; Academic Pres: Cambridge, MA, USA, 2009; pp. 11–17.
24. Schaller, O. *Illustrated Veterinary Anatomical Nomenclature*; Ferdinand Enke Verlag: Stuttgart, Germany, 1992; pp. 1–614.

25. Roston, R.A.; Lickorish, D.; Buchholtz, E.A. Anatomy and age estimation of an early blue whale (*Balaenoptera musculus*) fetus. *Anat. Rec.* **2013**, *296*, 709–722. [CrossRef]
26. Wilkie Tinker, S. *Whales of the World*; Brill, E.J., Ed.; Bess Pr Inc.: Leiden, The Netherlands, 1988; pp. 1–310.
27. Armfield, B.A.; Zheng, Z.; Bajpai, S.; Vinyard, C.J.; Thewissen, J. Development and evolution of the unique cetacean dentition. *PeerJ* **2013**, *1*, e24. [CrossRef]
28. Rommel, S.A.; Pabst, D.A.; McLellan, W.A. Skull anatomy. In *Encyclopedia of Marine Mammals*, 2nd ed.; Perrin, W.F., Würsig, B., Thewissen, J.G.M., Eds.; Academic Press: Cambridge, MA, USA, 2009; pp. 1033–1056.
29. Huggenberger, S.; Oelschläger, H.; Cozzi, B. *Atlas of the Anatomy of Dolphins and Whales*; Academic Press, Elsevier: London, UK, 2019; pp. 1–513.
30. Cozzi, B.; Huggenberger, S.; Oelschläger, H. *Anatomy of Dolphins. Insights into Body Structure and Function*; Academic Press: London, UK, 2017; pp. 1–438.
31. Sandoval, J. *Tratado de Anatomía Veterinaria. Tomo III: Cabeza y Sistemas Viscerales*; Imprenta Sorles: León, Spain, 2000; pp. 1–457.
32. Nickel, R.; Schummer, A.; Seiferle, E. *The Anatomy of the Domestic Animals. The Locomotor System of the Domestic Mammals*; Verlag Paul Parey: Berlin/Hamburg, Germany, 1986; Volume 1, pp. 1–499.
33. Werth, A.J. Adaptations of the cetacean hyolingual apparatus for aquatic feeding and thermoregulation. *Anat. Rec.* **2007**, *290*, 546–568. [CrossRef] [PubMed]
34. Heithaus, M.R.; Dill, L.M. Feeding strategies and tactics. In *Encyclopedia of Marine Mammals*, 2nd ed.; Perrin, W.F., Würsig, B., Thewissen, J.G.M., Eds.; Academic Press: San Diego, CA, USA, 2009; pp. 414–423.
35. Pyenson, N.D.; Goldbogen, J.A.; Vogl, A.W.; Szathmary, G.; Drake, R.L.; Shadwick, R.E. Discovery of a sensory organ that coordinates lunge feeding in rorqual whales. *Nature* **2012**, *485*, 498–501. [CrossRef] [PubMed]
36. Lambertsen, R.H.; Hintz, R.J. Maxillomandibular cam articulation discovered in North Atlantic minke whale. *J. Mammal.* **2004**, *85*, 446–452. [CrossRef]
37. Girard, G.; Lagacé, A.; Higgins, R.; Béland, P. Adenocarcinoma of the salivary gland in a beluga whale (*Delphinapterus leucas*). *J. Vet. Diagn. Invest.* **1991**, *3*, 264–265. [CrossRef] [PubMed]
38. Harper, C.J.; McLellan, W.A.; Rommel, S.A.; Gay, D.M.; Dillaman, R.M.; Pabst, D.A. Morphology on the melon and its tendinous connections to the facial muscles in bottlenose dolphins (*Tursiops truncatus*). *J. Morphol.* **2008**, *269*, 820–839. [CrossRef] [PubMed]
39. O´Corry-Crowe, G.M. Beluga whale. In *Encyclopedia of Marine Mammals*, 2nd ed.; Perrin, W.F., Würsig, B., Thewissen, J.G.M., Eds.; Academic Press: Cambridge, MA, USA, 2009; pp. 108–112.
40. Behrmann, G. Funktion und Evolution der Delphinnase. *Nat. Mus.* **1983**, *113*, 71–78.
41. Reidenberg, J.S.; Laitman, J.T. Sisters of the sinuse cetacean air sacs. *Anat. Rec.* **2008**, *291*, 1389–1396. [CrossRef]
42. Lawrence, B.; Schevill, W.E. The Functional Anatomy of the Delphinid Nose. *Bull. Mus. Comp. Zool.* **1956**, *114*, 103–150.
43. Schenkkan, E.J. The occurrence and position of the "connecting sac" in the nasal tract complex of small odontocetes (Mammalia, Cetacea). University of Amsterdam. *Beaufortia* **1971**, *246*, 37–43.
44. Heyning, J.E. Comparative facial anatomy of beaked whales (Ziphiidae) and a systematic revision among the families of extant *odontoceti*. *Nat. Hist. Mus. Los Angel. Cty.* **1989**, *405*, 1–59.
45. Cranford, T.W. Visualizing dolphin sonar signal generation using high speed video endoscopy. *J. Acoust. Soc. Am.* **1997**, *102*, 3123. [CrossRef]
46. Cranford, T.W.; Amundin, M.; Kenneth, S.N. Functional morphology and homology in the Odontocete nasal complex: Implications for sound generation. *J. Morph.* **1996**, *228*, 223–285. [CrossRef]
47. Reidenberg, J.S.; Laitman, J.T. Generation of sound in marine mammals. In *Handbook of Mammalian Vocalization—An Integrative Neuroscience Approach*; Brudzynski, S.M., Ed.; Academic Press/Elsevier: London, UK, 2010; pp. 451–465.
48. Whitlow, W.L. Echolocation. In *Encyclopedia of Marine Mammals*, 2nd ed.; Perrin, W.F., Würsig, B., Thewissen, J.G.M., Eds.; Academic Press: Cambridge, MA, USA, 2009; pp. 348–357.
49. Reidenberg, J.S.; Laitman, J.T. Cetacean prenatal development. In *Encyclopedia of Marine Mammals*, 2nd ed.; Perrin, W.F., Würsig, B., Thewissen, J.G.M., Eds.; Academic Press: Cambridge, MA, USA, 2009; pp. 220–230.
50. Racicot, A.; Berta, A. Comparative morphology of porpoise (Cetacea: Phocaenidae) pterygoid sinuses: Phylogenetic and functional implications. *J. Morphol.* **2013**, *274*, 49–62. [CrossRef] [PubMed]
51. Gatesy, J.; Geisler, J.H.; Chang, J.; Buell, C.; Berta, A.; Meredith, R.W.; Springer, M.S.; McGowen, M.R. A phylogenetic blueprint for a modern whale. *Mol. Phylogenet. Evol.* **2013**, *66*, 479–506. [CrossRef] [PubMed]

52. Macleod, C.C.; Reidenberg, J.S.; Weller, M.; Santos, M.B.; Herman, J.; Goold, J.; Piercel, G.J. Breaking symmetry: The marine environment, prey size, and the evolution of asymmetry in cetacean skulls. *Anat. Rec.* **2007**, *290*, 539–545. [CrossRef] [PubMed]
53. Kellogg, A.R. The history of the whales: Their adaptation to life in the water. *Q. Rev. Biol.* **1928**, *3*, 29–76. [CrossRef]
54. Fraser, F.C.; Purves, P.E. Hearing in cetaceans-evolution of the accessory air sacs and the structure and function of the outer and middle ears in recent cetaceans. *Bull. Br. Mus.* **1960**, *7*, 1–140.
55. Moran, M.M.; Numella, S.; Thewissen, J.G.M. Development of the skull of the pantropical spotted dolphin (*Stenella attenuata*). *Anat. Rec.* **2011**, *294*, 1743–1756. [CrossRef]
56. McGeady, T.A.; Quinn, P.J.; Fitzpatrick, E.S.; Ryan, M.T.; Kilroy, D.; Lonergan, P. *Veterinary Embryology*; Wiley Blackwell: Oxford, UK, 2017; pp. 1–400.
57. García Monterde, J.; Gil Cano, F. *Embriología Veterinaria: Un Enfoque Dinámico del Desarrollo Animal*; Inter-médica: Buenos Aires, Argentina, 2013; pp. 1–185.
58. Houser, D.S.; Finneran, J.; Carder, D.; Van Bonn, W.; Smith, C.; Hoh, C.; Mattrey, R.; Ridgway, S. Structural and functional imaging of bottlenose dolphin (*Tursiops truncatus*) cranial anatomy. *J. Exp. Biol.* **2004**, *207*, 3657–3665. [CrossRef]
59. Bouetel, V. Phylogenetic implications of skull structure and feeding behaviour in Balaenopterids (Cetacea, Mysticeti). *J. Mammal.* **2005**, *86*, 139–146. [CrossRef]
60. Field, D.J.; Campbell-Malone, R.; Goldbogen, J.A.; Shadwick, R.E. Quantitative computed tomography of humpback whale (*Megaptera novaeangliae*) mandibles: Mechanical implications for rorqual lunge-feeding. *Anat. Rec.* **2010**, *293*, 1240–1247. [CrossRef] [PubMed]
61. Reidenberg, J.S.; Laitman, J.T. Anatomy of the hyoid apparatus in Odontoceti (toothed whales): Specializations of their skeleton and musculature compared with those of terrestrial mammals. *Anat. Rec.* **1994**, *240*, 598–624. [CrossRef] [PubMed]
62. Kienle, S.S.; Ekdale, E.G.; Reidenberg, J.S.; Deméré, T.A. Tongue and Hyoid Musculature and Functional Morphology of a Neonate Gray Whale (Cetacea, Mysticeti, *Eschrichtius robustus*). *Anat. Rec.* **2015**, *298*, 660–674. [CrossRef] [PubMed]
63. Hampe, O.; Franke, H.; Hipsley, C.A.; Kardjilov, N.; Muller, J. Prenatal cranial ossification of the humpback whale (Megaptera novaeangliae). *J. Morph.* **2015**, *276*, 564–582. [CrossRef]

MDPI

Article

Comparative Anatomy of the Nasal Cavity in the Common Dolphin *Delphinus delphis* L., Striped Dolphin *Stenella coeruleoalba* M. and Pilot Whale *Globicephala melas* T.: A Developmental Study

Alvaro García de los Ríos y Loshuertos [1,2], Marta Soler Laguía [3], Alberto Arencibia Espinosa [4], Alfredo López Fernández [5,6], Pablo Covelo Figueiredo [7], Francisco Martínez Gomariz [1], Cayetano Sánchez Collado [1], Nuria García Carrillo [7] and Gregorio Ramírez Zarzosa [1,*,†]

1 Departamento de Anatomía y Anatomía Patológica Comparadas, Facultad de Veterinaria, Universidad de Murcia, 30100 Murcia, Spain; agrios@ceuta.es (A.G.d.l.R.yL.); f.gomariz@colvet.es (F.M.G.); scollado@um.es (C.S.C.)
2 Centro de Estudio y Conservación de Animales Marinos (CECAM), 51002 Ceuta, Spain
3 Departamento de Medicina y Cirugía Animal, Facultad de Veterinaria, Universidad de Murcia, 30100 Murcia, Spain; mtasoler@um.es
4 Departamento de Morfología, Anatomía y Embriología, Facultad de Veterinaria, Universidad de Las Palmas de Gran Canaria, Trasmontaña, Arucas, 35416 Las Palmas de Gran Canaria, Spain; alberto.arencibia@ulpgc.es
5 Departamento de Biología—CESAM, Campus Universitario de Santiago, Universidade de Aveiro, 3810-193 Aveiro, Portugal; a.lopez@ua.pt
6 Coordinadora para el Estudio de los Mamíferos Marinos (CEMMA), Ap. 15, Gondomar, 36380 Pontevedra, Spain
7 Servicio de Experimentación Animal, Área Científica-Técnica de Investigación, Edificio Centro de Experimentación e Investigación Biomédica (CEIB), Universidad de Murcia, 30100 Murcia, Spain; pablo_cov@yahoo.es (P.C.F.); ngc2@um.es (N.G.C.)
* Correspondence: grzar@um.es; Tel.: +34-868887546; Fax: +34-868884147
† Campus de Excelencia Internacional de Ámbito Regional (CEIR), Campus Mare Nostrum, Universidad de Murcia, 30003 Murcia, Spain.

Citation: García de los Ríos y Loshuertos, A.; Soler Laguía, M.; Arencibia Espinosa, A.; López Fernández, A.; Covelo Figueiredo, P.; Martínez Gomariz, F.; Sánchez Collado, C.; García Carrillo, N.; Ramírez Zarzosa, G. Comparative Anatomy of the Nasal Cavity in the Common Dolphin *Delphinus delphis* L., Striped Dolphin *Stenella coeruleoalba* M. and Pilot Whale *Globicephala melas* T.: A Developmental Study. *Animals* **2021**, *11*, 441. https://doi.org/10.3390/ani11020441

Academic Editor: Matilde Lombardero

Received: 16 December 2020
Accepted: 28 January 2021
Published: 8 February 2021

Publisher's Note: MDPI stays neutral with regard to jurisdictional claims in published maps and institutional affiliations.

Simple Summary: The developmental anatomy of the dolphin head has been studied mostly in single fetuses and few works have been made using a wide range of specimens. In this study, fetal specimens are the main subjects, but newborn, juvenile and adult specimens were also used. Our study analyzes the external nose and nasal cavities during pre- and postnatal development. The nose and nasal cavities were studied using a high-resolution endoscopy to analyze changes in the mucosa of fetal specimens, newborns and juveniles. Magnetic Resonance Imaging (MRI) was also used in fetuses to locate and identify significant structures. Computed Tomography (CT) allowed us to understand the development of the facial bones and the nasal cavity. The histological samples were compared with a horse, a terrestrial mammal with a complex nasal anatomy. Dissections and anatomical sections in two spatial planes were compared with MRI and CT studies. Endoscopy of the external nose showed interesting morphological changes as only two different diverticula (air sacs) were observed in the vestibular part and one recess in the respiratory and olfactory part. We conclude that nasal cavity development of the striped and common dolphins and the pilot whale is simpler than in the bottlenose dolphin and the melon is part of the nose both anatomically and functionally.

Abstract: Our goal was to analyze the main anatomical structures of the dolphin external nose and nasal cavity from fetal developmental stages to adult. Endoscopy was used to study the common development of the external nose and the melon, and nasal mucosa. Magnetic resonance imaging (MRI) and anatomical sections were correlated with anatomical sections. Computed tomography (CT) was used to generate 3D reconstructions of the nasal bones and nasal cavities to study its development. Dissections, histological and pathological studies were carried out on the nasal mucosa to understand its function. These results were compared with the horse. Endoscopy showed an external nose with two lips and the upper lip is divided by a groove due to the nasal septum and an obstruction of right nasal cavity was diagnosed in a newborn. Two diverticula (air sacs) were found in the nasal vestibule and an incisive recess (premaxillary sac) in the nasal cavity. These findings were corroborated by 3D reconstructions of the nasal cavities, MRI, anatomical sections and dissections. The presphenoid and ethmoid bones were fused at early stages of fetal development. The ethmoid is the last bone to ossify in the nasal cavity.

Keywords: striped dolphin (*Stenella coeruleoalba*); common dolphin (*Delphinus delphis*); pilot whale (*Globocephala melas*); fetal development; nose; nasal cavity; endoscopy; PET/SPECT/CT; MRI; sectional anatomy; dissection; 3D reconstruction; histology; ontogeny

1. Introduction

Through evolution, mammals that colonized the aquatic environment have undergone numerous adaptations in their anatomical structures, especially those of the cephalic region [1]. These adaptative changes are mainly seen in the asymmetry and telescoping of the skull [2–4] and have direct implications in feeding [5], mechanical protection of cephalic structures, echolocation, breathing and diving. The repositioning of the bones of the dorsal skull affect the position of the nasal opening [6], whose dorsal position enables breathing while the animal is on the surface. Along with the air sacs and sinuses which have respiratory, vocal and structural functions [7–10], these modifications in cephalic anatomy results in the most highly evolved nose of all mammals.

There are not many articles in the scientific literature about the cetacean upper respiratory system, and some of the existing studies differ in their descriptions of structures and the terminology used to identify these structures... Most of those articles are about adult specimens, either odontocetes [11–20] or mysticetes [21–24] and many of them focused on sound production and biosonar [14,25,26]. The few papers on fetal specimens [27–32] do not cover different stages of development and only show one fetus without comparison with other fetal stages, essential to the understanding of cetacean ontogeny and its application in the fields of biology and veterinary medicine [29]. The development of the bones and other nasal structures continues post-partum with opening of the blowhole, expansion of the lungs and independent breathing. In the current study, we analyze the nasal complex of 22 odontocetes belonging to three species: (striped dolphin *Stenella coeruleoalba*, common dolphin *Delphinus delphis* and pilot whale *Globicephala melas*) of all ages (16 fetuses of different stages, three newborn, two juvenile and three adults). We applied several diagnostic techniques: computed tomography (CT), magnetic resonance imaging (MRI), anatomical sections, dissections, 3D reconstructions, histology and histopathology. Endoscopy is commonly used in dolphin medicine especially to view the lower respiratory tract (lungs and bronchus) [33–37] but is not often used to visualize the nasal passages, a region of key importance, not only in live animals, but also during necropsies to confirm any pathology incompatible with a correct air passage and therefore likely to interfere with vital activities such as diving and feeding that will decrease the life expectancy of the animal.

A comparison with terrestrial mammals will serve also to describe the structures of the head following the Illustrated Veterinary Anatomical Nomenclature [38] and to better understand the actual function and position of the different hard and soft tissue structures that form the nasal cavity of small cetaceans.

2. Materials and Methods

2.1. Animals

A total of sixteen prenatal, three perinatal specimens, two juvenile and four adult dolphins; one foal fetus and two adult horses were used in this study (Table 1). Additional information is located in Table S1 (Supplementary Materials). The mothers of each fetus were stranded along the Spanish Atlantic coast. The youngest newborn specimen was stranded on the Spanish African coast and the two others on the Mediterranean coast. The juvenile and adult specimens were stranded on the Spanish Mediterranean coast. All stranded dolphin specimens were found dead, two horse cadaver heads were obtained from the Orihuela abattoir; consequently, ethics committee clearance was not necessary. Endoscopic studies were made between September 2019 and January 2020 in a Veterinary Clinic ("Bonafé"), La Alberca (Murcia). Fourteen fetuses, one newborn and one adult specimen were transported to the image analysis units to perform scans. Six fetuses, one newborn and one adult were transported to the dissection room for silicone injections. The silicone was injected through the blowhole towards the nasal cavity, filling the vestibule first. The amount of silicone went from 1 to 8 cm^3 in the fetuses, 16 cm^3 in the newborn and 20–28 cm^3 in the adult. After that, the blowhole was covered with a tight elastic band to avoid backflow of the silicone. Two adult specimens were used to obtain anatomical sections, one of them for histological analysis, one newborn for pathological study and two adults for dissection. One horse specimen was sectioned and the other dissected.

Table 1. Specimens of dolphin used in this study.

Study Code	Species, Sex [29,39,40]	Anatomical, Surgical/Imaging Diagnostic and 3D Reconstruction Techniques
dde1	*Delphinus delphis* L., male fetus	Endoscopy, MRI
dde2	*Delphinus delphis* L., male fetus	Endoscopy, MRI, PET/SPEC/CT, AMIRA®, silicone injection
dde3	*Delphinus delphis* L., female fetus	Endoscopy, MRI
scop1	*Stenella coeruleoalba* M., female fetus	Endoscopy, MRI, PET/SPECT/CT, AMIRA®, silicone injection
gma1	*Globicephala melas* T., male fetus	Endoscopy, MRI, PET/SPEC/CT
dde4	*Delphinus delphis* L., male fetus	Endoscopy
dde5	*Delphinus delphis* L., female fetus	Endoscopy, MRI, PET/SPEC/CT, AMIRA®
dde6	*Delphinus delphis* L., female fetus	Endoscopy, PET/SPEC/CT, AMIRA®, silicone injection
dde7	*Delphinus delphis* L., male fetus	Endoscopy, MRI
dde8	*Delphinus delphis* L., female fetus	Endoscopy
dde9	*Delphinus delphis* L., male fetus	Endoscopy, MRI, CT, AMIRA®, silicone injection
dde10	*Delphinus delphis* L., female fetus	Endoscopy, MRI
dde11	*Delphinus delphis* L., male fetus	Endoscopy, MRI
dde12	*Delphinus delphis* L., male fetus	Endoscopy, MRI
dde13	*Delphinus delphis* L., female fetus	Endoscopy, MRI, CT, AMIRA®, silicone injection
dde14	*Delphinus delphis* L., female fetus	Endoscopy, MRI, CT, AMIRA®, silicone injection
scoce1	*Stenella coeruleoalba* M., male newborn	Endoscopy, Head dissection, Anatomopathological study
scomu1	*Stenella coeruleoalba* M., female newborn	CT, AMIRA®, silicone injection
scomu2	*Stenella coeruleoalba* M., male newborn	Head coronal section
scomu3	*Stenella coeruleoalba* M., male juvenile	Head sagittal section
scomu4	*Stenella coeruleoalba* M., Male juvenile	Endoscopy
scomu5	*Stenella coeruleoalba* M., female adult	Head dissection, vascular latex injection
scomu6	*Stenella coeruleoalba* M., female adult	Head sagittal section, histological analysis
scomu7	*Stenella coeruleoalba* M., female adult	Head dissection

Table 1. *Cont.*

Study Code	Species, Sex [29,39,40]	Anatomical, Surgical/Imaging Diagnostic and 3D Reconstruction Techniques
scomu8	*Stenella coeruleoalba* M., female adult	CT, AMIRA®, silicone injection
ecal1	*Equus caballus* L., male, fetus	Head dissection
ecal2	*Equus caballus* L., adult	Head dissection
ecal3	*Equus caballus* L., adult	Head section
ecal4	*Equus caballus* L., adult	Head section

DDE: *Delphinus delphis* from Galicia, Spain; SCOP: *Stenella coeruleoalba* from Galicia, Spain; SCOCE: *Stenella coeruleoalba* from Ceuta, Spain; SCOMU: *Stenella coeruleoalba* from Murcia, Spain; MRI: Magnetic resonance imaging; CT: Computed Tomography, CEMMA: Coordinator Center for the study of the marine mammals, Galicia; CECAM: Center for the study and conservation of marine animals, Ceuta; CRFS: Wildlife rehabilitation Center, Murcia; ECAL: *Equus caballus* from Alicante, Spain; AMIRA® for FEI systems is a software platform for 3D and 4D data visualization, processing and analysis.

2.2. Endoscopy

A fixed endoscopy unit (Karl Storz Autocon 200, Tuttlingen, Germany) located at Clínica Veterinaria "Bonafé", La Alberca (Murcia), Spain with a camera processor (Storz image 1 hub, camera head Karl Storz Image 1 H3 HD, a Storz power led 175) was used to obtain endoscopic images. For the external nose we employed a forward telescope (0° enlarged view, diameter 3 mm, length 14 cm, with incorporated fiber optic light transmission) and for the nasal cavity we used a forward-oblique telescope 30°, diameter 2.7 mm, length 18 cm, with incorporated fiber optic light transmission.

The endoscopic procedure was performed on the fetuses and juvenile specimens following standard protocols. Each specimen was placed in ventral recumbency and facing the endoscopist. Initial image acquisition was of the external nose morphology, using an optic of 4 mm diameter and 0° vision angle. Following this, the nasal cavity was examined by introducing an optic of 2.7 mm and 30° vision angle protected by a 3 mm sheath forming an irrigation channel, visualizing first the vestibule, then turning the optic to obtain a complete exposition of this part of the nasal cavity. After visualizing the nasal plugs, observation and study of the respiratory tract within the nasal cavity was conducted. For this purpose, and with the specimens in a lateral position, the endoscope was introduced through the vestibule towards this respiratory part, which allowed us to observe structures such as the incisive recess and the choanae, while irrigating physiological serum through the endoscopy irrigation channel to clean cavities and to obtain good images. All endoscopic images were stored in external and internal hard disks at the CVB (Bonafé Veterinary Clinic) and at the Department of Anatomy and Embryology, Facultad de Veterinaria, Universidad de Murcia, Spain.

2.3. Magnetic Resonance Imaging

Magnetic Resonance images were obtained with a high-field MR apparatus (General Electric Sigma Excite, Schenectady, NA, USA; Centro Veterinario de Diagnóstico por Imagen de Levante, Ciudad Quesada, Alicante, Spain), 1.5 Tesla using a human quad knee coil (dde1 to 3, 5, 7, 9 to 12, gma1), wrist coil (scop1, dde1) and head coil (dde13 and 14). All dolphin specimens were positioned in ventral recumbency. The MR images were transferred to a DICOM workstation. MR images were analyzed with Radiant DICOM viewer. MRI parameters used are in Table S2 (Supplementary Materials). All MR images were stored in external and internal hard disks at the CVDIL (Veterinary Center for Imaging Diagnosis), Ciudad Quesada, Alicante, Spain and at the Department of Anatomy and Embryology, Facultad de Veterinaria, Universidad de Murcia, Spain.

2.4. Computed Tomography and 3D Reconstruction of Bony Nasal Cavity

Four fetuses were scanned with Positron Emission Tomography (PET), Single Photon Emission Computed Tomography (SPECT)—Computed Tomography (CT) (PET/SPECT/CT AlbiraTM Systems, Valencia, Spain; Centro de Investigación Biomédica, Universidad de Murcia, Murcia, Spain). The following parameters were used: single-slice: 1 detector arrays; type of acquisition: helical; thickness: 0.125 mm; image reconstruction interval or index: 0.0125 mm; pitch: 0; tube rotation time: 0.12 s; mA: 0.4; Kv: 45; FOV: 68 cm; Matrix dimensions: 2240 × 2360; reconstruction algorithm: FBP filtered back projection; WW: 600/WL: 300).

Three fetuses (one newborn and one adult) were scanned with a CT (General Electric Medical Systems-HiSpeed dual; Hospital Clínico Veterinario, Universidad de Murcia, Spain). All dolphin specimens were positioned in ventral recumbency. The following parameters were used: multislice: two detector arrays; type of acquisition: helical; thickness: 1 mm; image reconstruction interval or index: 2.5 mm; pitch: 0.35; tube rotation time: 1 s; mA: 45 (dde2), 50 (dde6), 55 (scog1, gma1), 60 (dde9,dde13), 75 (scomu1), 80 (dde14), 150 (scomu5); Kv: 120; image field of view: 40 cm; acquisition matrix: 512 × 512; reconstruction algorithm: standard; WW: 400/WL: 40) (Table 1).

All CT images were transferred to a DICOM workstation and CT images were analyzed with AMIRA for Fei Systems 5.6 (Thermo Fisher Sci and Zuse Institute, Berlin, Germany). Volume rendering was generated to obtain 3D renderings of internal anatomy. All CT images were stored in external and internal hard disks at CEIB (Experimental Center of Biomedical Research) building, SAI (Support Research Facility) building and at the Department of Anatomy and Embryology, Facultad de Veterinaria, Universidad de Murcia, Spain.

2.5. Computed Tomography and 3D Reconstruction of Nasal Cavity Spaces

The nasal cavities of several fetal, newborn and adult specimens were injected with Silicone Xiameter © RTV-4230-E Base, 2 to 30 mL depending on size of specimen, (Dow Corning Co, Midland, MI, USA; Dissection room, Facultad de Veterinaria, Murcia, Spain) to obtain nasal endocasts and to enhance CT contrast properties. Injected specimens were scanned with CT and DICOM images were used to obtain 3D reconstructions of the nasal cavities. Three-dimensional reconstructions of the nasal cavity were obtained using AMIRA for Fei Systems 5.6. All Dicom images were stored in external and internal hard disks at SAI (Support Research Facility) building and at the Department of Anatomy and Embryology, Facultad de Veterinaria, Universidad de Murcia, Spain.

2.6. Anatomic Evaluation: Sectional Anatomy and Dissection Techniques

One newborn and one juvenile striped dolphin (*Stenella coeruleoalba*) were frozen at −20 °C prior to obtaining coronal and sagittal sections of the head. One adult striped dolphin (*Stenella coeruleoalba*) was frozen at −46 °C prior to obtaining sagittal sections. All specimens were cut with a band saw (Anatomical Lab., Department of Anatomy and Embryology, Universidad de Murcia, Murcia, Spain), obtaining 0.5–0.7 cm thick slices. Head sections and slices were immersed in 10% formaldehyde for preservation and then stored in a cooling chamber (5 °C) at the Department of Anatomy and Embryology, Facultad de Veterinaria, Universidad de Murcia, Spain.

Fourteen fetuses were preserved by immersion in formaldehyde (10%) and two fetuses were fixed with embalming solution injected into the umbilical arteries and veins. In a striped dolphin (*Stenella coeruleoalba*) (scomu5), the external jugular vein and the left and right atria were injected with embalming solution using an electrical pump. After 48 h, the arteries and veins were injected with red and blue latex, respectively. A deep head dissection of one ill newborn (scoce1) was made to observe abnormal anatomy of the vestibule and nasal cavity and an adult specimen was dissected to observe the normal anatomy of nasal cavity (Table 1). All specimens used were stored in a cooling and freezer chambers at the Department of Anatomy and Embryology, Facultad de Veterinaria, Universidad de Murcia, Spain.

2.7. Histological Analysis

The mucosa of the nasal cavity (vestibule and respiratory and olfactory parts) was histologically analyzed in one adult dolphin specimen. Elongated rectangles of nasal mucosa were removed at different levels of the nasal cavity. Samples were oriented perpendicular to the paraffin block base and then processed using a special saw. Paraffin blocs were cut to obtain slices. Routine histological processing was carried out and sections were stained with Haemotoxylin and Eosin. Samples were then photographed with a computed light microscope (Axioskop 40, Zeiss, Jena, Germany) with an incorporated Insight 2 Axiocam 105 color camera. Histological sections were stored at the Department of Anatomy and Embryology, Facultad de Veterinaria, Universidad de Murcia, Spain.

3. Results

3.1. Endoscopic Study of the External Nose and Nasal Cavity

The endoscopic figures are described by columns, observing in the left column the external nose closed in its natural position to avoid the entrance of amniotic liquid into the lungs.

3.1.1. External Nose

In the youngest fetus (dde1), a common dolphin *Delphinus delphis*, a very small protuberance was observed between the forehead and the melon. It was seen between a semi-circular line (U-shaped) (rima naris), slightly concave right to angulus naris, closing it hermetically. We observed two lips, the upper (caudal) close to the forehead and the lower (rostral) close to the primordial melon. The lower lip was a little bit prominent compared to the upper (Figure 1A). In the next fetus (dde2), a common dolphin (*Delphinus delphis*), both lips are prominent, the upper due to the fact that it underlies the vestibular fold, muscle fibers and the nasal bones mesenchyme and the lower lip, due to the proximity to the developing melon. The lips line (rima) could be now observed as slightly convex (Figure 1C). In the third common dolphin (*Delphinus delphis*) fetus (dde3) (4.5 months gestation) the epidermis of the external nose differs from the forehead epidermis, with a different colour (light pink) and with a clear separation edge of the upper lip and the forehead skin. This lip, now prominent, was clearly divided vertically by a supranasal groove. The lower lip was undivided and elevated right to the contact with the melon, which has a similar tone to that of the nose. The rima naris was slightly convex (Figure 1E). In the next fetus (scop1), a striped dolphin (*Stenella coeruleoalba*), the external nose maintained its change in colour with respect to adjacent areas. The rima naris shows a slight concavity in both angulus. The upper lip is prominent, compared to the common dolphin (*Delphinus delphis*) fetuses with a light groove dividing them in two while the lower lip descends rostrally, forming a valley, extending rostrally to the melon prominence (Figure 1G).

In the pilot whale (*Globocephala melas*) fetus (gma1), the external nose presented a whitish pigmentation extending to the melon, while the pigmentation of the rest of the head is dark grey. The rima naris is slightly convex and differs from the prominent upper lip divided by the supranasal groove. The rostral lip is visible (Figure 2A). The three next fetuses of the common dolphin (*Delphinus delphis*), more developed than the pilot whale (*Globocephala melas*) fetus, showed the same features as the common dolphin (*Delphinus delphis*) fetus (dde3), with a separation area holding the nose and melon, though the depigmentation can only be observed at the upper lip. The convexity in the rima naris is very slight (Figure 2E,I,M).

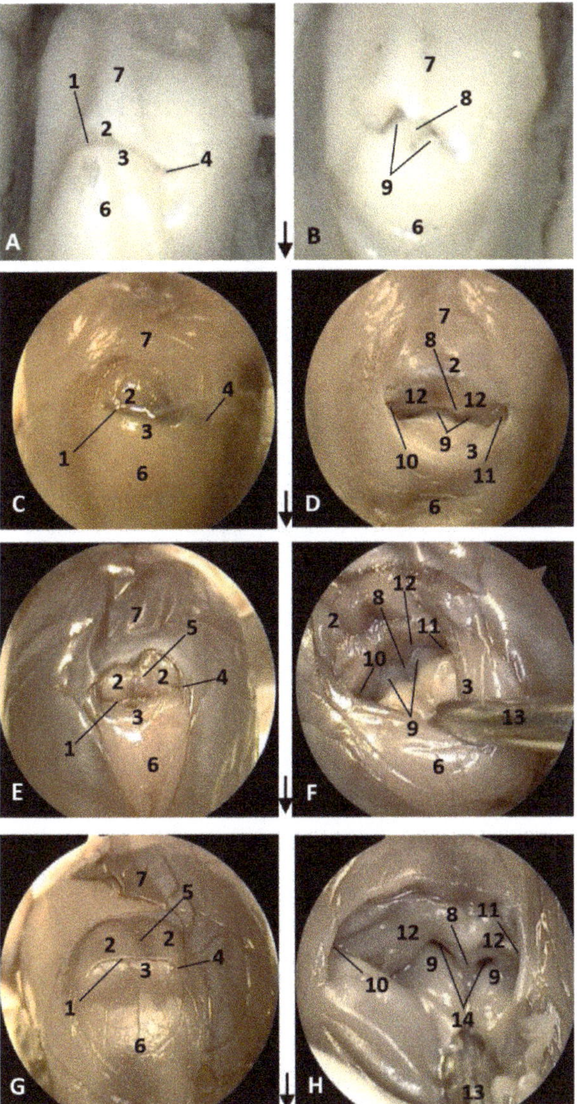

Figure 1. Endoscopic images of the external nose and nasal cavity at the level of nasal vestibule. Images are oriented so that the left side of the head is to the right of the image and rostral is at the bottom (**arrow**). (**A**,**B**) 1.5 months, dde1; (**C**,**D**) 3.5 months, dde2; (**E**,**F**) 4 months, dde3; (**G**,**H**) 4.5 months, scop1. 1, Rima naris; 2, Upper lip; 3, Lower lip; 4, Angulus naris; 5, Nasal groove; 6, Melon; 7, Forehead; 8, Nasal septum: membranous part; 9, Nasal plugs; 10. Vestibule: right diverticulum (vestibular sac); 11, Vestibule: left diverticulum (vestibular sac); 12, Vestibular folds; 13, Delicate hook; 14, Nasal cavity: respiratory part.

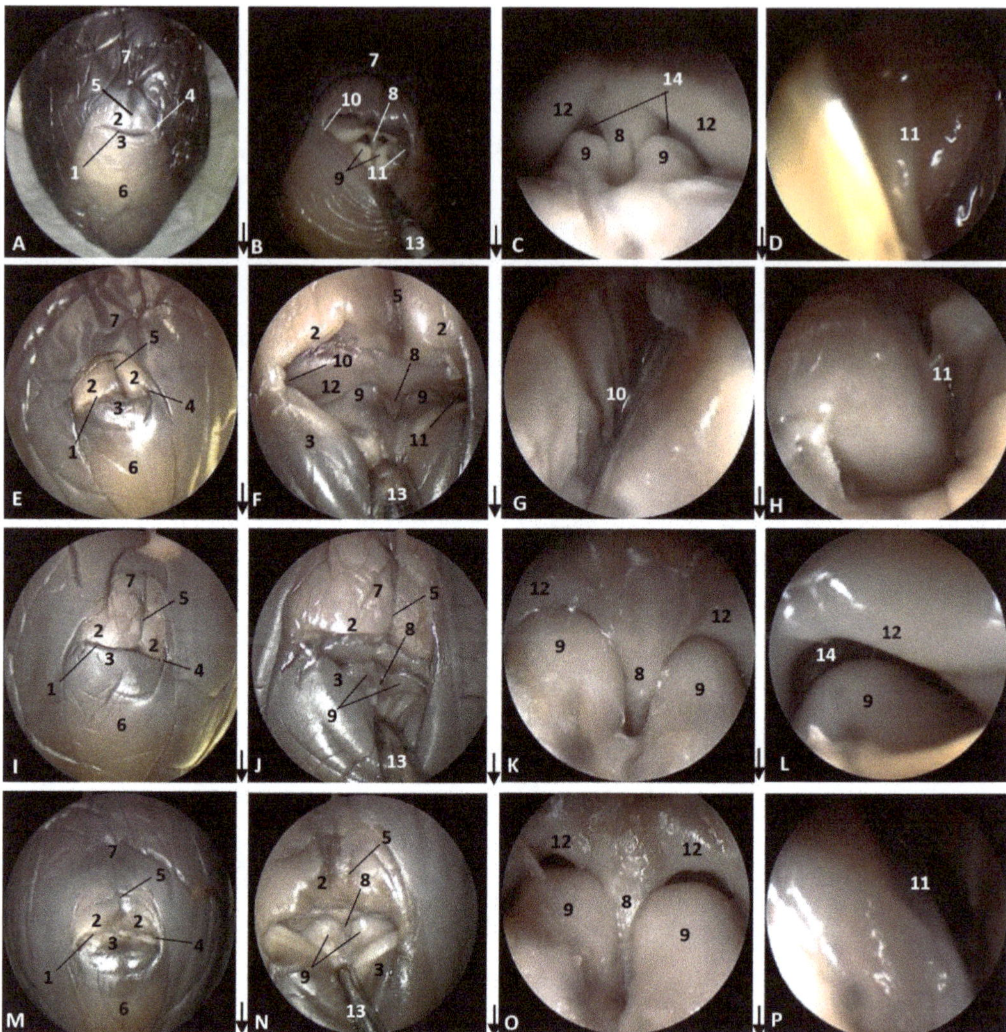

Figure 2. Endoscopic images of the external nose and nasal cavity at level of the vestibule. Images are oriented so that the left side of the head is to the right of the image and rostral is at the bottom (**arrow**). Detail of monkey or phonic lips from now on (**K**,**L**,**O**). (**A–D**) 5 months, gma1; (**E–H**) 5.5 months, dde4; (**I–L**) 6 months, dde8; (**M–P**) 7 months, dde9. 1, Rima naris; 2, Upper lip; 3, Lower lip; 4, Angulus naris; 5, Nasal groove; 6, Melon; 7, Forehead; 8, Nasal septum: membranous part; 9, Nasal plugs; 10. Vestibule: right diverticulum; 11, Vestibule: left diverticulum; 12, Vestibular folds; 13, Delicate hook; 14, Nasal cavity: respiratory part.

The four next common dolphin (*Delphinus delphis*) fetuses, from 7 to 9 months of gestation, mantain the external features described in the last three, though the midline rima naris forms a triangle with respect to the supranasal groove, and it appears as a horizontal line (Figure 3A,E,I). In the last fetus a light concavity is seen (Figure 3M).

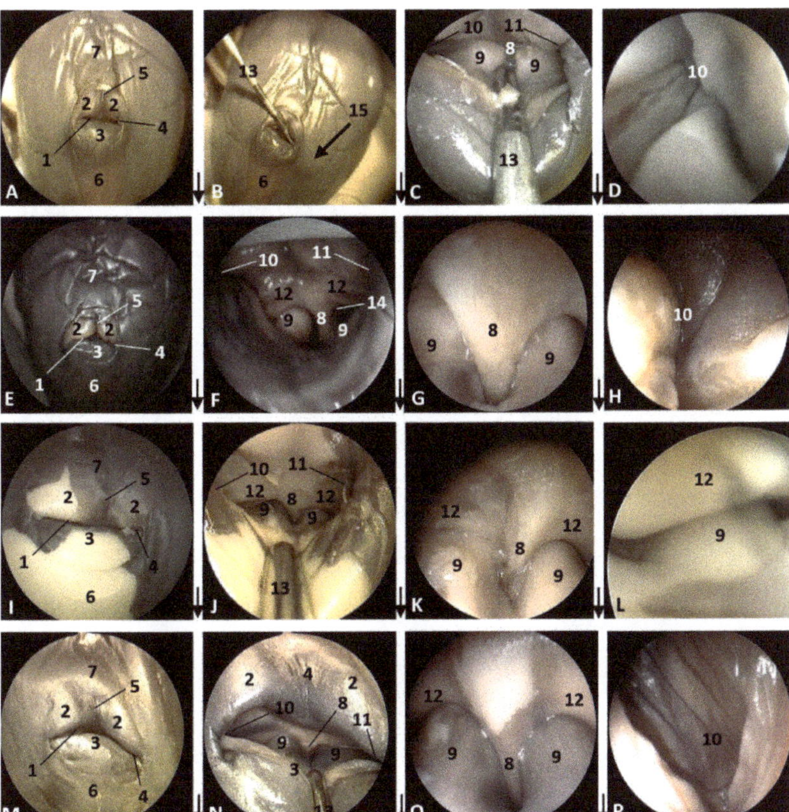

Figure 3. Endoscopic images of the external nose and nasal cavity at the level of vestibule. Images are oriented so that the left side of the head is to the right of the image and rostral is at the bottom (**arrow**). (**A–D**) 7.5 months, dde10; (**E–H**) 8 months, dde11; (**I–L**) 8.5 months, dde12; (**M–P**) 9 months, dde13 1, Rima naris; 2, Upper lip; 3, Lower lip; 4, Angulus naris; 5, Nasal groove; 6, Melon; 7, Forehead; 8, Nasal septum: membranous part; 9, Nasal plugs; 10. Vestibule: right diverticulum; 11, Vestibule: left diverticulum; 12, Vestibular folds; 13, Delicate hook; 14, Nasal cavity: respiratory part; 15, Diverticulum extension.

In the last common dolphin (*Delphinus delphis*) fetus, the depigmentation of the nose and upper lip remains and misses the delimitation of the nose and melon with respect to the rest of the head. The rima naris keeps the concavity towards the angulus naris, commissures or nose angles (Figure 4A). In the next specimen, a newborn striped dolphin (*Stenella coeruleoalba*), showed a hyperkeratosis between the two lips and the right part of upper lip preventing the access of the endoscope (Figure 4E). In the last specimen studied, a juvenile striped dolphin (*Stenella coeruleoalba*), the nose was tightly closed and uniform pigmentation was seen. The upper lip is divided in two by the supranasal groove. The rima naris was curved towards the commissures or angulus naris and presented many folds sinking to the rima. The lower lip is undivided and prominent remaining close to the melon (Figure 4I). Between both nasal lips, there are striations and a double internal fold hermetically sealing the lips (Figure 4J,K,L).

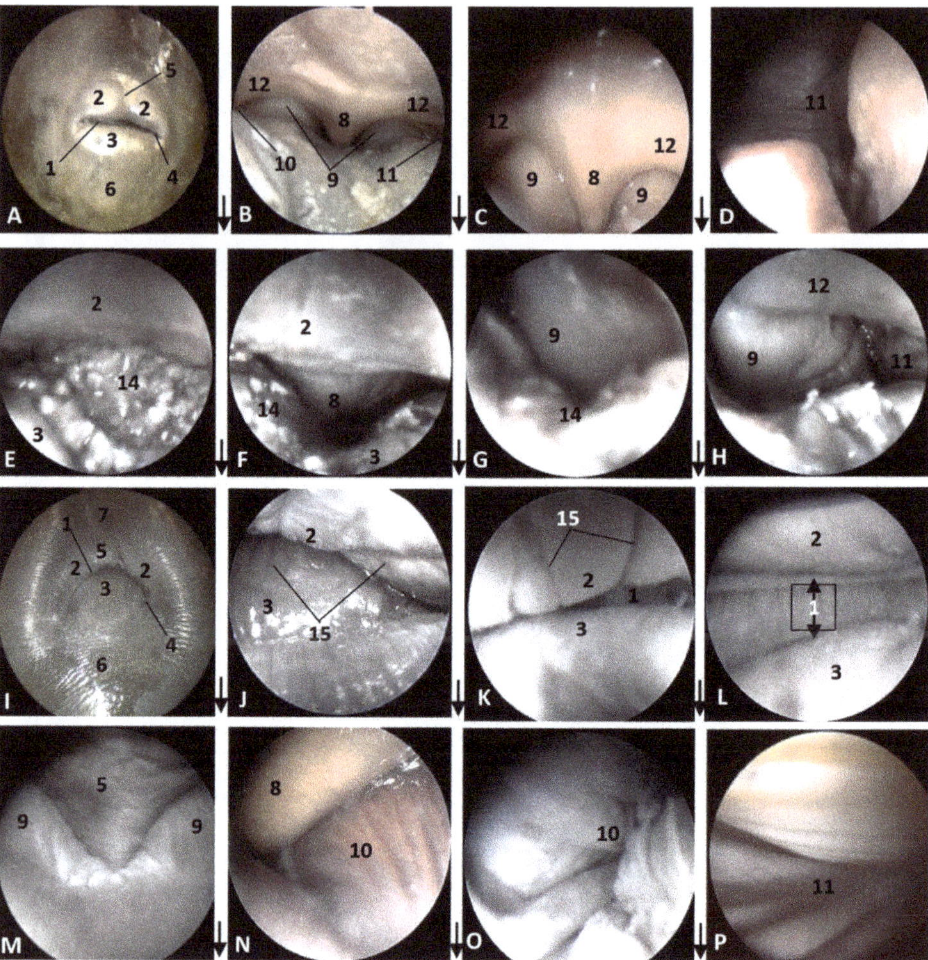

Figure 4. Endoscopic images of the external nose and nasal cavity at the level of vestibule. Images are oriented so that the left side of the head is to the right of the image and rostral is at the bottom (**arrow**). (**A–D**) 10 months, dde14; (**E–H**) newborn, scoce1; (**I–P**) juvenile, scomu4. 1, Rima naris; 2, Upper lip; 3, Lower lip; 4, Angulus naris; 5, Nasal groove; 6, Melon; 7, Forehead; 8, Nasal septum: membranous part; 9, Nasal plugs; 10. Vestibule: right diverticulum; 11, Vestibule: left diverticulum; 12, Vestibular folds; 13, Delicate hook, 14. Hyperkeratosis; 15, Striation marks.

3.1.2. Nasal Cavity: Vestibule

In the second, third and fourth columns the external nose was opened allowing us to see the first part of the nasal cavity, the vestibule (Figures 1–4).

After caudal retraction of the skin of the forehead region of the least developed common dolphin (*Delphinus delphis*) fetus (dde1), the nose was opened to expose the vestibule. Only the membranous part of the nasal septum was seen as were the two nasal plugs. The mucosa is of a similar colour of that of the epidermis, a little clearer at the plugs (Figure 1B). In the next specimen (dde2), we could also see the vestibular folds dorsal to the nasal plugs and laterally the entry to the nasal diverticula (Figure 1D). In the third common dolphin (*Delphinus delphis*) fetus (dde3), the lower lip was pulled rostrally in order to observe the vestibule. The nasal plugs could be seen between the nasal septum. The vestibular folds along with the membranous part of the nasal septum stick to the plugs, closing the meatus tightly. The nasal diverticula (vestibular sacs) (Figure 1F) of the striped dolphin (*Stenella coeruleoalba*) fetus (scop1), already more developed, showed a mucosa with a whitish tone which was uniform throughout the entire vestibule. Also, the meatus was opened between the 'monkey lips' (both the nasal plugs and the vestibular folds) (Figure 1H). In the pilot whale (*Globocephala melas*) fetus (gma1), the mucosal colour of the vestibule remains the same (Figure 2B,C), though it differs from the greyish tone inside the lips (Figure 2B). The mucosa of the nasal diverticulum remained dark and had folds (Figure 2D). In the three next common dolphin (*Delphinus delphis*) fetuses, the vestibular mucosa goes from pink, to grey to white (Figure 2F,J,N), shown in detail (Figure 2K,L,O). The mucosa of the diverticula has greyish folds arranged in a dorsoventral direction (Figure 2G,H,P). The next group of common dolphin (*Delphinus delphis*) fetuses had a more advanced gestation time. A clamp was used to mark the internal extension of one diverticulum (Figure 3B). We subsequently observed prominent nasal plugs, the nasal septum and well-defined folds and a greyish mucosa, different from the earlier fetal periods (Figure 3C,F,J,N,G,L,O and Figure 4B,C). The mucosa of both left and right diverticula was growing in anfractuosity and the longitudinal folds more abundant (Figure 3D,H,P and Figure 4D).

The newborn (scoce1) was diagnosed with epithelial inflammation with local hyperkeratosis probably caused by a congenital alteration. This obstructed the right nasal cavity (Figure 4E). This specimen had a thickened membranous part of the nasal septum so the nasal plugs are barely visible (Figure 4F). Only the left nasal plug and the dilated mucosa of left diverticulum were visible (Figure 4G,H).

In the juvenile striped dolphin (*Stenella coeruleoalba*), the mucosa of vestibule was wrinkled with striations and was tightly adhered to the plug, septum and vestibular folds (Figure 4M,N). The mucosa of the nasal diverticula presented numerous longitudinal folds (Figure 4O,P).

3.1.3. Nasal Cavity: Respiratory and Olfactory Part

In this part of the nasal cavity, nasal cavities (left and right) were analyzed from superficial to deep regions after passing the nasal plugs. Each row corresponds with each studied specimen, with the first row corresponding to the specimen with the earliest fetal development. This part of the study started analyzing more developed specimens in which the nasal cavity diameter allowed easy passage of the endoscope. In the nasal cavity, we could distinguish four walls: the caudal, once the endoscope passed the nasal plug; the rostral wall, ventral to the nasal plug and containing the incisive recess (premaxillary sac). The lateral wall is the external wall of both cavities and finally, the median wall, built by the vomer bone, separating both choanae.

The study starts in a common dolphin (*Delphinus delphis*) fetus of approximately 5.5 months of gestation. We had observed, caudal to the nasal plug, how the mucosa of longitudinal folds was placed vertically towards the floor of the right nasal cavity (Figure 5A). On turning the endoscope rostrally, we observed the incisive recess (Figure 5B). The mucosa of the deepest part of the recess in the left nasal cavity contained small vesicles. Also, at the bottom of the nasal cavity we could see the choanae and the bony portion of the nasal septum (Figure 5C). Unlike the vestibule, the mucosa of the respiratory and olfactory part shows a pinkish colour in every specimen studied. The presence of vesicles in the mucosa and longitudinal folds was clearly seen (Figure 5D–L). Small mucosal fossae close to the choanae were observed mainly in the common dolphin (*Delphinus delphis*) fetus 8.5 months of gestation (Figure 5L).

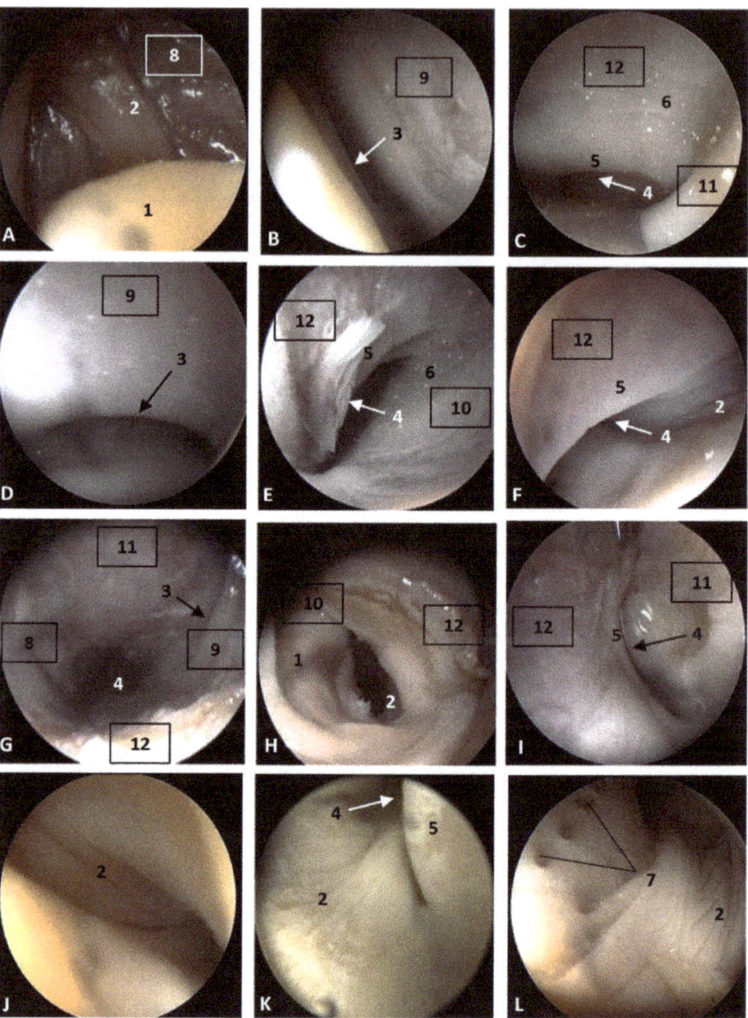

Figure 5. Endoscopic images of the nasal cavity at the level of respiratory and olfactory parts. Arrows show the entry to the incisive recess (3) and to the nasopharynx (4) (**A**,**B**,**H**) Right nasal cavity, (**C**–**G**,**I**–**L**) Left nasal cavity, (**A**–**C**) 5.5 months, dde4; (**D**–**F**) 6 months, dde7. (**G**–**I**) 7 months, dde9; (**J**–**L**) Detailed images. 8.5 months, dde12. Walls indicate orientation. Vertical view. 1. Nasal plug; 2, Longitudinal folds; 3, Incisive recess (premaxillary sac); 4, Choanae; 5, Nasal septum: bony part; 6, Small vesicles; 7, Small fossae; 8, Caudal wall; 9, Rostral wall; 10, Right lateral wall; 11, Left lateral wall; 12, Medial wall.

In the last two common dolphin (*Delphinus delphis*) fetuses closest to birth, we highlight that the incisive recess is more visible and widened and the mucosa had few folds, a flat aspect and a pinkish colour (Figure 6A–F). In the newborn (scoce1), we could only study the left respiratory part of the nasal cavity since the right one was blocked and was impossible to introduce the endoscope due to hyperkeratosis. The mucosa was greyish in colour and there was inflammation of the nasal mucosa folds. The incisive recess could be observed (Figure 6G–I).

In the last specimen studied with the endoscope, the striped juvenile dolphin, we observed the pinkish mucosa and thick folds, the incisive recess and the choanae (Figure 6J–L). Also, small fossae could be seen close to the choanae (Figure 6L).

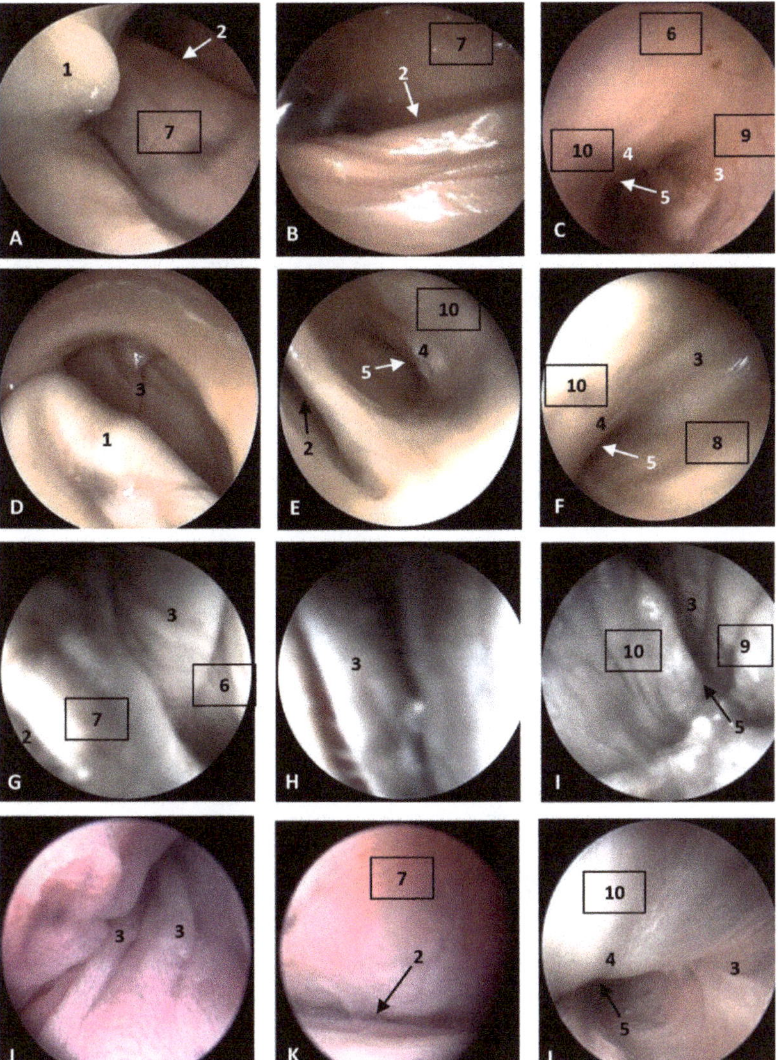

Figure 6. Endoscopic images of the nasal cavity at the level of respiratory and olfactory parts. Arrows show the entry to the incisive recess (2) and to the nasopharynx (5) 9 months, dde13. (**A–C**,**L**) Left nasal cavity. (**D–K**) Right nasal cavity. (**A–C**) 9 months, dde13; (**D–F**) 10 months, dde14; (**G–I**) Hypertrophied longitudinal folds. Newborn, scoce1; (**J–L**) juvenile, scomu4. Walls indicate orientation. Vertical view. 1, Nasal plug; 2, Incisive recess; 3, Nasal folds; 4, Nasal septum: bony part. 5, Choanae; 6, Caudal wall; 7, Rostral wall; 8, Right lateral wall; 9, Left lateral wall; 10, Median wall.

3.2. MRI Study of the External Nose and Nasal Cavity

MRI were first used in the sagittal plane in order to have an overall vision of the whole nasal cavity and afterwards in three coronal planes at the level of vestibule, the incisive recess and finally at the choanal region.

The least developed specimen of common dolphin (*Delphinus delphis*) dde1, corresponds to first stage of fetal development. The sagittal MRI allowed us to identify the external nose with two lips and the rima naris. The melon is beginning to develop and was observed hyperintense in both sagittal images. The nasal cavity from the vestibule to the choanae is shown as a mass of mesenchymal tissue slightly hypointense in T1 and hyperintense in T2 (Figure 7A,B). The coronal MRI at the vestibule level allows us to identify the nasal septum slightly hypointense at level of the vestibule (L1) and hyperintense at the respiratory part (L2). Nasal cavities were observed always hypointense (Figure 7C,D).

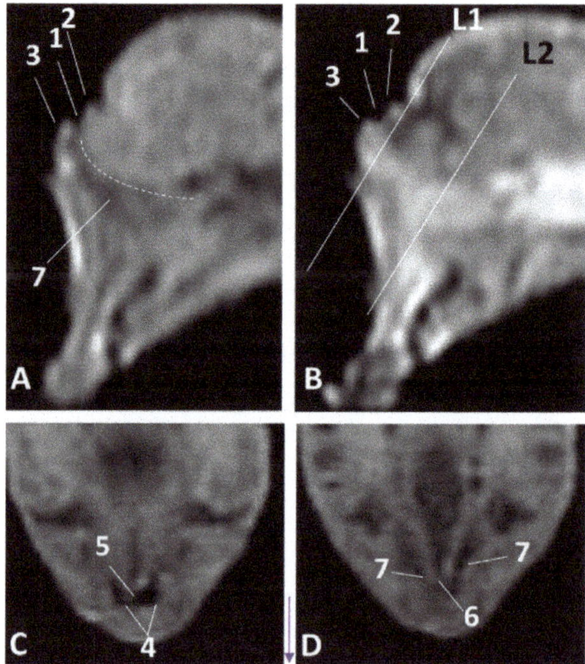

Figure 7. Magnetic Resonance Imaging (MRI) of the external nose and nasal cavity. MR sagittal images (**A**,**B**) are oriented so that the rostral is to the left and the caudal to the top. MR coronal images (**C**,**D**) are oriented so that the rostral is to the bottom and the caudal to the top. (**A**) T1-weighted Spin echo (SE) sagittal plane, (**D**) T2-weighted Fast Recovery Fast Spin Echo (FrFSE) sagittal plane. (**C**,**D**) T1-weighted SE coronal plane. (**C**) Level 1. (**D**) Level 2. 1.5 months, dde1. 1, Rima naris; 2, Upper lip; 3, Lower lip; 4, Nasal cavity: vestibule; 5, Nasal septum: membranous part; 6, Nasal septum: bony part; 7. Nasal cavity: respiratory part.

In the fetus of four months' gestation, dde3, the sagittal sections clearly showed the external nose. The nasal cavity vestibule section does not show the developing cavities, although slightly hypointense (diverticula) were noted. The nasal plug could be seen as slightly hyper/hypointense in T1 and T2, and the respiratory part of the nasal cavity was hypointense in T1 and T2. The main bones which form the walls are more distinct within the mesenchyme especially the presphenoid, but the ethmoid bone was not clearly identified (Figure 2A,B). In the coronal sections we could see the nasal plugs hypointense in

T1 and slightly hyperintense in T2. The nasal septum was observed with different intensity in T1 and T2 as well as the vestibule and respiratory parts of the nasal cavity (Figure 8C,H).

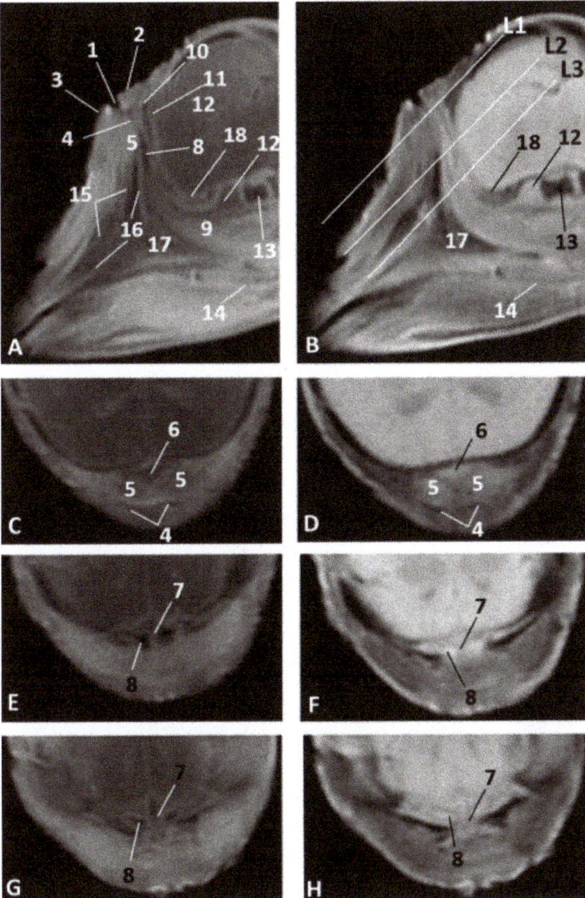

Figure 8. MRI of the external nose and nasal cavity. MR sagittal images (**A**,**B**) are oriented so that the rostral is to the left and the dorsal to the top. MR coronal images (**C**,**H**) are oriented so that the rostral is to the bottom and the caudal to the top. (**A**) T1-weighted SE sagittal plane, (**B**) T2-weighted FrFSE sagittal plane. (**C**,**E**,**G**) T1-weighted SE coronal plane (**D**,**F**,**H**) T2-weighted FrFSe coronal plane. (**C**,**D**) Level 1, (**E**,**F**) Level 2, (**G**,**H**) Level 3. 4 months, dde3. 1. Rima naris; 2, Upper lip; 3, Lower lip; 4, Nasal cavity: vestibule (left and right diverticula); 5, Nasal plugs; 6, Nasal septum: membranous part; 7, Nasal septum: bony part; 8, Nasal cavity: respiratory part; 9, Choanae: 10, Nasal bone; 11, Frontal bone; 12, Presphenoid bone; 13, Basisphenoid bone; 14, Pterygoid bone; 15, Incisive bone (premaxilla); 16, Maxillary bone; 17, Vomer bone; 18, Ethmoidal mesenchyme.

In a similar development stage (striped dolphin *Stenella coeruleoalba* fetus, scop1), we could better identify the lumen of the respiratory part of the nasal cavity, but the bones were less defined (Figure 9A–H).

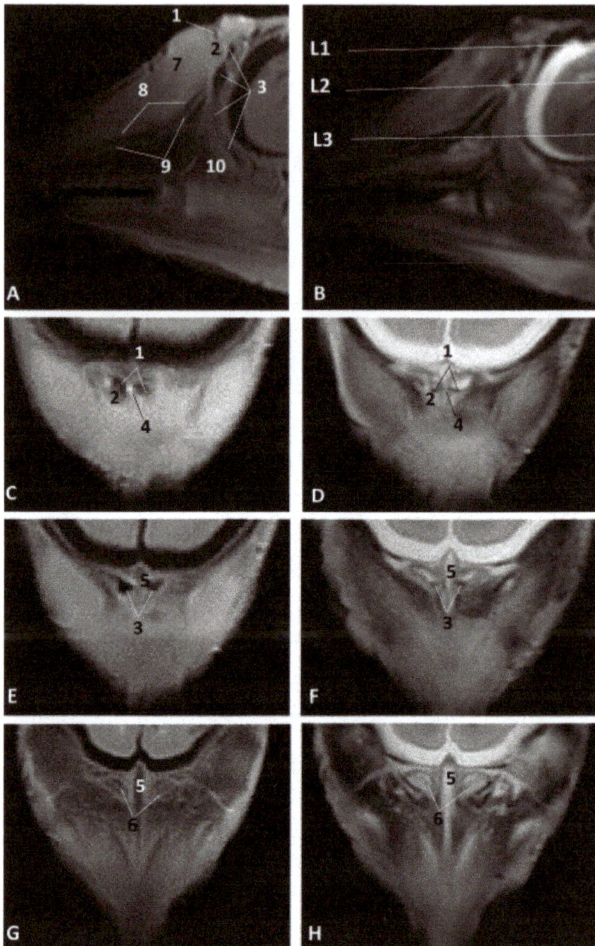

Figure 9. Images of the external nose and nasal cavity. MR sagittal images (**A**,**B**) are oriented so that rostral is to the left and dorsal to the top. MRI coronal images (**C**–**H**) are oriented so that the rostral is to the bottom and the caudal to the top. (**A**) T1-weighted SE sagittal plane. (**B**) T2-weighted FrFSE sagittal plane. (**C**,**E**,**G**) T1-weighted SE coronal plane, (**D**,**F**,**H**) T2-weighted FrFSE coronal plane. (**C**,**D**) Level 1. (**E**,**F**) Level 2. (**G**,**H**) Level 3. 4.5 months, scop1. 1, Nasal cavity: vestibule; 2, Nasal plug; 3, Nasal cavity: respiratory part; 4, Nasal septum: membranous part; 5, Nasal septum: bony part; 6, Choanae; 7, Melon; 8, Incisive bone; 9, Maxillary bone; 10, Vomer bone.

In the pilot whale (*Globocephala melas*) fetus, gma1, we found better definition of the nasal plugs, the vestibule and respiratory parts of the nasal cavity in the coronal sections and most of the bony nasal cavity. At this stage, the perpendicular lamina of the ethmoid bone was seen clearly using MRI (Figure 10A–D).

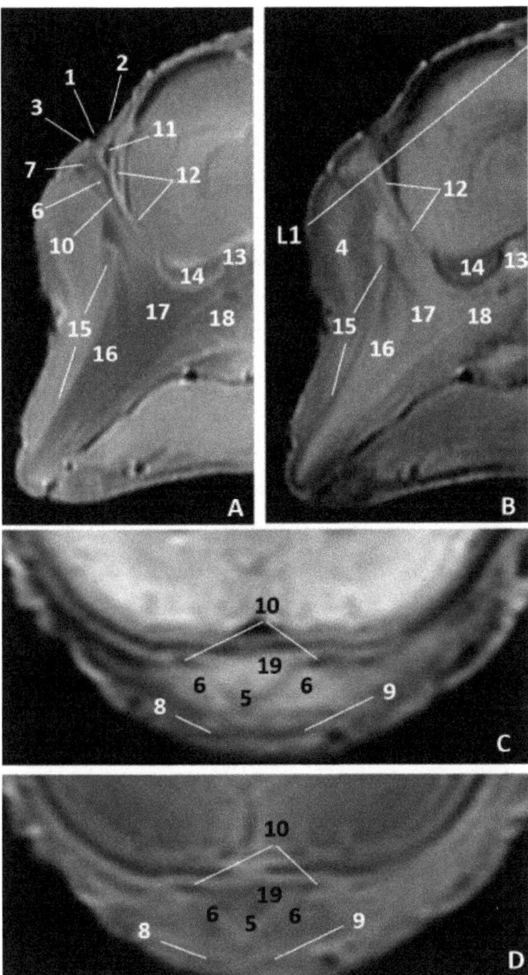

Figure 10. Images of the external nose and nasal cavity. MR sagittal images (**A**,**B**) are oriented so that the rostral is to the left and the dorsal to the top. MR coronal images (**C**,**D**) are oriented so that rostral is to the bottom and caudal to the top. (**A**) T1-weighted SE sagittal plane, (**B**) T2-weighted FrFSE sagittal plane. (**C**) T1-weighted SE coronal plane. (**D**) T2-weighted FrFSE coronal plane. (**C**,**D**) Level 1. 5 months, gma1. 1, Rima naris; 2, Upper lip; 3, Lower lip; 4, Melon; 5, Nasal septum: membranous part; 6, Nasal plugs; 7, Nasal cavity: vestibule; 8, Vestibule: right diverticulum; 9, Vestibule: left diverticulum; 10, Nasal cavity: respiratory part; 11, Nasal bone; 12, Frontal bone; 13, Presphenoid bone; 14, Ethmoid bone: perpendicular lamina; 15, Incisive bone; 16, Maxillary bone; 17, Vomer bone, 18, Pterygoid bone; 19, Vestibular folds.

In the common dolphin (*Delphinus delphis*) fetus of six months' gestation, dde7, the melon was observed as more developed and we identified the incisive recess using MRI, as well as the respiratory part of the nasal cavity, as a hypointense cavity in T1 and hyperintense in T2 (Figure 11A,B). In the three anatomical sections we also identified the nasal plugs, the nasal septum (membranous and bony parts) and the mesorostral cartilage (Figure 11C–H).

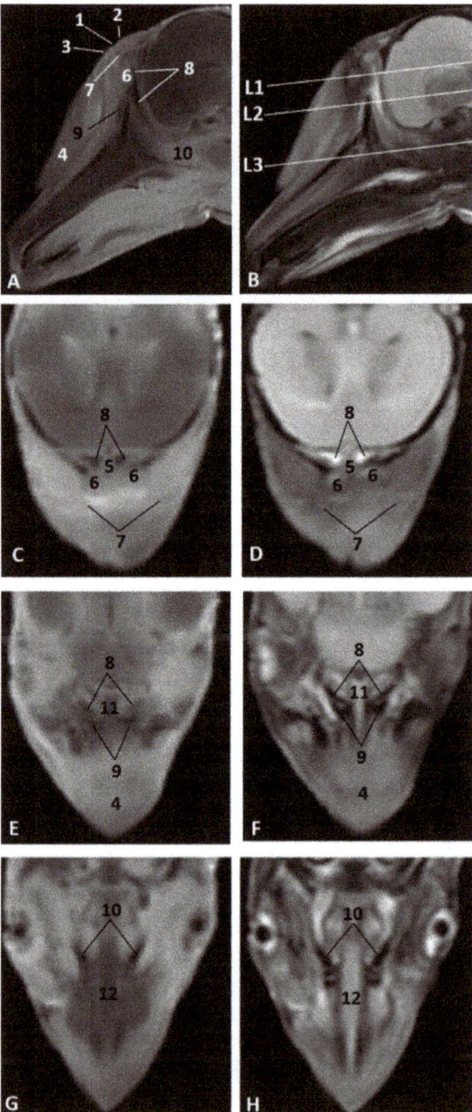

Figure 11. Images of the external nose and nasal cavity. MR sagittal images (**A**,**B**) are oriented so that the rostral is to the left and the dorsal to the top. MR coronal images (**C**,**H**) are oriented so that the rostral is to the bottom and caudal to the top. (**A**) T1-weighted SE sagittal plane, (**B**) T2-weighted FrFSE sagittal plane. (**C**,**E**,**G**) T1-weighted SE coronal plane. (**D**,**F**,**H**) T2-weighted FrFSE coronal plane. (**C**,**D**) Level 1. (**E**,**F**) Level 2. (**G**,**H**) Level 3. 6 months, dde7. 1, Rima naris; 2, Upper lip; 3, Lower lip; 4, Melon; 5, Nasal septum: membranous part; 6, Nasal plugs; 7, Nasal cavity: vestibule; 8, Nasal cavity: respiratory part; 9, Nasal cavity: incisive recess; 10, Choanae; 11, Nasal septum: bony part; 12, Mesorostral cartilage.

In an advanced stage of fetal development all the anatomical structures of the nasal cavity could be seen perfectly defined (Figure 12A–H). We identified the perpendicular lamina of the ethmoid bone in the common dolphin (*Delphinus delphis*) fetus of 7.5 month's gestation (dde10) as hyperintense in T1 and hypointense in T2. Between the frontal and ethmoid bones mesenchyme still is observed (not bony contact) (Figure 12A,B).

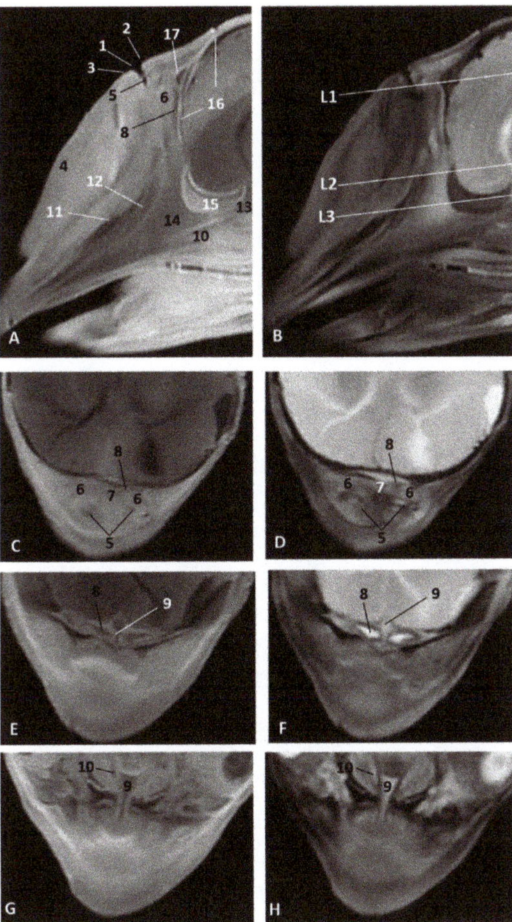

Figure 12. Images of the external nose and nasal cavity. MR sagittal images (**A**,**B**) are oriented so that the rostral is to the left and the dorsal to the top. MR coronal images (**C**–**H**) are oriented so that the rostral is to the bottom and caudal to the top. (**A**) T1-weighted SE sagittal plane, (**B**) T2-weighted FrFSE sagittal plane. (**C**,**E**,**G**) T1-weighted SE coronal plane. (**D**,**F**,**H**) T2-weighted FrFSE coronal plane. (**C**,**D**) Level 1. (**E**,**F**) Level 2. (**G**,**H**) Level 3. 7.5 months, dde10. 1, Rima naris; 2, Upper lip; 3, Lower lip; 4, Melon; 5, Nasal cavity: vestibule; 6, Nasal plugs; 7, Nasal septum: membranous part; 8, Nasal cavity: respiratory part; 9, Nasal septum: bony part; 10, Choanae; 11, Incisive bone; 12, Maxillary bone; 13, Presphenoid bone; 14. Vomer; 15, Ethmoid bone: perpendicular lamina; 16, Frontal bone; 17, Nasal bone.

In the common dolphin (*Delphinus delphis*) fetus, ten months old, we could see the lumen of the vestibule and of the nasal (vestibular sacs) and accessory diverticula (nasofrontal sacs), all hypointense in T1 and T2. Also, the nasal plugs, the vestibular folds and their associated muscles are moderately hyperintense in T1 and hypointense in T2 (Figure 13A–D). The respiratory part and incisive recess are hypointense in T1 and T2. The incisive recess is surrounded rostrally by the nasal plug muscles and caudally by the incisive and maxillary bones (Figure 13A–D). This recess could also be observed in the coronal slices extending to the choanae (Figure 13E–H). The majority of the bones forming the nasal cavity have been identified (Figure 13A,B,E,H).

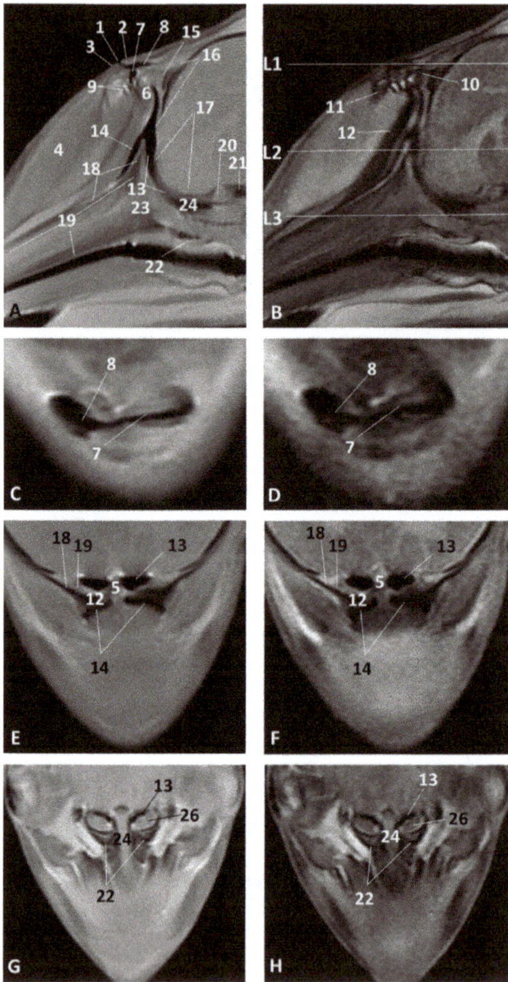

Figure 13. Images of the external nose and nasal cavity. MR sagittal images (**A**,**B**) are oriented so that the rostral is to the left and the dorsal to the top. MR coronal images (**C**,**H**) are oriented so that the rostral is to the bottom and caudal to the top. (**A**) T1-weighted SE sagittal plane, (**B**) T2-weighted FrFSE sagittal plane. (**C**,**E**,**G**) T1-weighted SE coronal plane. (**D**,**F**,**H**) T2-weighted FrFSE coronal plane. (**C**,**D**) Level 1. (**E**,**F**) Level 2. (**G**,**H**) Level 3. 10 months, dde14. 1, Rima naris; 2, Upper lip; 3, Lower lip; 4, Melon; 5, Nasal septum: bony part; 6, Nasal plug; 7, Nasal cavity: vestibule; 8, Vestibule: nasal diverticulum; 9, Vestibule: nasal accessory diverticulum (nasofrontal sac); 10, Caudal vestibular muscles; 11, Rostral vestibular muscles; 12, Nasal plug muscles; 13, Nasal cavity: respiratory part; 14, Nasal cavity: incisive recess; 15, Nasal bone; 16, Frontal bone; 17, Ethmoid bone: perpendicular lamina; 18, Incisive bone; 19, Maxillary bone; 20, Presphenoid bone; 21, Basisphenoid bone; 22, Pterygoid bone; 23, Vomer bone; 24, Mesorostral cartilage; 25, Choanae; 26, Pharyngeal muscles.

3.3. Computed Tomography and 3D Reconstruction of Bony Nasal Cavity

The 3D reconstruction of the skull bones has been performed in order to show nasal cavity development from early fetal stages where the mesenchymal tissue is abundant and the bones of the skull are being formed by intramembranous and endochondral ossification.

In the less developed second fetus (dde2), we could see the main bones forming the rostral part of the bony nasal cavity such as the incisive and maxillary bones and the vomer groove dividing both cavities (Figure 14A,B). The presphenoid (body) and the ethmoid (body) bones are in close contact from an early developmental stage, and could only be distinguished as separate using MRI (Figure 10). In a dorsal view, we observed the bones of the braincase but not the bones separating the nasal and the cranial cavities (Figure 14C). The choanae were observed formed by the palatine and pterygoid bones and divided ventrally by the vomer crest (Figure 14D).

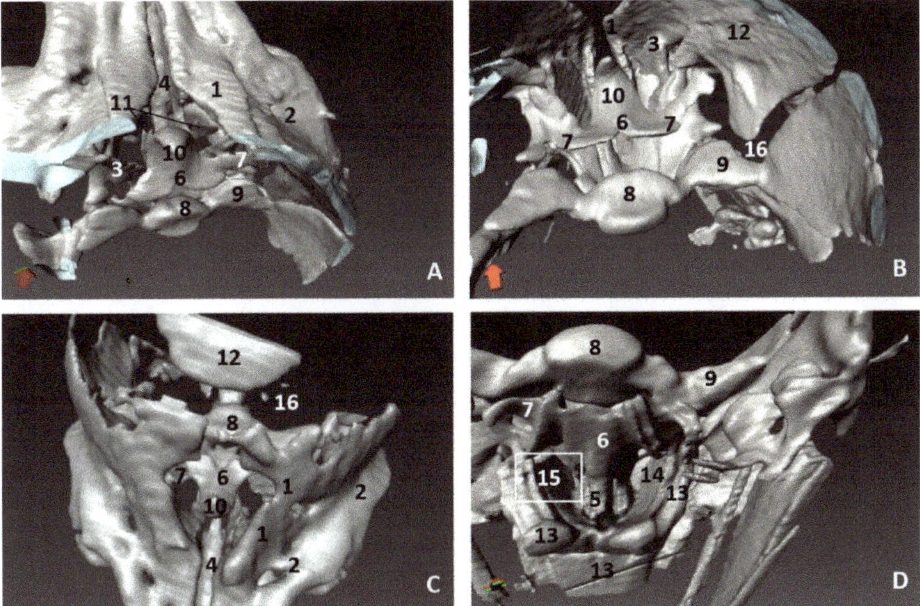

Figure 14. 3D reconstruction images of bony nasal cavity using AMIRA and PET/SPEC/CT. 3.5 months, dde2. (**A**) *Dorsal view.* (**B**) *Caudal view.* (**C**) *Dorsal view.* (**D**) *Ventral view.* 1, Incisive bone; 2, Maxillary bone; 3, Nasal face of maxillary bone; 4, Vomer bone: groove; 5, Vomer bone: ventral crest; 6, Presphenoid bone: body; 7, Presphenoid bone: wings; 8, Basisphenoid bone: body; 9, Basisphenoid bone: wings; 10, Ethmoid bone; 11, Maxillary bone (nasal face): ossification nuclei; 12, Frontal bone: cerebral face; 13, Pterygoid bone; 14 Palatine bone; 15, Choanae; 16, Fontanelles.

A more developed striped dolphin (*Stenella coeruleoalba*) fetus, scop1, showed the frontal bone growing towards the incompletely-ossified ethmoid bone (Figure 15A). The rostral wall of the nasal cavity, mainly formed by the maxillary bone, is closing towards the vomer. The ethmoid bone (ethmoidal fossa) remains unossified and the bones at the base of the skull are quite separated (Figure 15B).

At five and a half months, the fetus (dde5) shows an unossified ethmoidal fossa and bony projections from the frontal bone towards the ethmoid bone were observed (Figure 16A). The choanae are clearly seen, but there are still areas lacking ossification (Figure 16B).

In the next common dolphin (*Delphinus delphis*) fetus of almost 6 months' gestation, dde6, the rostral wall of the nasal cavity is almost closed. The maxilloincisive fissure is almost formed and fontanelles are decreasing in size (Figure 17A,B).

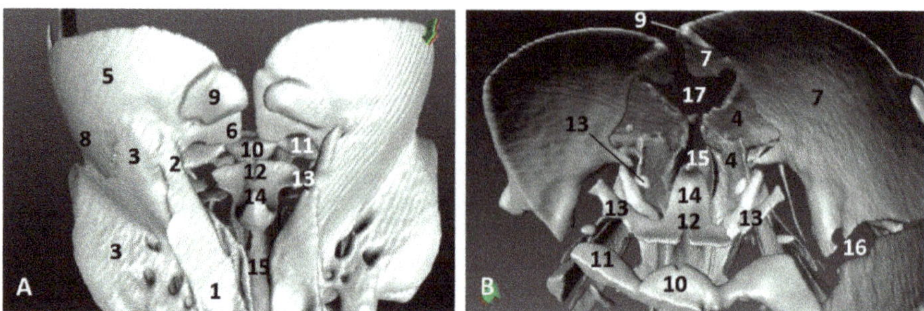

Figure 15. 3D reconstruction images of bony nasal cavity using AMIRA and PET/SPEC/CT. 4.5 months, scop1. (**A**) *Rostral view.* (**B**) *Caudal view.* 1, Incisive bone; 2, Incisive bone: nasal process; 3, Maxillary bone: external face; 4, Maxillary bone: nasal face; 5, Frontal bone: external face; 6, Frontal bone: nasal face; 7, Frontal bone: cerebral face; 8, Temporal bone; 9, Nasal bone; 10, Basisphenoid bone: body; 11, Basisphenoid bone: wings; 12, Presphenoid bone: body; 13, Presphenoid bone: wings; 14, Ethmoid bone; 15, Vomer bone: groove; 16, Fontanelles; 17, Maxilloincisive fissure (bony naris opening).

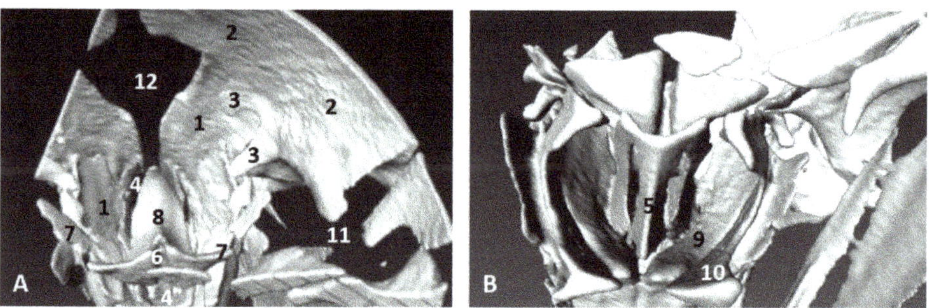

Figure 16. 3D reconstruction images of bony nasal cavity using AMIRA and PET/SPEC/CT. 5.5 months, dde5. (**A**) *Caudal view.* (**B**) *Ventrocaudal view.* 1, Maxillary bone: nasal face; 2, Frontal bone: cerebral face; 3, Frontal bone: wall projections; 4, Vomer bone: groove; 5, Vomer: ventral crest; 6, Presphenoid bone: body; 7, Presphenoid bone: wings; 8, Ethmoid bone: crista galli; 9, Palatine bone; 10, Pterygoid bone; 11, Fontanelles; 12, Maxilloincisive fissure (bony naris opening).

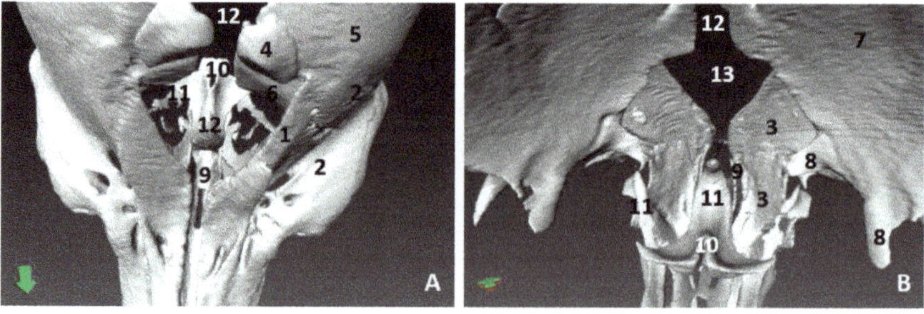

Figure 17. 3D reconstruction images of bony nasal cavity using AMIRA and PET/SPEC/CT. 5.8 months, dde6. (**A**) *Dorsal view.* (**B**) *Caudal view.* 1, Incisive bone: external face; 2, Maxillary bone: external nasal face; 3, Maxillary bone: nasal face; 4, Nasal bone; 5, Frontal bone: external face; 6, Frontal bone: nasal faces; 7, Frontal bone: cerebral face; 8, Frontal bone: wall projections; 9, Vomer bone: groove; 10, Presphenoid bone: body; 11, Presphenoid bone: wings; 12, Ethmoid bone; 12, Fontanelles; 13, Maxilloincisive fissure (bony naris opening).

In the common dolphin (*Delphinus delphis*) fetus of 9 months' gestation, dde13, the walls of the nasal cavities, near to the choanae, are closing as is the rostral wall, though there are still mesenchymal areas (Figure 18A,B).

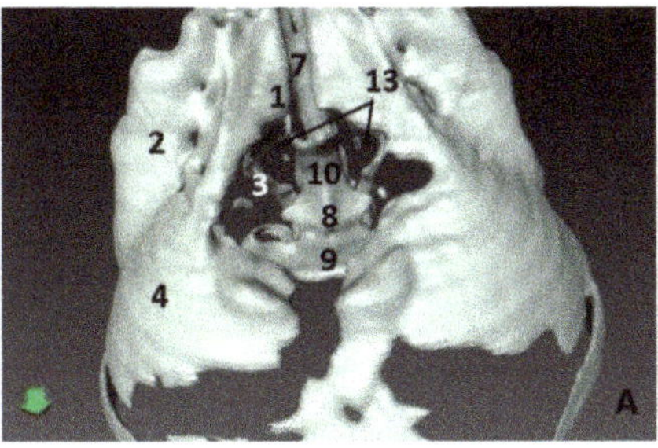

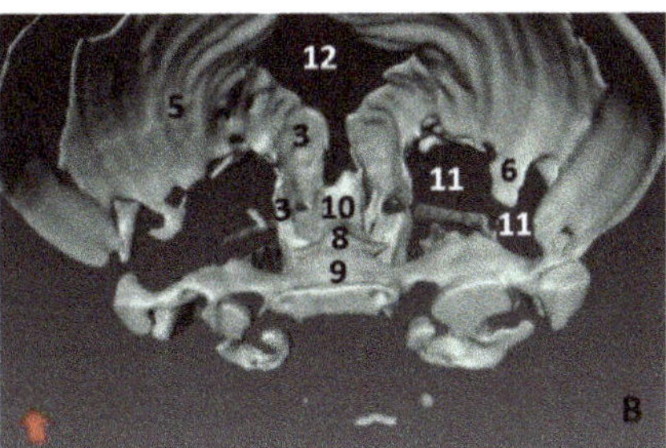

Figure 18. 3D reconstruction images of bony nasal cavity using AMIRA and CT. 9 months, dde13. (**A**) *Dorsal view.* (**B**) *Caudal view.* 1, Incisive bone; 2, Maxillary bone: external nasal face; 3, Pterygoid bone; 4, Frontal bone: external face; 5, Frontal bone: cerebral face; 6, Frontal bone: wall projections; 7, Vomer bone: groove; 8, Basisphenoid bone; 9, Presphenoid bone; 10, Ethmoid bone: crista galli; 11, Fontanelles; 12, Maxilloincisive fissure (bony naris opening); 13, Nasal cavities.

In the common dolphin (*Delphinus delphis*) fetus, 10 months of gestation, dde14, the rostral wall of the nasal cavity is totally closed by the maxillary bone, forming the bony walls in both cavities at both sides of the ethmoid bone. The lamina cribosa and the vomer wings are incomplete. The choanae are also seen well delimited (Figure 19A,B).

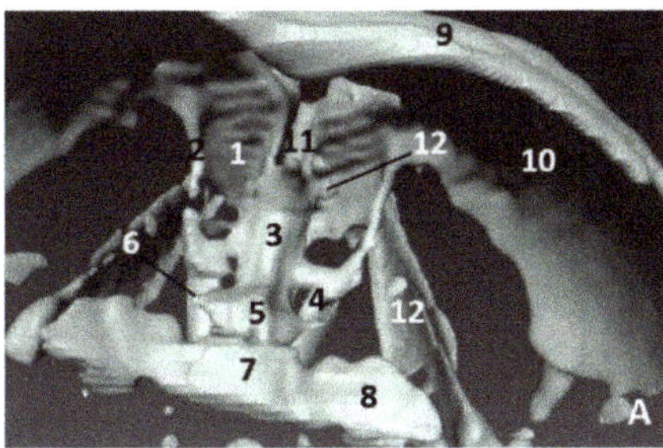

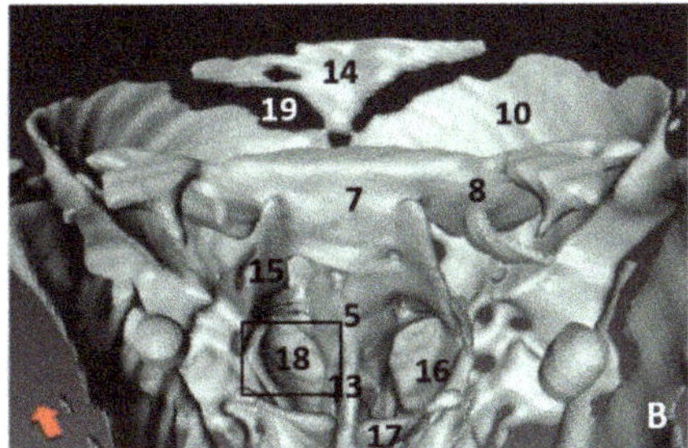

Figure 19. 3D reconstruction images of bony nasal cavity using AMIRA and CT. 10 months, dde14. (**A**) *Caudal view.* (**B**) *Ventrocaudal view.* 1, Maxillary bone: nasal face (rostral nasal wall); 2, Maxillary bone: lateral nasal projections; 3, Ethmoid bone; 4, Ethmoid bone: ossification area (lamina cribosa); 5, Presphenoid bone: body; 6, Presphenoid bone: wings; 7, Basisphenoid bone: body; 8, Basisphenoid bone: wings; 9, Frontal bone: external face; 10, Frontal bone: cerebral face; 11, Vomer bone: groove; 12, Vomer bone: wings; 13, Vomer bone: ventral crest; 14, Interparietal bone; 15, Basisphenoid bone: pterygoid crest; 16, Palatine bone; 17, Pterygoid bone; 18, Choanae; 19, Fontanelles.

The newborn striped dolphin (*Stenella coeruleoalba*), scomu1, shows the caudal wall of the nasal cavity almost closed, with bony projections from the frontal and maxillary bones towards the ethmoid bone. (Figure 20A,B). The lamina cribosa of the ethmoid bone is closing the wall of right and left nasal cavities (Figure 20C,D).

In the adult striped dolphin (*Stenella coeruleoalba*), three sections of the nasal cavity have been performed (Figure 21A). We observed the bony nasal septum formed by vomer and perpendicular lamina of ethmoid bone and the closed caudal wall of the nasal cavity (Figure 21B) formed by the ethmoidal and cerebral fossae (Figure 21C). In the last section we could see the completely defined caudal region of the nasal cavities (Figure 21D). Several vestigial perpendicular fissures were observed in the ethmoidal fossae and a small cribriform area in the nasal aspect of the ethmoid bone.

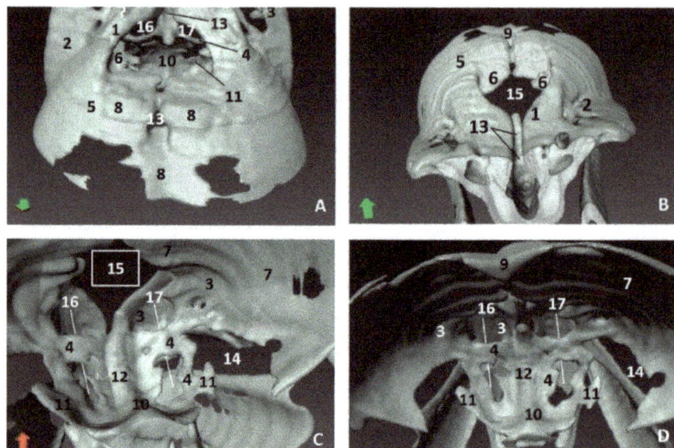

Figure 20. 3D reconstruction images of bony nasal cavity using AMIRA and CT. newborn, scomu1. (**A**) *Dorsal view.* (**B**) *Frontal view.* (**C**,**D**) *Caudal view.* 1, Incisive bone; 2, Maxillary bone: external face; 3, Maxillary bone: nasal face; 4, Ethmoid bone: ossification nuclei (lamina cribosa); 5, Frontal bone: external face; 6, Frontal bone: nasal face; 7, Frontal bone: cerebral face; 8, Nasal bone; 9, Interparietal bone; 10, Presphenoid bone: body; 11, Presphenoid bone: wings; 12, Ethmoid bone: crista galli; 13, Ethmoid bone: lamina perpendicular; 14, Fontanelles; 15, Maxilloincisive fissure (bony naris opening); 16, Left nasal cavity; 16, Right nasal cavity.

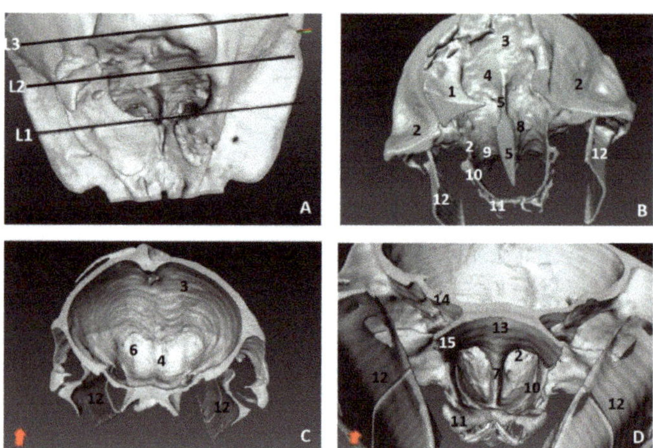

Figure 21. 3D reconstruction images of bony nasal cavity using AMIRA and CT. newborn, scomu1. (**A**) Level sections. *Dorsal view*; (**B**) Level 1 (L1), *Rostral view*; (**C**) Level 2 (L2), *Caudal view*; (**D**) Level 3 (L3), *Ventrocaudal view*. 1, Incisive bone; 2, Maxillary bone; 3, Frontal bone; 4, Ethmoid bone; 5, Ethmoid bone: perpendicular lamina; 6, Ethmoidal fossa: lamina cribosa; 7, Vomer bone: ventral crest; 8, Vomer bone: wings; 9, Presphenoid bone: wings; 10, Palatine bone; 11, Pterygoid bone; 12, Mandibles; 13, Basisphenoid bone: body; 14, Basisphenoid bone: wings; 15, Basisphenoid bone: pterygoid crest.

3.4. Computed Tomography and 3D Reconstruction of Nasal Cavity Spaces

In order to give a better perspective of the respiratory cavity, 3D casts were observed in both lateral and oblique views along with the 3D reconstructions of the skull bones. In the second less developed common dolphin (*Delphinus delphis*) fetus, dde2, we injected a small amount of silicone, with similar properties to CT iodinate contrast medium, to enable us to obtain an endocast and dilate the cavities to make the 3D reconstruction. Within the vestibule we could already distinguish the bilateral nasal diverticula. The accessory diverticulum was not clearly seen. We could observe the endocast of respiratory and olfactory part of the nasal cavity (Figure 22A). A more developed striped dolphin (*Stenella coeruleoalba*) fetus, scop1, showed the incisive recesses extending to the snout base overlapping the incisive bone (Figure 22B). In the pilot whale (*Globocephala melas*) fetus of five months' gestation, gma1, we could observe the expansion of the incisive recesses as the fetus is developing (Figure 22C). In common dolphin (*Delphinus delphis*) fetus of 5.5 months' gestation, dde5, the left lateral view allowed us to distinguish the accessory nasal diverticulum, which was very small and oriented rostrally (Figure 22D). In the last studied fetus, dde14, we observed the accessory nasal diverticulum ventral to the left nasal diverticulum (Figure 22E). In the adult striped dolphin (*Stenella coeruleoalba*) specimen, scomu1, we saw well-dilated left and right diverticula, the left accessory diverticulum, left incisive recess and parts of the left respiratory and olfactory parts of the nasal cavity. The right accessory diverticulum and right incisive recess were not dilated by air or the injected silicone, so it could not be reconstructed (Figure 22F).

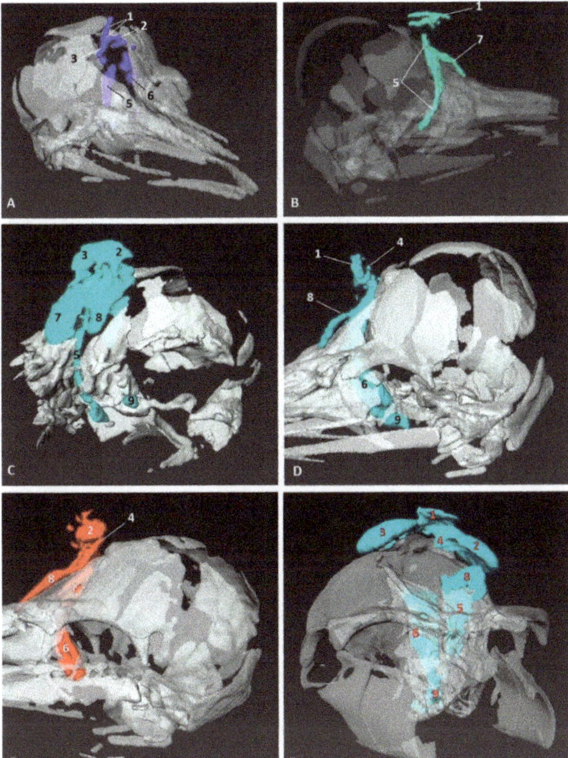

Figure 22. Amira 3D reconstructions of nasal cavity spaces after injecting silicone. Hiperattenuated CT images and air spaces were used to obtain the internal endocast. (**A**) *Right rostral aspect*. 3.5 months, dde2. (**B**) *Right lateral aspect*. 4.5 months, scop1. (**C**) *Left rostral aspect*. 5 months, gma1. (**D**) *Left lateral aspect*. 5.8 months, dde6. (**E**) *Left lateral aspect*. 10 months, dde14. (**F**) *Left rostral aspect*. Adult, scomu5. 1, Nasal cavity: vestibule; 2, Vestibule: left diverticulum; 3, Vestibule: right diverticulum; 4, Vestibule: left accessory diverticulum; 5, Nasal cavity: right respiratory part; 6, Nasal cavity: left respiratory part; 7, Nasal cavity: right incisive recess; 8, Nasal cavity: left incisive recess; 9, Choanae.

3.5. *Study of the Nasal Cavity Using Sectional Anatomy and Dissection Techniques*

3.5.1. Sectional Anatomy

The sagittal sections were performed in both a juvenile and an adult striped dolphin (*Stenella coeruleoalba*). The coronal sections were performed in a newborn striped dolphin (*Stenella coeruleoalba*) covering the whole nasal cavity. The two first sections are placed in the vestibule of the nasal cavity while the other four are in the respiratory and olfactory part of the nasal cavity.

At the level of the vestibule we could see many vestibular folds, together with the nasal plugs forming the "monkey lips" involved in sound production. We also observed muscles of the external nose, which enable the opening of its caudal lip. We also could see the accessory diverticulum and the vestibule (Figure 23A,B).

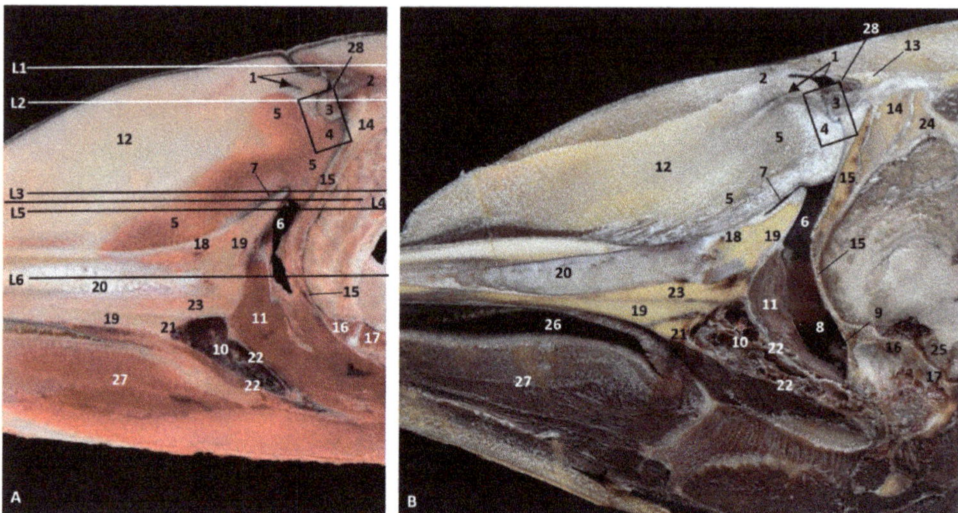

Figure 23. (**A**,**B**) Sagittal sections of nasal cavity. Sagittal sections images are oriented so that the rostral is to the left and the caudal to the right. (**A**) Level sections (L1–L6), Fresh section, juvenile, scomu3. (**B**) Fixed section, adult, scomu6. 1, Nasal cavity: vestibule (line) and (arrow) accessory nasal diverticulum; 2, Upper lip and vestibular fold muscles; 3, Vestibular fold; 4, Nasal plug; 5, Nasal plug and lower lip muscles; 6, Nasal cavity: respiratory part; 7, Nasal cavity: incisive recess; 8, Choanae; 9, Aditus laryngis; 10, Pterygopalatine recess: pharyngeal recess of pterygoid and palatine bones; 11, Palatopharyngeal muscles (sectioned); 12, Melon; 13, Nasal bone; 14, Frontal bone; 15, Ethmoid bone; 16, Presphenoid bone; 17, Basisphenoid bone; 18, Incisive bone; 19, Maxillary bone; 20, Mesorostral cartilage; 21, Palatine bone; 22, Pterygoid bone; 23, Vomer bone; 24, Interparietal bone; 25, Hypophysis; 26, Oral cavity; 27, Tongue; 28, "Monkey or phonic lips".

The coronal sections at the vestibule level showed the left and right diverticulum and nasal plugs close to the vestibular fold (Figure 24A,B).

3.5.2. Dissections

Dolphin Specimens

The dissections were performed in a newborn (scocel) stranded in the Mediterranean coast of Ceuta and two adults (scomu5 and scomu7) stranded in the Mediterranean coast of Murcia.

After removing the external nose in the newborn, we observed dorsally the right nasal diverticulum and ventrally the accessory nasal diverticulum caudal to the right nasal plug (Figure 25A). The right nasal plug was clogged and the left one was normal (Figure 25B). We also observed two functional positions of the left nasal plug, one opened and two closed, and the nasal cavities divided by the nasal septum (Figure 25C,D).

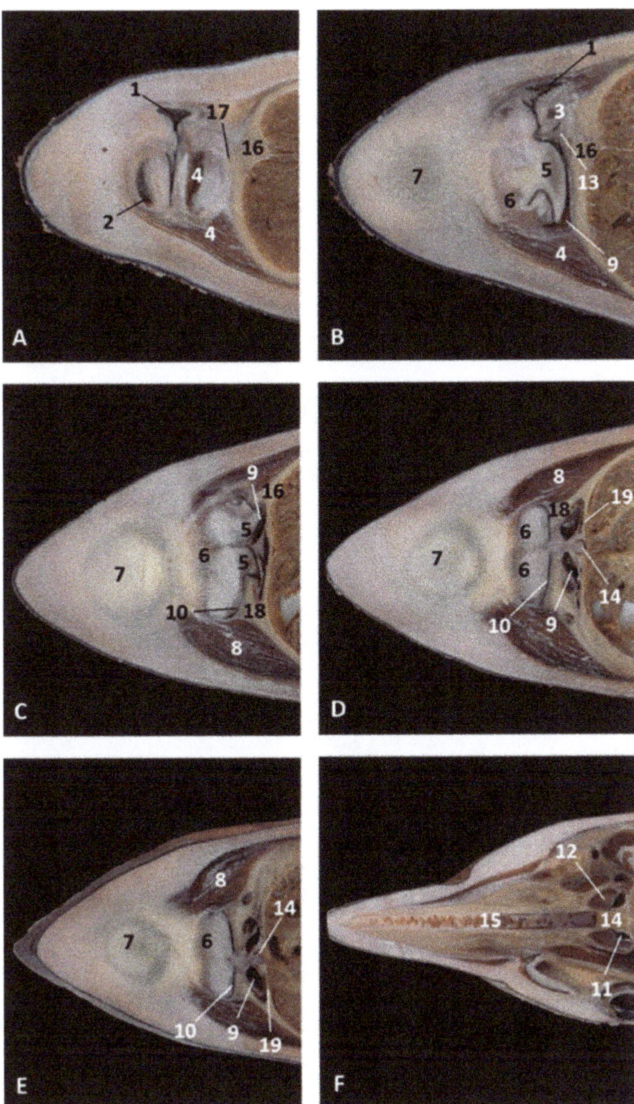

Figure 24. Coronal sections of nasal cavity. Coronal sections images are oriented so that the rostral is to the left and the caudal to the right. (**A**,**B**) Level 1 and 2. Nasal cavity: vestibule. (**C**–**F**) Levels 3 to 6. Nasal cavity: respiratory and olfactory parts. Newborn, scomu2. 1, Vestibule: right diverticulum; 2, Vestibule: left diverticulum; 3, Vestibular folds; 4, External nose muscles; 5, Nasal plugs; 6, Nasal plug muscles and connective tissue; 7, Melon; 8, Melon muscles; 9, Nasal cavity: respiratory part (nasal mucosae); 10, Nasal cavity: incisive recess; 11, Choanae; 12, Pharyngeal muscles; 13, Nasal septum: membranous part; 14, Nasal septum: bony part (vomer); 15, Mesorostral cartilage; 16, Frontal bone; 17, Nasal bone; 18, Incisive and maxillary bones; 19, Ethmoid bone.

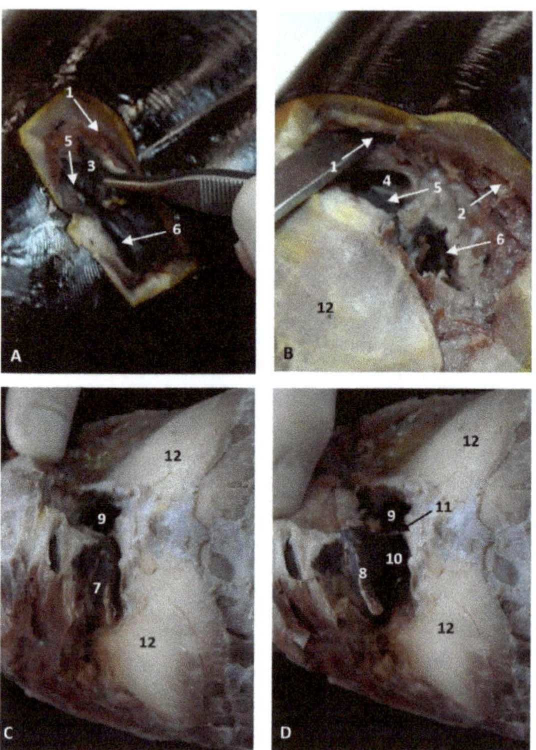

Figure 25. Dissection of nasal cavity. (**A**,**B**) Head is oriented so that the rostral side of the head is to the right up corner and caudal is to the left down corner. (**C**,**D**) Head is oriented so that the rostral to the left and the caudal to the right. scoce1. (**A**) Right nasal cavity: vestibule. Clogged right nasal cavity (hyperkeratosis). (**B**) Nasal cavity: vestibule after removed melon skin. (**C**,**D**) Nasal plug opened and closed. Newborn, scoce1. 1, Nasal cavity: right diverticulum; 2, Nasal cavity: left diverticulum; 3, Nasal cavity: accessory diverticulum opened; 4, Accessory diverticulum opened and mucosa partly removed; 5, Right nasal plug clogged; 6, Left nasal plug normal but hypertrophied; 7, Left nasal plug closed; 8, Left nasal plug opened; 9, Right nasal cavity: respiratory part; 10, Left nasal cavity: respiratory part; 11, Nasal septum: bony part (lamina perpendicular ethmoid bone); 12, Frontal bones.

In a preserved female striped dolphin (*Stenella coeruleoalba*), scomu5, the right nasal diverticulum was dilated to better understand its size and form (Figure 26A). In subsequent images, the rostral wall of diverticulum has been removed to show the dark mucosa and ventrally the closed nasal plug (Figure 26B,C). The final images show in detail the closed nasal plug and the access to the accessory diverticulum, first closed (Figure 26D) and then opened showing a portion of the less pigmented mucosa (Figure 26E).

In a female fresh adult striped dolphin (*Stenella coeruleoalba*), scomu7, we could identify the right and the accessory diverticula. The mucosa of both diverticula is dark in colour. The right nasal plug is tightly closed (Figure 27A). The nasal plug has been displaced rostrally showing the respiratory part of the nasal cavity (Figure 27B). The vestibular fold and nasal plug have been displaced caudally and rostrally, respectively, and inside the respiratory part of the nasal cavity we could observe the dark nasal mucosa (Figure 27C,D). Finally, we examined in detail the closed entrance to the left nasal plug and the respiratory part of the right nasal cavity which was patent after removal of the nasal plug (Figure 27D).

Figure 26. Dissection of nasal cavity: vestibule. Detailed images of a nasal diverticulum. Adult, scomu5. (**A**) Nasal cavity: vestibule. Right diverticulum dilated. *Right lateral aspect*. (**B**) Nasal cavity: vestibule. Right diverticulum partially sectioned. *Rostral view*. (**C**) Right diverticulum opened. *Rostral view*. (**D**) Right diverticulum opened. *Caudal view*. (**E**) Right diverticulum border displaced dorsally. *Caudal view*. 1, Nasal cavity: right diverticulum (dilated); 2, Right diverticulum: mucosa; 3, Nasal plug; 4, Accessory diverticulum.

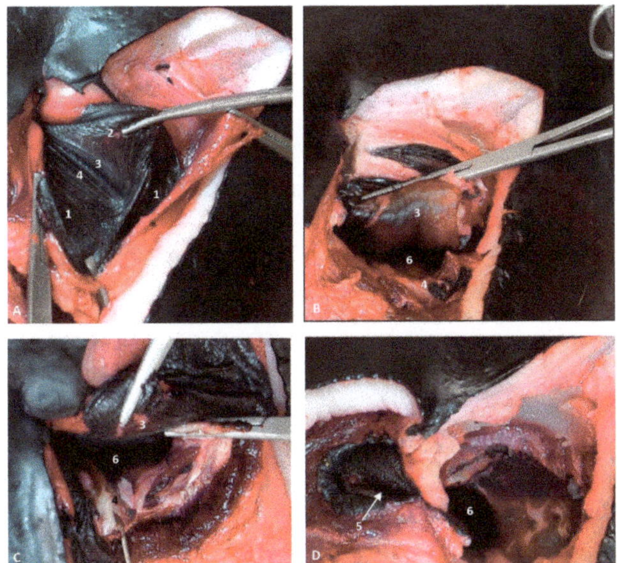

Figure 27. Dissection of nasal cavity: vestibule and respiratory part. (**A,D**) Head is oriented so that the rostral side of the head is to the top and caudal is at the bottom. Adult, scomu7. (**A**) Right nasal cavity: vestibule open. (**B**) Right nasal cavity: nasal plug displaced rostrally. (**C**) Right nasal cavity: nasal plug displaced rostrally and vestibular fold caudally. (**D**) Right nasal plug has been removed and left is observed closed. 1, Nasal cavity: right diverticulum; 2, Accessory diverticulum opened; 3, Right nasal plug; 4, Vestibular fold; 5, Left nasal cavity: left nasal plug; 6, Right nasal cavity: respiratory part.

Horse Specimens

The dissections were performed in one foal fetus (ecal1) and two adult (ecal2 and 3) horses from the slaughter house (Alicante).

The fetal vestibule was observed in its anatomical position showing the alar fold and nostril (false nostril is the portion of the nostril continuous with the nasal diverticulum) which is the orifice of nasal vestibule (Figure 28A–C). Dissections showed the differences between the mucosa of nasal diverticulum, vestibule, alar fold and respiratory part of the nasal cavity (Figure 28B–D).

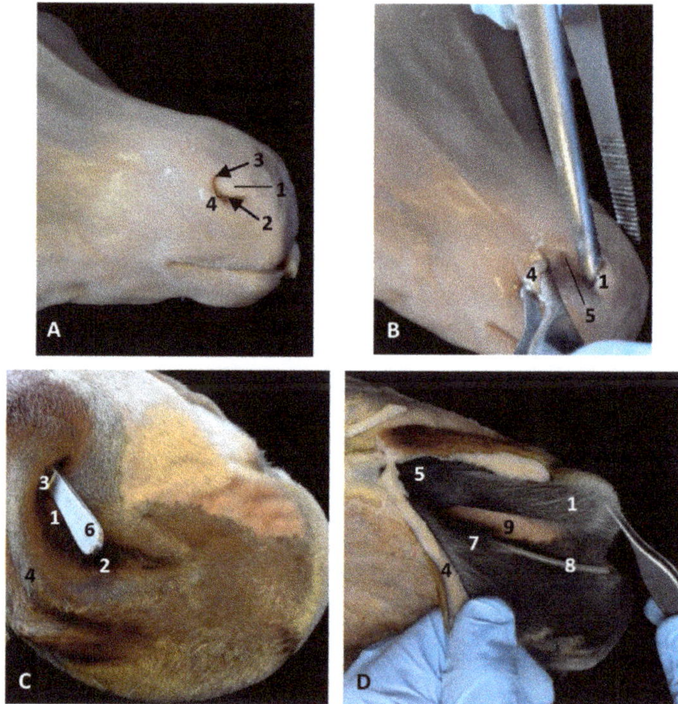

Figure 28. Nose dissections at right nasal vestibule level in a foal fetus (**A**,**B**) and two adult horses. (**C**,**D**). (**B**) Wing of the nostril has been retracted laterally to observe the nasal diverticulum. (**C**) Forceps are inside the nasal diverticulum. (**D**) The nose has been dissected to observe pigmentation differences between mucosa of the vestibule and the respiratory part of the nasal cavity. (**A–D**) *Right lateral view*. 1, Alar fold; 2, Nostril; 3, False nostril; 4, Wing of the nostril; 5, Nasal diverticulum; 6, Forceps; 7, Nasal vestibule: mucosa; 8, Nasolacrimal orifice (plastic tube); 9, Nasal cavity: respiratory part.

3.6. Histological Study of the Nasal Cavity

3.6.1. Dolphin

The vestibular mucosa, specifically its diverticulae, is a stratified squamous epithelium, both pigmented and keratinized. The papillary layer is wide. The connective tissue base is normal and contains small vessels (Figure 29A). The vestibular folds show a connective tissue base with cartilage and striated muscle (Figure 29B). The nasal plug has a stratified epithelium, narrow and pigmented with a papillary layer wide and flattened. It has stratified musculature in discontinued fascicles. Its connective tissue base is strong with muscular fascicles (Figure 29C). The mucosa in the respiratory part has a pseudostratified arrangement but cilia were not observed (10 layers). A deep and narrow papillary layer is observed. A dense connective tissue base with a few muscular fibers and blood vessels was seen (white areas enlarged and empty) (Figure 29D). The mucosa of the olfactory part is a pseudostratified epithelium but cilia were not seen and it was more rounded and wider than the respiratory part (15–16 layers). The papillary layer is narrow and vascularized with a connective tissue base (Figure 29E). A stratified squamous epithelium lines the incisive recess. The papillary layer is wide with little vascularization. The connective tissue base containing some nerves was seen (Figure 29F).

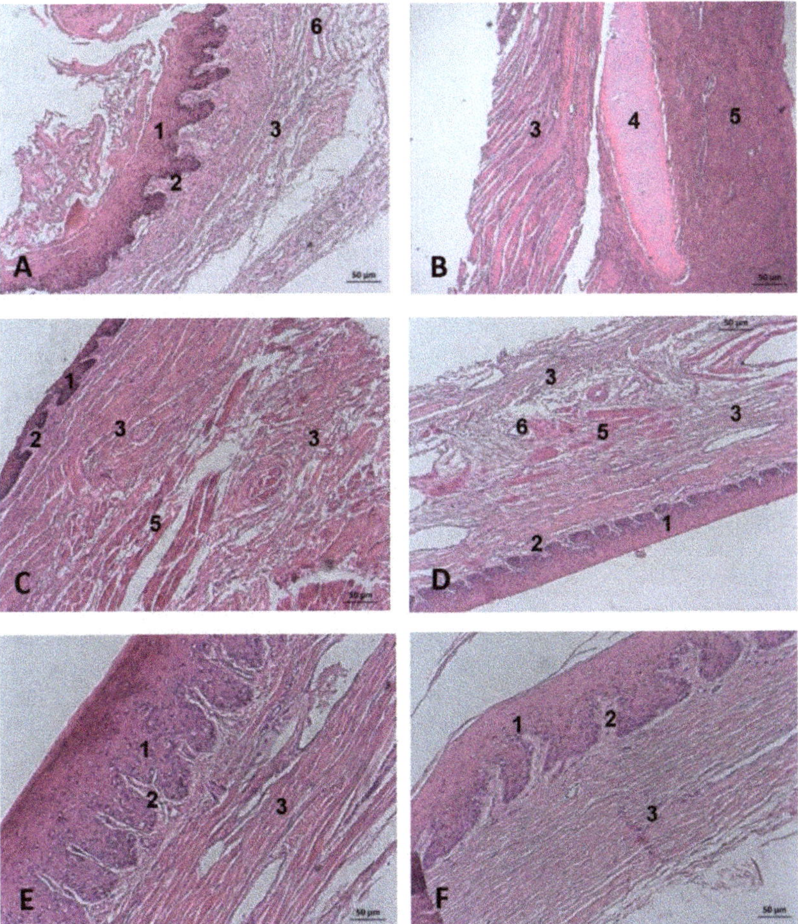

Figure 29. Histological study of nasal mucosa: vestibule, respiratory and olfactory parts. H-E staining technique. Adult, scomu6. (**A**) Left diverticulum, 10×. (**B**) Vestibular fold, 10×. (**C**) Nasal plug, 10×. (**D**) Respiratory part, 10×. (**E**) Olfactory part, 20×. (**F**) Incisive recess, 20×. 1, Epithelium: stratified squamous keratinized and pigmented; 2, Papillary layer; 3, Connective tissue base; 4, Cartilage; 5, Striated muscular base; 6, Vessels.

3.6.2. Horse

The vestibular mucosa of the nasal cavity, specifically its diverticulae, is a stratified squamous epithelium, both pigmented and keratinized with hair and with associated sebaceous glands. The papillary layer is not well defined. The connective tissue base contains some adipose tissue (Figure 30A). The alar fold has a stratified squamous epithelium which is not keratinized. The connective tissue base is large and dense. There are groups of glands between the sebaceous tissue (Figure 30B). The respiratory part of the nasal cavity has a pseudostratified epithelium, forming a cylindrical mucosa with cilia and a cartilaginous base (Figure 30C). The mucosa of the olfactory part is a pseudostratified epithelium but olfactory cilia were not clearly seen (Figure 30D).

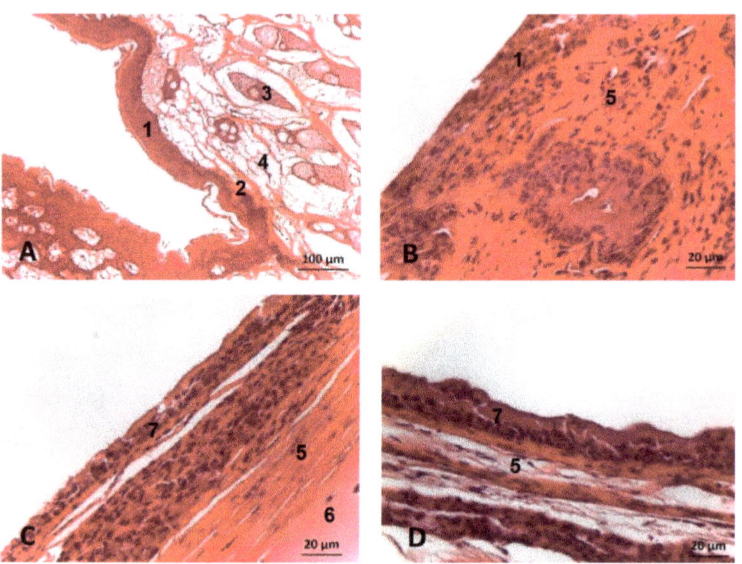

Figure 30. Histological study of nasal mucosa: vestibule, respiratory and olfactory parts. H_E stain technique. Adult Ecal4 (**A**) Nasal vestibule 10×. (**B**) Alar fold 40×. (**C**) Respiratory part. 40×. (**D**) Olfactory part. 20×. 1, Epithelium: stratatified squamous keratinized and pigmented; 2, Papillary stratum; 3, hair; 4, Fat tissue; 5, Connective tissue base; 6, Cartilage; 7, Respiratory epithelium.

4. Discussion

4.1. Anatomical, Comparative and Functional Commentaries

4.1.1. External Nose

The external nose of adult odontocetes and mysticetes is well described [10,11,15,16,19,21,25,26,41–44] but the information is sometimes confusing, often due to differing terminology. Our study goes from the first development stages of the fetus to the adult stage.

The endoscopy technique to analyze the nasal cavity has allowed us to observe the external nose and nasal cavity morphology caudal to the choanae. In mammals, the external nose is part of the face rostral to the frontal region and dorsal to the infraorbital, buccal and oral region [38]. We have observed the highly characteristic morphology of the cetacean nose with the endoscopy images. The presence of nasal lips (rostral and caudal) and its anatomical position, closed from the beginning of development to avoid the entrance of water and salt into the nasal cavity is one of the first image proofs of the development of the two prominences in the caudal lip and their division by a median groove from the nasal septum to the caudal lip. During early stages of fetal development the external nose and the melon were observed as a common anatomical area delimited by a line and covered by an epidermis paler than the rest of the head. Together with the function of the melon in the projection of sounds produced by "phonic lips" [6,43,44], this supports our idea that the melon is forming part of the external nose both anatomically and functionally, and not merely as a part of the nasal complex.

These changes in morphology will be helpful in determining the stages of fetal development.

The MRI and dissection images allowed us to observe, as well as other authors [19,25,26,41] that both the caudal and the rostral lips present retractor muscles which act to open the nose for breathing when the animal rises above the water. Also, we have observed that the most retracted lip was the caudal one.

We also discovered that closing of the lips becomes more airtight as development progresses to adulthood, whose particular anatomy forces these lips to double their size with an increased number of striations, rendering them more waterproof.

4.1.2. Nasal Cavities

The nasal cavity of odontocetes is similar to that of mysticetes except in the sperm whale, a species with a unique morphology [13,18,42].

The odontocete nasal cavity has only one nostril (naris) and one vestibule with paired diverticula, and two respiratory and olfactory parts divided by a nasal septum and ending caudally at the choanae. The equine nose has two nostril, two vestibules of the nose and a cutaneous blind sac (diverticulum nasi) [38]. Except for the sperm whale, the cetacean nasal cavity has a vertical orientation and the external nose and melon are located dorsally [6] with the choanae ventrally, and as described before, the nasal bones are retracted towards the frontal bones [1–3,15]. Therefore, we suggest that the cetacean nasal cavity should be more properly called the maxillary cavity.

Vestibule

The nasal cavity vestibule is a space described in all the terrestrial mammals as the hall of the nasal cavity rostral to therespiratory and olfactory part. Though theoretically it is a simple part of this cavity, most of the studies define it in the bottlenose dolphin as a very complex area [9,15] and in our opinion it is not fully explained in previous publications. The presence of two diverticula of the nasal vestibule is clearly shown by endoscopy, 3D reconstructions, sections and dissection studies. These diverticula present a pigmented mucosa as in other species [18] and also in the dissections, sections and histology of our study. The stratified squamous epithelium is similar to that of the equine nose and corroborates our idea about its function, which is protection from external agents and water [17,28]. The vestibular fold shows a very different mucosa from that of the respiratory region, probably related to is proposed function in sound production. Endoscopy permitted us to observe the morphology of the right and left diverticula as spaces with many folds and an anfractuous arrangement as in other cetacean species [17]. The nasal endocast and the volumetric reconstructions enabled us to study its morphology and to locate another accessory nasal diverticulum that we confirmed by dissections. Though we know the bottlenose dolphin vestibule morphology as very complex [9,15,26], our developmental study shows that the vestibule in common and striped dolphins *Stenella coeruleoalba* and pilot whale *Globocephala melas* is less complex.

Another important aspect of our study is the endoscopic examination of the "monkey or phonic lips" which are responsible for sound production [6,43,44]. According to our study, the dorsal parts of the sound production mechanism were the vestibular folds placed under the upper lip and the ventral parts were the nasal plugs divided by the membranous part of the nasal septum. According to our anatomical idea, the "monkey or phonic lips" are solely in charge of producing sound projected towards the melon. Diverticula, wrongly named as air sacs [7–10] because they are only present in birds in order to decrease the total weight during flight. The bronchi extend outside the lung in a form of thin-walled transparent chambers called air sacs [45] extending through the celomic cavity and bones. Diverticula avoid the entrance of possible water filtrations through the blowhole that could have entered during sound production, diving or under a stressful situation such as vessel crash or an underwater shark attack, since they are very far from the phonic lips. It is interesting to observe that vestibular folds lack an epithelium and present a connective tissue and muscular base explaining its function as sound generator.

Respiratory and Olfactory Parts of the Nasal Cavity

This region delivering air towards the lungs is characterized by the absence of nasal conchae or cornettes and paranasal sinuses except for the maxillary sinus [28] which is absent in the adult. The maxiloturbinate, well developed in terrestrial mammals, is vestigial in odontocetes, though there are differences between mysticetes and odontocetes [31]. If the function of heating air at this level of the nasal cavity is the main goal in the domestic mammals due to the presence of cornettes and sinuses. In cetaceans it is not the case. Also, it seems that these structures lack the olfactory function in odontocetes, as our histological analysis did not find olfactory epithelium.

Bony Nasal Cavity

The tomographic study of the fetal nasal cavity permitted the analysis of the interior of the cavities Volumetric reconstruction of the nasal cavity allowed us to determine the stages of ossification of the nasal cavity The maxillary bones forming the rostral wall are first to form followed by the frontal bones closing slowly towards the cranium base, but they only slowly ossify and connect with the lamina cribosa of the ethmoid bone. The final bone to ossify is the ethmoid bone, which indicates that it is to allow the passage of the olfactory nerves and form the lamina cribosa, even though the olfactory bulb is vestigial in cetaceans and persists as a remnant of phylogeny [21,30]. The vomer bone is growing rostrally slowly and we can see the groove on which the mesorostral cartilage will be placed. The bones closing the ventral floor of nasal cavity are the pterygoid and the palatine [31]. The ethmoid bones take longer to ossify and close the caudal wall of the nasal cavity which was visible in the volumetric reconstructions. The mold of the respiratory and olfactory parts showed two cavities bending caudally towards the nasopharynx surrounded by the aforementioned bones.

Nasal Mucosa

The endoscopic study of the nasal cavity shows at the early developmental stage a smooth mucosa which forms longitudinal folds as development progresses. We have not seen this described by other authors using other techniques, including endoscopy [34,35]. The endoscope allowed us to examine the incisive recess (premaxillary sac according to other authors [8,12,14]) that our study suggests serves to store, along with the vestibular sac [19], water escaping from the nasal vestibule which is expelled when the animal emerges to breathe [28]. The extent of the recesses could be seen in the nasal casts, as was their growth during fetal development, findings supported by the MRI data, anatomical sections and even dissection. As seen in other endoscopy studies [46] small mucosal fossae were detected close to the choanae whose function is uncertain. The nasal folds are very abundant in the adult and the lumen was narrow. Histological analysis of the nasal vestibule produced similar findings to the nasal diverticulum of the horse [7] (Figure 29A and Figure 30A) and the respiratory part had a respiratory mucosa not so similar to that of the horse), but we have found few histological studies at these levels [34,42,46]. The spherical nuclei of the sensorial cells were not observed in the olfactory part confirming the absence of olfactory function (Figure 29E).

Pathological Findings

The endoscopic study detected an obstruction in the right nasal cavity vestibule in the newborn from Ceuta stranded with its mother and dying afterwards. Post mortem pathologic analysis confirmed this diagnosis. Many stranded calves with respiratory problems are not fully examined due to the expense of the procedure and insufficient equipment. Using endoscopy, parasites were detected, belonging mainly to two families (*Pseudaliidae* and *Prassicaudidae*) and four genera (*Halocercus*, *Pharurus*, *Pseudalius* and *Stenurus*) [47,48]. This often undiagnosed infestation, along with the obstructions could lead to stranding of neonatal cetaceans. Except for rare cases where it is used during necropsies [46], endoscopy is restricted to live animals to examine of the lower respiratory tract (bronchis) and to diagnose the respiratory pathologies of parasitic, bacterial or fungal kind, either by direct visualization, bronchial flushes, or taking biopsies [36,37]. When necropsies are performed, we think that endoscopy of the nasal cavity should be included in necropsy protocols, despite the difficulty of the technique.

5. Conclusions

Fetal anatomical endoscopic study allowed us to observe the simultaneous development of the melon and the external nose. Also, we have also seen the form and function of the external nose showing a closing formed by two lips very simple during fetal development and very sophisticated at adult stage. The vestibule showed us the "monkey lips", two diverticula and two incisive recesses. Longitudinal mucosal folds were seen in the respiratory and olfactory parts. 3D reconstructions of nasal spaces and nasal skull gave us a spatial representation of nasal cavity development and confirmed our endoscopic observations. MRI data of sagittal and coronal sections using T1 and T2 MRI sequences were important in confirming that bony anatomical structures were correctly identified such as the presphenoid and ethmoid bone. Also, sectional anatomy and dissections aided our identification of structures seen in endoscopic, CT and MRI studies.

The histological analysis has confirmed the similarity of the dolphin nasal mucosa compared with horse nasal mucosa but cilia were not observed in the respiratory part of the nasal mucosa in odontocetes. Pathological findings which showed hyperkeratosis within the nasal vestibule should be taken into account during necropsies in stranded specimens.

The retraction of the nasal bones, the vertical position of the nasal cavities and the special bony walls leads us to recommend that "nasal cavity" should be referred to as "maxillary cavity" as these bones close the respiratory space in dolphins as nasal bones do in domestic mammals.

The nasal vestibule contains different cavities such as the nasal diverticulum which is described in horses. The cetacean diverticulum functions as a water reservoir and protection against external agents. Vestibular folds are comparable to the modified alar folds described in the horse.

The fusion between the ethmoid and presphenoid bones was seen in the second early fetal specimen.

Finally, we conclude that the lamina cribosa of the ethmoid bone is the last bone to ossify in the nasal cavity in a newborn specimen. It shows a small perpendicular fissure at both sides of the crista galli and a small cribriform area in the nasal aspect of the ethmoid bone. Olfactory cells were not detected during histological analysis of the nasal mucosa.

Supplementary Materials: The following are available online at https://www.mdpi.com/2076-2615/11/2/441/s1, Table S1: Other parameters observed in this study, Table S2: MRI parameters used in this study.

Author Contributions: Conceptualization, A.G.d.l.R.yL. and G.R.Z.; Formal analysis, A.A.E. and N.G.C.; Investigation, A.G.d.l.R.yL.; Methodology, A.A.E., M.S.L., F.M.G., A.L.F., P.C.F. and C.S.C.; Resources, F.M.G., A.L.F., P.C.F., N.G.C. and C.S.C.; Supervision, A.G.d.l.R.yL., A.A.E. and G.R.Z.; Writing—original draft, A.G.d.l.R.yL. and G.R.Z.; Writing—review & editing, A.G.d.l.R.yL., A.A.E., M.S.L. and G.R.Z. All authors have read and agreed to the published version of the manuscript.

Funding: CT and MRI acquisitions were financed by Departamento de Anatomía y Anatomía Patológica Comparadas. Facultad de Veterinaria. Universidad de Murcia. Spain. The Galician stranding network is supported by the regional government Xunta de Galicia—Dirección Xeral de Patrimonio Natural. CESAM/FCT: thanks are due to FCT/MCTES for the financial support to CESAM (UIDP/50017/2020+UIDB/50017/2020), though national funds. Norma transitoria. Alfredo López is funded by national funds (OE), through FCT—Fundaçâo para a Ciência e a Tecnologia, I.P. in the scope of the framenetwork contract foreseen in numbers 4, 5 and 6 of the article 23, of the Decree-Law 57/2016, of 29 August, changed by Law 57/2017, of 19 July.

Institutional Review Board Statement: Not applicable.

Data Availability Statement: Not applicable.

Acknowledgments: We are grateful to Consejería de Sanidad, Ceuta, Spain. Especial thanks to CESAM and Xunta de Galicia, Spain. We thank CEMMA volunteers that helped with necropsies and sample collection. We are grateful to Miguel Ángel Gómez Sánchez and Serafín Gómez Cabrera for the Histological and Anatomopathological analysis, respectively at the Departamento de Anatomía y Anatomía Patológica Comparadas, Facultad de Veterinaria, Murcia, Spain. We are also thankful to image technician Oscar Blázquez Pérez for the MRI scan performed at Centro Veterinario de Diagnostico por Imagen del Levante, Ciudad Quesada, Rojales Alicante, Spain. We give special thanks to Mª José Gens Abujas (Oficina de impulso Socioeconómico del Medio Ambiente, Dirección General de Medio Natural, Consejería de Empleo, Universidades, Empresas y Medio Ambiente, Región de Murcia, Spain). Thank you very much to the CRFS Veterinary Team, in a special way, Fernando Escribano Cánovas, Luisa Lara Rosales and Alicia Gómez de Ramón Ballesta, El Valle, Murcia, Spain, for allowing us to have access to the carcasses stranded in their regional area.

Conflicts of Interest: The authors of this manuscript have no conflict of interest to declare.

References

1. Thewisen, J.G.M.; Cooper, L.N.; George, J.C.; Bajpai, S. From land to water: The origin of whales, dolphins, and porpoises. *Evol. Educ. Outreach* **2009**, *2*, 272–288. [CrossRef]
2. Berta, A.; Sumich, J.L.; Kovacs, K.M. Cetacean evolution and systematics. In *Marine Mammal Evolutionary Biology*, 2nd ed.; Academic Press: San Diego, CA, USA, 2005; pp. 165–209.
3. Roston, R.A.; Roth, V.L. Cetacean skull telescoping brings evolution of cranial sutures into focus. *Anat. Rec.* **2019**, *302*, 1055–1073. [CrossRef] [PubMed]
4. Armfield, B.A.; George, J.C.; Vinyard, C.J.; Thewissen, J.G.M. Allometric patterns of fetal head growth in mysticetes and odontocetes: Comparison of *Balaena mysticetus* and *Stenella attenuata*. *Mar. Mammal Sci.* **2011**, *27*, 819–827. [CrossRef]
5. Macleod, C.C.; Reidenberg, J.S.; Weller, M.; Santos, M.B.; Herman, J.; Goold, J.; Piercel, G.J. Breaking symmetry: The marine environment, prey size, and the evolution of asymmetry in cetacean skulls. *Anat. Rec.* **2007**, *290*, 539–545. [CrossRef]
6. Berta, A.; Ekdale, E.; Cranford, T.W. Review of the cetacean nose: Form, function, and evolution. *Anat. Rec.* **2014**, *297*, 2205–2215. [CrossRef] [PubMed]
7. Browers, M.E.L.; Kaminga, C. The use of computed tomography in cetacean research air sac determination of *Lagenorhyncus albirostris*. Part 1. *Aquat. Mamm.* **1990**, *16*, 145–155.
8. Reidenberg, J.S.; Laitman, J.T. Sisters of the sinuses: Cetacean air sacs. *Anat. Rec.* **2008**, *291*, 1389–1396. [CrossRef]
9. Houser, D.S.; Finneran, J.; Carder, D.; Van Bonn, W.; Smith, C.; Hoh, C.; Mattrey, R.; Ridgway, S. Structural and functional imaging of the bottlenose dolphin (*Tursiops truncatus*) cranial anatomy. *J. Exp. Biol.* **2004**, *207*, 3657–3665. [CrossRef] [PubMed]
10. Schenkkan, E.J. The occurrence and position of the "connecting sac" in the nasal tract complex of small odontocetes (Mammalia, Cetacea). University of Amsterdam. *Beaufortia* **1971**, *246*, 37–43.
11. Cave, A.J.E. The os maxilloturbinare in the Cetacea. *Investig. Cetacea* **1988**, *21*, 41–50.
12. Huggenberger, S.; Rauschmann, M.A.; Volg, T.J.; Oelschläger, H. Functional morphology of the nasal complex in the harbor porpoise (*Phocoena phocoena* L.). *Anat. Rec.* **2009**, *292*, 902–920. [CrossRef] [PubMed]
13. Raven, H.C.; Gregory, W.K. The spermaceti organ and nasal passages of the Sperm Whale (*Physeter catodon*) and other odontocets. *Am. Mus. Novitatis.* **1993**, *677*, 1–18.
14. Cranford, T.W.; Amundin, M.; Norris, K.S. Functional morphology and homology in the odontocete nasal apparatus: Implications for sound generation. *J. Morph.* **1996**, *228*, 223–228. [CrossRef]
15. Mead, J.G. Anatomy of the external nasal passages and facial complex in the delphinidae (Mammalia: Cetacea). *Smithson. Contr. Zool.* **1975**, *207*. [CrossRef]
16. Heyning, J.E.; Mead, J. Evolution of the nasal anatomy of cetaceans. In *Sensory Abilities of Cetaceans*; Thomas, J., Kastelain, R., Eds.; Plenum Press: New York, NY, USA, 1990; pp. 67–79.
17. Schenkkan, E.J. On the comparative anatomy and function of the nasal tract in odontocetes (Mammalia, Cetacea). *Bijdr Dierkd* **1973**, *43*, 127–159. [CrossRef]
18. Schenkkan, E.J.; Purves, P.E. The comparative anatomy of the nasal tract and the function of the spermaceti organ in the physeteridae (Mammalia, Odontoceti). *Bijdr Dierkd* **1973**, *43*, 93–112. [CrossRef]
19. Lawrence, B.; Schevill, W.E. The functional anatomy of the delphinid nose. *Bull. Mus. Comp. Zool.* **1956**, *114*, 103–151.
20. Heyning, J.E. *Comparative Facial Anatomy of Beaked Whales (Ziphiidae) and a Systematic Revision among the Families of Extant Odontoceti*; Natural History Museum of Los Angeles: Los Angeles, CA, USA, 1989; Volume 405, pp. 1–59.
21. Cave, A.J.E. Note on olfactory activity in mysticetes. *J. Zool.* **1988**, *214*, 307–311. [CrossRef]
22. Honigmann, H. Bau und entwicklung des knorpelschadels vom Buckelwal. *Zoology* **1917**, *69*, 1–87.
23. De Burlet, H.M. Zur entwicklungsgeschichte des walsch€adels III. Das primordialcranium eines embryo von *Balaenoptera rostrata* (105 mm). *Morph. Jb.* **1914**, *49*, 119–178.
24. Buono, M.R.; Fernández, M.S.; Fordyce, R.E.; Reidenberg, J.S. Anatomy of nasal complex in the southern right whale, *Eubalaena australis* (Cetacea, Mysticeti). *J. Anat.* **2015**, *226*, 81–92. [CrossRef] [PubMed]

25. Dormer, K.J. The mechanism of sound production and recycling in delphinids: Cineradigraphic evidence. *J. Acoust. Soc. Am.* **1979**, *65*, 229. [CrossRef]
26. Green, R.F.; Ridgway, S.H.; Evans, W.E. Functional and descriptive anatomy of the bottlenosed dolphin nasolaryngeal system with special reference to the musculature associated with sound production. In *Animal Sonar Systems*; Busnel, R.G., Fish, J.F., Eds.; Plenum Press: New York, NY, USA, 1980; pp. 199–238.
27. Moran, M.M.; Nummela, S.; Thewissen, J.G.M. Development of the skull of the pantropical spotted dolphin (*Stenella attenuata*). *Anat. Rec.* **2011**, *294*, 1743–1756. [CrossRef]
28. García de los Ríos, A.; Arencibia, A.; Soler, M.; Gil, F.; Martínez, F.; López, A.; Ramírez, G. A study of the head during prenatal and perinatal development of two fetuses and one newborn Striped Dolphin (*Stenella coeruleoalba*, Meyen 1833) using dissections, sectional anatomy, CT, and MRI: Anatomical and functional implications in cetaceans and terrestrial mammals. *Animals* **2019**, *9*, 1139.
29. Reidenberg, J.S.; Laitman, J.T. Prenatal development in cetaceans. In *Encyclopedia of Marine Mammals*, 2nd ed.; Perrin, W.F., Würsig, B., Thewissen, J.G.M., Eds.; Academic Press: San Diego, CA, USA, 2009; pp. 220–230.
30. Klima, M. Cetacean phylogeny and systematics based on the morphogenesis of the nasal skull. *Aquat. Mamm.* **1995**, *21*, 79–89.
31. Klima, M. Development of the cetacean nasal skull. *Adv. Anat. Embryol. Cell Biol.* **1999**, *149*, 1–143.
32. Klima, M. Development of the cetacean nasal skull. *Mar. Mammal Sci.* **2001**, *17*, 425–426.
33. Chaplin, M.; Kamolnick, T.; Van Bonn, W.; Carder, D.; Ridgway, S.; Cranford, T. Conditioning *Tursiops truncatus* for nasal passage endoscopy. In Proceedings of the 27th International Association for Aquatic Animal Medicine Conference (IAAAM), Chattanooga, TN, USA, 11–15 May 1996; pp. 11–15.
34. Dover, S.R.; Van Bonn, W. Flexible and rigid endoscopy in marine mammals. In *CRC Handbook of Marine Mammals Medicine*, 2nd ed.; Dierauf, L., Gulland, M.D., Eds.; CRC Press LLC: Boca Raton, FL, USA, 2001; Volume 27, pp. 621–642.
35. Harrel, J.H.; Reiderson, T.H.; McBain, J.; Sheetz, H. Bronchoscopy of the bottlenose dolphin. In Proceedings of the 27th International Association for Aquatic Animal Medicine Conference (IAAAM), Chattanooga, TN, USA, 11–15 May 1996; pp. 11–15.
36. Tsang, W.K.; Kinoshita, R.; Rouke, N.; Yuen, Q.; Hu, W.; Lam, W.K. Bronchoscopy of Cetaceans. *J. Wildl. Dis.* **2002**, *38*, 224–227. [CrossRef]
37. Reiderson, T.H.; McBain, J.; Harrel, J.H. The use of bronchoscopy and fungal serology to diagnose *Aspergillus fumigatus* lung infection in a bottlenose dolphin. In Proceedings of the 27th International Association for Aquatic Animal Medicine Conference (IAAAM), Chattanooga, TN, USA, 11–15 May 1996; pp. 11–15.
38. Schaller, O. *Illustrated Veterinary Anatomical Nomenclature*; Ferdinand Enke Verlag: Stuttgart, Germany, 1992; pp. 1–614.
39. Roston, R.A.; Lickorish, D.; Buchholtz, E.A. Anatomy and age estimation of an early blue whale (*Balaenoptera musculus*) fetus. *Anat. Rec.* **2013**, *296*, 709–722. [CrossRef]
40. Wilkie Tinker, S. *Whales of the World*; Brill, E.J., Ed.; Bess Pr Inc.: Leiden, The Netherlands, 1988; pp. 1–310.
41. Lawrence, B.; Schevill, W.E. Gular musculature in delphinids. *Bull. Mus. Comp. Zool.* **1965**, *133*, 1–65.
42. Huggenberger, S.; Ridgway, S.H.; Oelschläger, H.H.; Kirschenbauer, I.; Vogl, T.J.; Klima, M. Histological analysis of the nasal roof cartilage in a neonate sperm whale (*Physeter macrocephalus*—Mammalia, Odontoceti). *Zool. Anz.* **2006**, *244*, 229–238. [CrossRef]
43. Pouchet, M.G.; Beauregard, H. Sur "l'organe des spermaceti". *Comptes Rendus Soc. Biol.* **1885**, *11*, 343–344.
44. Cranford, T.W.; Elsberry, W.R.; Bonn, W.G.W.; Jeffress, J.A.; Chaplin, M.S.; Blackwood, D.J.; Carder, D.A.; Kamolnick, T.; Todd, A.; Ridgway, S.H. Observation and analysis of sonar signal generation in the bottlenose dolphin (*Tursiops truncatus*): Evidence for two sonar sources. *J. Exp. Mar. Biol. Ecol.* **2011**, *407*, 81–96. [CrossRef]
45. McLelland, J. *A Colour Atlas of Avian Anatomy*; Wolfe Publishing Ltd.: Aylesbury, UK, 1990; p. 113.
46. Rowles, T.K.; Frances, M.; Dolah, V.; Hohn, A.A. Gross necropsy and specimen collection protocols. In *CRC Handbook of Marine Mammals Medicine*, 2nd ed.; Dierauf, L., Frances, M., Gulland, M.D., Eds.; CRC Press LLC: Boca Raton, FL, USA, 2001; Volume 21, pp. 449–470.
47. Dailey, M.D. Parasitic diseases. In *CRC Handbook of Marine Mammals Medicine*, 2nd ed.; Dierauf, L.A., Gulland, F.M.D., Eds.; CRC Press LLC: Boca Raton, FL, USA, 2001; pp. 357–379.
48. Raga, J.A.; Fernández, M.; Balbuena, J.A.; Aznar, F.J. Parasites. In *Encyclopedia of Marine Mammals*, 2nd ed.; Perrin, W.F., Würsig, B., Thewissen, J.G.M., Eds.; Academic Press: Cambridge, MA, USA, 2009; pp. 821–830.

MDPI

Article

Endoscopic Study of the Oral and Pharyngeal Cavities in the Common Dolphin, Striped Dolphin, Risso's Dolphin, Harbour Porpoise and Pilot Whale: Reinforced with Other Diagnostic and Anatomic Techniques

Álvaro García de los Ríos y Loshuertos [1,2], Marta Soler Laguía [3], Alberto Arencibia Espinosa [4], Francisco Martínez Gomariz [1], Cayetano Sánchez Collado [1], Alfredo López Fernández [5,6], Francisco Gil Cano [1], Juan Seva Alcaraz [1] and Gregorio Ramírez Zarzosa [1,*,†]

1 Departamento de Anatomía y Anatomía Patológica Comparadas, Facultad de Veterinaria, Universidad de Murcia, 30100 Murcia, Spain; agrios@ceuta.es (Á.G.d.l.R.y.L.); f.gomariz@colvet.es (F.M.G.); scollado@um.es (C.S.C.); cano@um.es (F.G.C.); jseva@um.es (J.S.A.)
2 Centro de Estudio y Conservación de Animales Marinos (CECAM), 51001 Ceuta, Spain
3 Departamento de Medicina y Cirugía Animal, Facultad de Veterinaria, Universidad de Murcia, 30100 Murcia, Spain; mtasoler@um.es
4 Departamento de Morfología, Anatomía y Embriología, Facultad de Veterinaria, Universidad de Las Palmas de Gran Canaria, Trasmontaña, Arucas, 35416 Las Palmas de Gran Canaria, Spain; alberto.arencibia@ulpgc.es
5 Departamento de Biología—CESAM, Campus Universitario de Santiago, Universidade de Aveiro, 3810-193 Aveiro, Portugal; a.lopez@ua.pt
6 Coordinadora para el Estudio de los Mamíferos Marinos–CEMMA, Ap. 15, Gondomar, 36380 Pontevedra, Spain
* Correspondence: grzar@um.es; Tel.: +34-868-887-546; Fax: +34-868-884-147
† Current address: Campus de Excelencia Internacional de Ámbito Regional (CEIR), Campus Mare Nostrum, Universidad de Murcia, 30100 Murcia, Spain.

Citation: García de los Ríos y Loshuertos, Á.; Soler Laguía, M.; Arencibia Espinosa, A.; Martínez Gomariz, F.; Sánchez Collado, C.; López Fernández, A.; Gil Cano, F.; Seva Alcaraz, J.; Ramírez Zarzosa, G. Endoscopic Study of the Oral and Pharyngeal Cavities in the Common Dolphin, Striped Dolphin, Risso's Dolphin, Harbour Porpoise and Pilot Whale: Reinforced with Other Diagnostic and Anatomic Techniques. *Animals* **2021**, *11*, 1507. https://doi.org/10.3390/ani11061507

Academic Editors: Matilde Lombardero and Mar Yllera Fernández

Received: 31 March 2021
Accepted: 19 May 2021
Published: 22 May 2021

Simple Summary: Developmental studies of the dolphin oral cavity have been scarce and were mostly carried out on adult specimens dealing with teeth and lingual development. Moreover, the adult pharyngeal cavity has been mentioned in cetacean monographic encyclopedias and handbooks. In this work, prenatal and perinatal studies of both the oral and pharyngeal cavities were performed on juvenile and adult specimens to better understand these anatomical structures. Our study analyzes these cavities using high-resolution endoscopy to observe changes in the mucosa and to compare these findings with terrestrial mammals. Even though endoscopy was the main technique used, our study was reinforced with Magnetic Resonance Imaging (MRI), anatomical techniques and fetal histology to locate and identify significant structures. Endoscopy of the oral cavity showed some interesting morphological changes. The incisive papilla, teeth, tongue papillae and lateral sublingual recesses and folds were observed in different development stages. The three different parts of the pharynx (oropharynx, laryngopharynx, and nasopharynx) were examined using endoscopy. The histological study helps us to understand the function of the pharyngeal cavity. The nasopharynx contained important structures such as the orifice of the auditory tube and its expansion, the pharyngeal diverticula of the auditory tubes. This special anatomical area was studied using MRI, serial sections and dissections. Some functional considerations are made about both cavities in the five species of odontocetes studied.

Abstract: In this work, the fetal and newborn anatomical structures of the dolphin oropharyngeal cavities were studied. The main technique used was endoscopy, as these cavities are narrow tubular spaces and the oral cavity is difficult to photograph without moving the specimen. The endoscope was used to study the mucosal features of the oral and pharyngeal cavities. Two pharyngeal diverticula of the auditory tubes were discovered on either side of the choanae and larynx. These spaces begin close to the musculotubaric channel of the middle ear, are linked to the pterygopalatine recesses (pterygoid sinus) and they extend to the maxillopalatine fossa. Magnetic Resonance Imaging (MRI), osteological analysis, sectional anatomy, dissections, and histology were also used to better understand the function of the pharyngeal diverticula of the auditory tubes. These data were then compared with the horse's pharyngeal diverticula of the auditory tubes. The histology revealed that a vascular plexus inside these diverticula could help to expel the air from this space to the nasopharynx. In the oral cavity, teeth remain inside the alveolus and covered by gums. The marginal papillae of the tongue differ in extension depending on the fetal specimen studied. The histology reveals that the incisive papilla is vestigial and contain abundant innervation. No ducts were observed inside lateral sublingual folds in the oral cavity proper and caruncles were not seen in the prefrenular space.

Keywords: Striped dolphin (*Stenella coeruleoalba*); Common dolphin (*Delphinus delphis*); Pilot whale (*Globicephala melas*); Risso's dolphin (*Grampus griseus*); Harbour porpoise (*Phocoena phocoena*); fetal development; mouth; buccal; oral; pharyngeal; cavity; endoscopy; sectional anatomy; dissection; histology; ontogeny; MRI

1. Introduction

The anatomy of the cephalic region of marine mammals has undergone many evolutionary changes [1]. Amongst all those adaptations in cetaceans, the complete independence of the digestive and respiratory system is one of the most important and has direct implications in the colonization of the aquatic environment. Even though the process of elongation of the skull and the migration of certain flat bones mainly affected the nasal and cranial cavities, the oral cavity also underwent a change in comparison with terrestrial animals, and the repositioning of these bones and the vertical placement of the nasal structures meant that the nasal bones and orifice disappeared from the face.

Inside the cavities surrounded by these bones, the muscles, organs and other anatomical structures also adapted to the new aquatic environment. The tongue, which occupies most of the space within the oral cavity, has the most important role (followed by the teeth) in maintaining the thermoregulatory and feeding [2] functions of terrestrial mammals, but also in acquiring the new function of expelling water from the mouth after feeding, a feature especially seen in Mysticetes.

The oral cavity is not the most studied region in the cetacean head, when compared, for example, with the nasal cavity [3–8]. The snout is a sensitive anatomical structure, the first one to have contact with the environment and it performs specific interactions, such as courtship (epicritic tactile sensibility) and defence (protopathic tactile sensibility). There are few articles in the scientific literature about the oral cavity and pharynx, and some of the existing studies may need to be clarified. Most of these articles are about adult specimens, [9–12], and the few papers on development, both in mysticetes [13,14] and odontocetes [15,16], focus mainly on the tongue of neonates, and do not cover different stages of development, being restricted to only one species [17].

Throughout gestation, bones and other oral structures are in the process of development and their formation continues after birth. Endoscopy is commonly used in dolphin medicine, especially to view the lower respiratory tract (lungs and bronchus) [18–23]. It is not often used to visualize the oropharyngeal tract, a region of key importance in stranded dolphins where pathological changes (such as tooth infections or oral obstructions) can be the cause of death.

We include the pharyngeal cavity, as food goes towards the oesophagus and stomach via the laryngopharynx. A comparison with terrestrial mammals will also serve to describe the structures of the head, following the Illustrated Veterinary Anatomical Nomenclature [24] and will aid in our understanding of the function and position of the different structures forming the buccopharyngeal cavity of our small cetaceans.

2. Materials and Methods

2.1. Animals

In the current study, we analysed the oral complex of 24 odontocetes belonging to five species (*Stenella coeruleoalba*, *Delphinus delphis*, *Globicephala melas*, *Grampus griseus* and *Phocoena phocoena)* of all ages (17 fetuses of different stages, three newborn, two juveniles and two adults) carrying out several diagnostic techniques on all specimens: namely, endoscopy, magnetic resonance imaging (MRI), dissections and histology (Table 1). Additional information is contained in Table S1 (Supplementary Materials). The mothers of each fetus were stranded along the Spanish Atlantic and Mediterranean coasts. The newborn and adult specimens were stranded along the Spanish Mediterranean coast. All stranded specimens were found dead and ethics committee clearance was not necessary. An endoscopic study was carried out, transporting sixteen fetuses and one juvenile to Veterinary Clinic "Bonafé", La Alberca, Murcia, (Spain) during November 2020. Eight fetuses were transported to the MRI unit to perform scans. One newborn, one juvenile and one adult specimen were used to obtain anatomical sections, the same adult and two fetuses for histological analysis and one newborn for dissection.

Table 1. Specimens of dolphin used in this study. dde: *Delphinus delphis* (Linnaeus 1758) from Galicia, Spain; scop: *Stenella coeruleoalba* (Meyen 1833) from Galicia, Spain; gma: *Globicephala melas* (Traill 1809) from Galicia, Spain; scoce: *Stenella coeruleoalba* (Meyen 1833) from Ceuta, Spain; scomu: *Stenella coeruleoalba* (Meyen 1833) from Murcia, Spain; phog: *Phocoena phocoena* (Linnaeus 1758) from Galicia, Spain; grgr: *Grampus griseus* (Cuvier 1912) from Valencia, Spain; MRI: Magnetic resonance imaging.

Study Code	Specie, Sex Stage [25–27].	Anatomical, Surgical/Imaging Diagnostic Techniques
dde1	*Delphinus delphis* L, male fetus	Endoscopy
dde2	*Delphinus delphis* L, male fetus	Endoscopy
dde3	*Delphinus delphis* L, female fetus	Endoscopy, MRI
scop1	*Stenella coeruleoalba* M, female fetus	Endoscopy
gma1	*Globicephala melas* T, male fetus	Endoscopy, MRI
dde5	*Delphinus delphis* L, female fetus	Endoscopy, MRI
dde6	*Delphinus delphis* L, female fetus	Endoscopy
dde7	*Delphinus delphis* L, male fetus	Photography
dde8	*Delphinus delphis* L, female fetus	Endoscopy. MRI
dde9	*Delphinus delphis* L, male fetus	Endoscopy
dde10	*Delphinus delphis* L, female fetus	Endoscopy, histological analysis
dde11	*Delphinus delphis* L, male fetus	Endoscopy, MRI
dde12	*Delphinus delphis* L, male fetus	Endoscopy
dde13	*Delphinus delphis* L, female fetus	Endoscopy, MRI
phog1	*Phocoena phocoena* L, female fetus	Osteology
dde14	*Delphinus delphis* L, female fetus	Endoscopy, MRI, histological analysis
scoce1	*Stenella coeruleoalba* M, male newborn	Head dissection
scomu1	*Stenella coeruleoalba* M, female newborn	Oral cavity analysis
scomu2	*Stenella coeruleoalba* M, male newborn	Head coronal section
grgr1	*Grampus griseus* C, female fetus	MRI
scomu3	*Stenella coeruleoalba* M, male juvenile	Head sagittal section
scomu4	*Stenella coeruleoalba* M, Male juvenile	Endoscopy
scomu6	*Stenella coeruleoalba* M, male adult	Head sagittal section, histological analysis
dde15	*Delphinus delphis* L, adult	Osteology

2.2. Endoscopy

A fixed endoscopy unit (Karl Storz Autocon 200, Tuttlingen, Germany) located at Clínica Veterinaria "Bonafé", La Alberca (Murcia), Spain with a camera processor (Storz image 1 hub, camera head Karl Storz Image 1 H3 HD, a Storz power led 175) was used to obtain endoscopic images. For the oral, oropharyngeal and laryngopharyngeal cavities, we employed a forward telescope (0° enlarged view, diameter 4 mm, length 14 cm, with incorporated fiber optic light transmission) and for the nasopharyngeal cavity, we used a forward-oblique telescope 30°, diameter 2.7 mm, length 18 cm, with incorporated fiber optic light transmission.

The endoscopic procedure was performed on the fetuses and juvenile specimens following standard protocols. Each specimen was placed in ventral recumbency and facing the endoscopist. Initial image acquisition was of the oral, oropharyngeal and laryngopharyngeal cavities using an optic of 4 mm diameter and 0° vision angle. Following this, the pharyngeal cavity was examined, first visualizing the oropharynx, then turning the optic to obtain a complete exposition of this part of the pharyngeal cavity. After visualizing the epiglottic mucosa, the endoscope was introduced through the left pyriform recess towards the oesophageal vestibule to reach the laryngoesophageal limit and the oesophageal mucosa.

The nasopharyngeal cavity was examined by introducing an optic of 2.7 mm and 30° vision angle protected by a 3 mm sheath forming an irrigation channel, visualizing first the vestibule, then the nasal plugs to the nasal cavity and finally the nasopharynx. All endoscopic images were stored in external and internal hard disks at the CVB (Bonafé Veterinary Clinic) and at the Department of Anatomy and Embryology, Facultad de Veterinaria, Universidad de Murcia, Spain.

2.3. Magnetic Resonance Imaging

Magnetic Resonance (MR) images were obtained with a high-field MR apparatus (General Electric Sigma Excite, Schenectady, NA, USA; Centro Veterinario de Diagnóstico por Imagen de Levante, Ciudad Quesada, Alicante, Spain), 1.5 Tesla using a human quadknee coil (dde3, 5, 8, 11,12, gma1) and head coil (dde13-14, grgr1). Images were used to analyze the oral and pharyngeal cavities, and for a special study of the pharyngeal diverticulum of the auditory tube during fetal development. All dolphin specimens were positioned in ventral recumbency. The MR images were transferred to a DICOM workstation. MR images were analyzed with Radiant DICOM viewer. MRI parameters used are in Table S2 (Supplementary Materials). All MR images were stored in external and internal hard disks at the CVDIL (Veterinary Center for Imaging Diagnosis), Ciudad Quesada, Alicante, Spain and the Department of Anatomy and Embryology, Facultad de Veterinaria, Universidad de Murcia, Spain.

2.4. Anatomic Evaluation: Sectional Anatomy and Dissection Techniques

One newborn, one juvenile and one adult *Stenella coeruleoalba* were frozen at −20 °C prior to obtaining coronal and sagittal sections of the head. One adult *Stenella coeruleoalba* was frozen at −46 °C prior to obtaining sagittal sections. All specimens were cut with a band saw (Anatomical Lab, Department of Anatomy and Embryology, Universidad de Murcia, Murcia, Spain), obtaining 0.5–0.7 cm thick slices. Head sections and slices were immersed in 10% formaldehyde for preservation and then stored in a cooling chamber (3 °C) at the Department of Anatomy and Embryology, Facultad de Veterinaria, Universidad de Murcia, Spain.

Sixteen fetuses were preserved by immersion in formaldehyde (10%). Two fetuses and one adult were fixed with embalming solution injected into the fetal umbilical artery and vein, and the adult external jugular vein and left auricule. In a *Stenella coeruleoalba* (scomu5), the external jugular vein and the auricles were injected with embalming solution using an electrical pump. Head coronal and sagittal sections and a deep head dissection of one newborn were made to observe the pharyngeal diverticulum of the auditory tube. One newborn specimen was used to study the normal anatomy of the oral and nasal cavities and one fetus and one adult specimen were used to inspect the bony anatomy of the oral and pharyngeal cavities (Table 1).

All specimens used were stored in a cooling and freezer chambers at the Department of Anatomy and Embryology, Facultad de Veterinaria, Universidad de Murcia, Spain.

2.5. Histological Analysis

The mucosa of the oral and pharyngeal cavities was histologically analyzed in two fetuses, one *Delphinus delphis* and one adult *Stenella coeruleoalba*. Elongated rectangles of nasal mucosa were removed at different levels of these cavities. Samples were oriented perpendicular to the paraffin block base and then processed using a special saw. Paraffin blocks were cut to obtain slices. Routine histological processing was carried out and sections were stained with Haematoxylin and Eosin. Samples were then photographed with a computed light microscope (Zeiss Axioskop 40, Jena, Germany) with Camera Insight 2 Axiocam 105 color incorporated. Histological sections were stored at the Department of Anatomy and Embryology, Facultad de Veterinaria, Universidad de Murcia, Spain.

3. Results

3.1. The Oral Cavity

The oral region is closed during fetal development. We have opened it to view the different parts, starting with the vestibule placed between the lips and the teeth. Beyond the vestibule is the oral cavity proper, which extends from the rostral part of the palatoglossal arch or folds to the lingual aspect of the incisive teeth. The dorsal limit is the oral cavity's roof, and the ventral limit is defined by the tongue and the proper oral cavity and the prefrenular space (Figure 1).

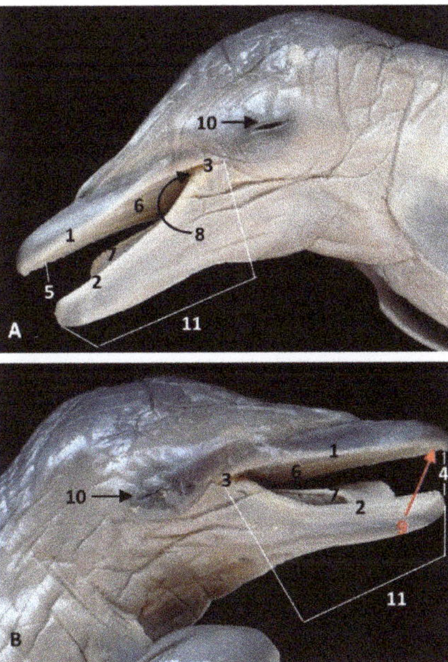

Figure 1. Head of a dolphin fetus showing the oral cavity opened. (**A**) Left aspect; (**B**) Right aspect. 6 months, dde7. 1, Upper lip; 2, Lower lip; 3, Angulus oris; 4, Rima oris; 5, Oral vestibule; 6, Oral cavity roof; 7, Tongue and oral cavity floor: 8, Palatoglossal arches or folds; 9, Incisive teeth (not erupted yet); 10, Eyelids (closed); 11, Oral region.

The bony part of the oral cavity's roof is composed of the palatine processes of the maxillary and incisive bones as seen in the *Phocoena phocoena* (phog1). The bony ventral limit of the oral cavity is represented by the left and right mandibles. In this specimen, close to birth, teeth have a conic shape. The intermandibular symphysis can be seen. Four incisive teeth were observed (Figure 2).

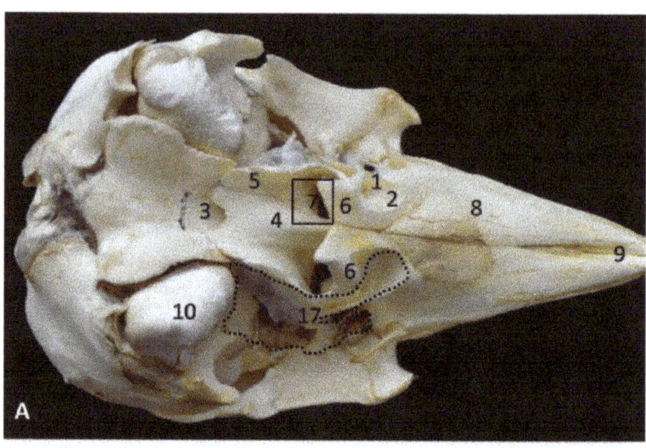

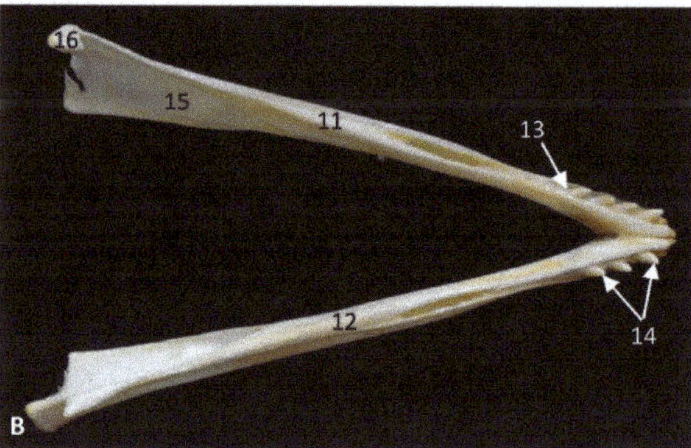

Figure 2. Bony anatomical base of the oral and pharyngeal cavities in a skull. (**A**) Ventral view. Right and left mandibles. (**B**) Dorsal view. Dots show the extension and form of the right pharyngeal diverticulum of the auditory tube. Photography Francisco Gil Cano. 9 months, phop1. 1, Palatine bone: perpendicular lamina; 2, Palatine bone: horizontal lamina; 3, Basisphenoid bone: body; 4, Vomer bone; 5, Pterygoid bone: medial lamina; 6, Pterygoid bone: lateral lamina (incomplete); 7, Choana; 8, Maxillary bone: palatine process; 9, Incisive bone: palatine process; 10, Temporal bone: petrous and tympanic part; 11; Left mandible: body; 12, Right mandible: body; 13, Dental alveolus; 14, Incisive teeth; 15, Mandibular channel; 16, Mandible: condyle; 17, Bony area of pharyngeal diverticulum of the auditory tube.

3.1.1. Endoscopic Study

In the least developed fetus a *Delphinus delphis* (dde1), the lips are proportionally large and immobile. The oral rim, like the blowhole, is tightly closed. A frenulum of the superior and inferior lips is absent. The buccal vestibule is not well-defined at this stage of development. The gums of the mandibles are wider than those of the maxilla. A hard palate is forming. An incisive papilla was not observed. The tongue is short, wide and without papillae (Figure 3).

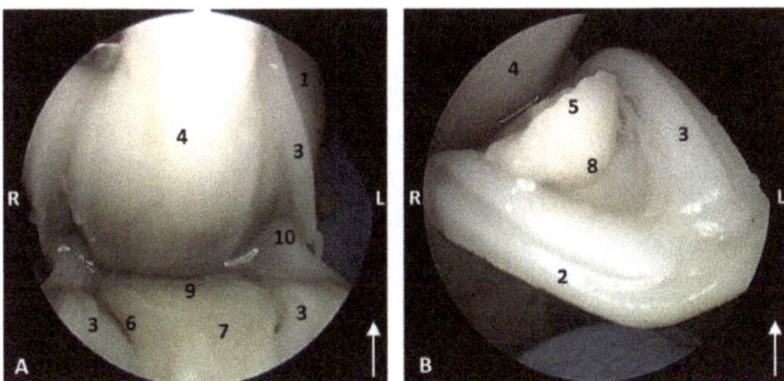

Figure 3. Endoscopic images of the oral cavity. The arrows point to the tip of the mouth. **L** (Left) **R** (Right). (**A**) Oral cavity open. (**B**) Oral cavity proper. 1,5 months, dde1. 1, Upper lip; 2, Lower lip; 3, Gums; 4, Hard palate; 5, Tongue: tip; 6, Tongue: border; 7, Tongue: dorsum; 8, Tongue: ventral part; 9, Tongue: root; 10, Angulus oris.

In the next fetuses (dde2, dd3), *Delphinus delphis*, a labial vestibule between the gums and lips was observed (not a buccal vestibule). The hard palate shows a middle palatine raphe in all species studied, and only in *Delphinus delphis* are both lateral palatine grooves parallel to the middle hard palate. No transverse ridges of the mucosa of hard palate were observed. We observed the incisive papilla at the tip of hard palate. Marginal papillae extend from the edges to the middle of the tongue. The dorsum of the tongue is smooth with an indistinguishable central groove. In the oral cavity proper, there are two lateral sublingual recesses, with a lateral sublingual fold which thickened rostrally (Figure 4).

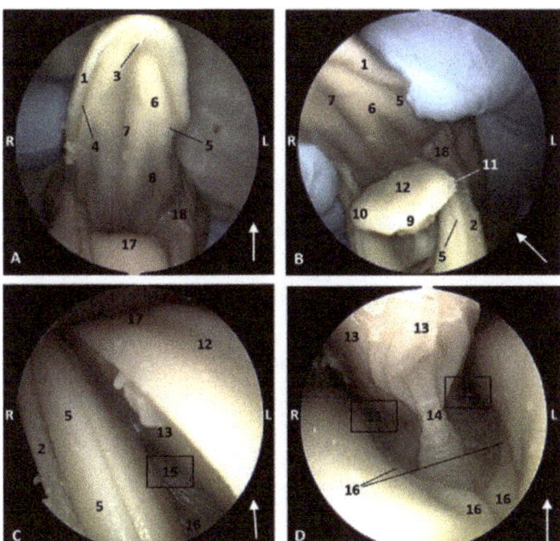

Figure 4. Endoscopic images of the oral cavity. The arrows show the tip of the mouth. **L** (Left) **R** (Right). (**A**,**B**) Oral cavity open. (**C**) Tongue: lateral part. (**D**) Oral cavity proper. 3,5 months, dde2. 1, Upper lip; 2, Lower lip; 3, Incisive papilla; 4, Labial vestibule; 5, Gums; 6, Hard palate; 7, Palatine raphe; 8, Greater palatine groove; 9, Tongue: tip; 10, Tongue: border; 11, Marginal papilla; 12, Tongue: dorsum; 13, Tongue: ventral part; 14, Lingual frenulum; 15, Lateral sublingual recesses; 16, Lateral sublingual folds; 17, Tongue: root; 18, Angulus oris.

In dde3, the lateral sublingual folds are very thin (Figure 5).

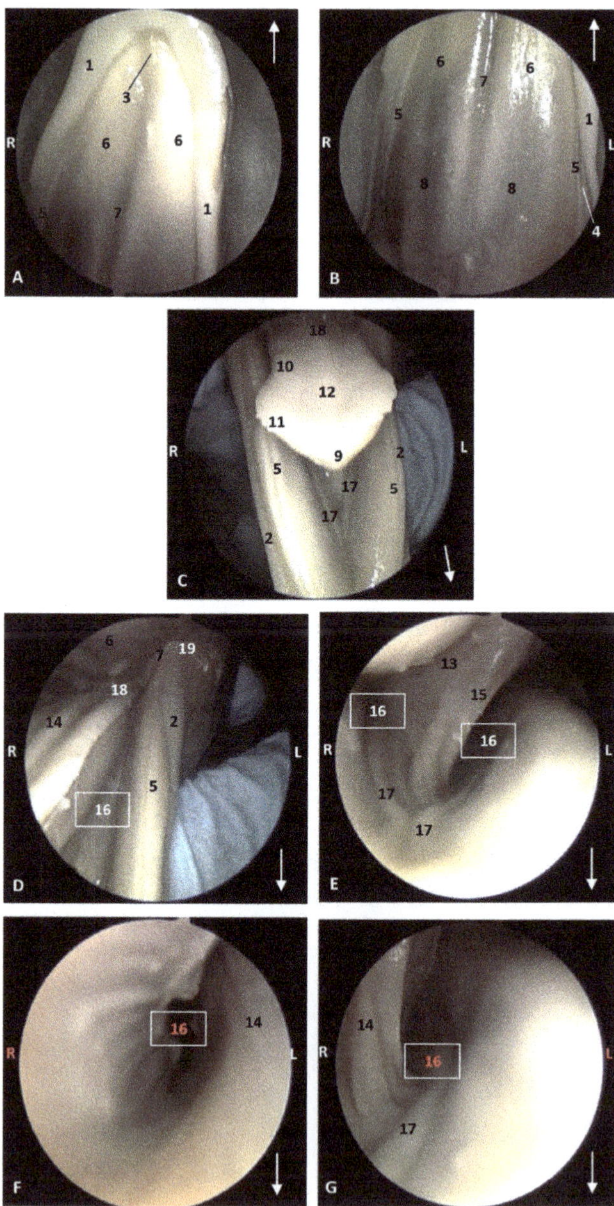

Figure 5. Endoscopic images of the oral cavity. In the different images, the arrows point to the tip of the mouth. **L** (Left) **R** (Right). (**A**,**B**) Hard palate. (**C**) Tongue1. (**D**–**G**) Oral cavity proper. 4 months, dde3. 1, Upper lip; 2, Lower lip; 3, Incisive papilla; 4, Labial vestibule; 5, Gums; 6, Hard palate; 7, Palatine raphe; 8, Greater palatine groove; 9, Tip of the tongue; 10, Tongue: border; 11, Marginal papilla; 12, Tongue: dorsum; 13, Tongue: ventral part; 14, Tongue: longitudinal prominence; 15, Lingual frenulum; 16, Lateral sublingual recess; 17, Lateral sublingual folds; 18, Tongue: root; 19, Angulus oris.

The ventral part of the tongue shows a simple lingual frenulum. In the pre-frenular space no sublingual caruncles were observed. Gums are long and smooth (Figures 4 and 5).

In the *Stenella coeruleoalba* (scop1), one month older than *Delphinus delphis* (dde3), the hard palate shows a palatine raphe and the incisive papilla is a little larger than in *Delphinus delphis* (dde2). The lateral sublingual folds are thin (Figure 6).

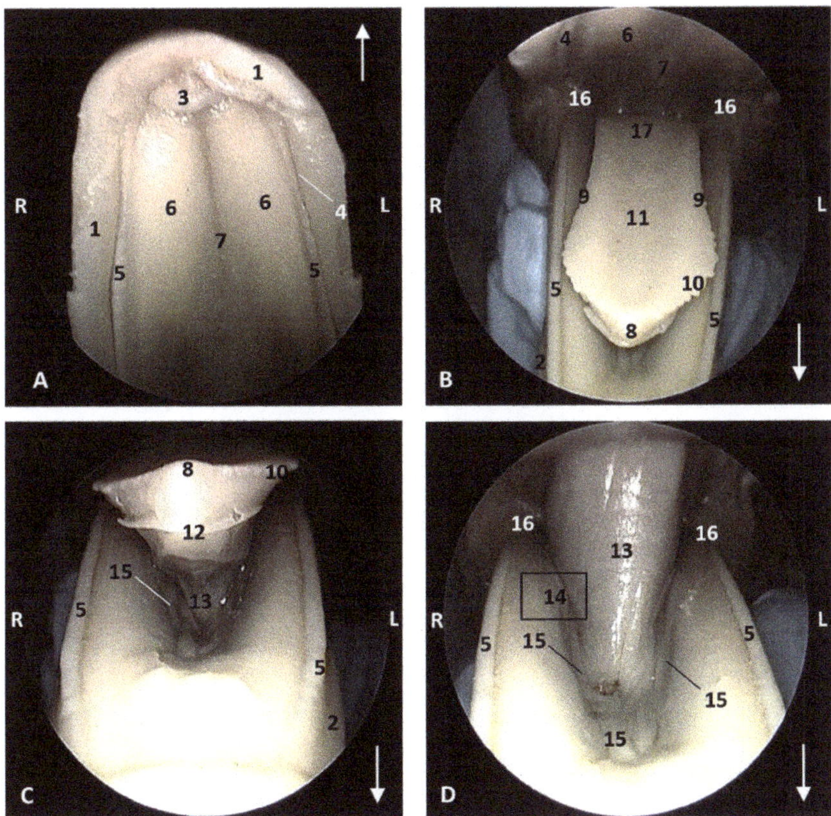

Figure 6. Endoscopic images of the oral cavity. The arrows show the tip of the mouth. **L** (Left) **R** (Right). (**A**) Hard palate. (**B**,**C**) Tongue. (**D**) Oral cavity proper. 4,5 months, scop1. 1, Upper lip; 2, Lower lip; 3, Incisive papilla; 4, Labial vestibule; 5, Gums; 6, Hard palate; 7, Palatine raphe; 8, Tongue: tip; 9, Tongue: border; 10, Marginal papilla; 11, Tongue: dorsum; 12, Tongue: ventral part; 13, Lingual frenulum; 14, Lateral sublingual recess; 15, Lateral sublingual folds; 16, Angulus oris; 17, Tongue: root.

In the *Globicephala melas* (gma1), a little older than *Stenella coeruleoalba* (scop1), the hard palate has a deep palatine raphe and the incisive papilla is larger than in other fetuses. The marginal papillae extend from the tip of the tongue almost to the root. The lingual frenulum is thinner than in other small fetuses. The gums are covering the teeth in both superior and inferior arcades. Teeth are well differentiated in this *Globicephala melas* at this stage (5 months). Also, the sublingual lateral folds are thin but their extremities are starting to thicken (Figure 7).

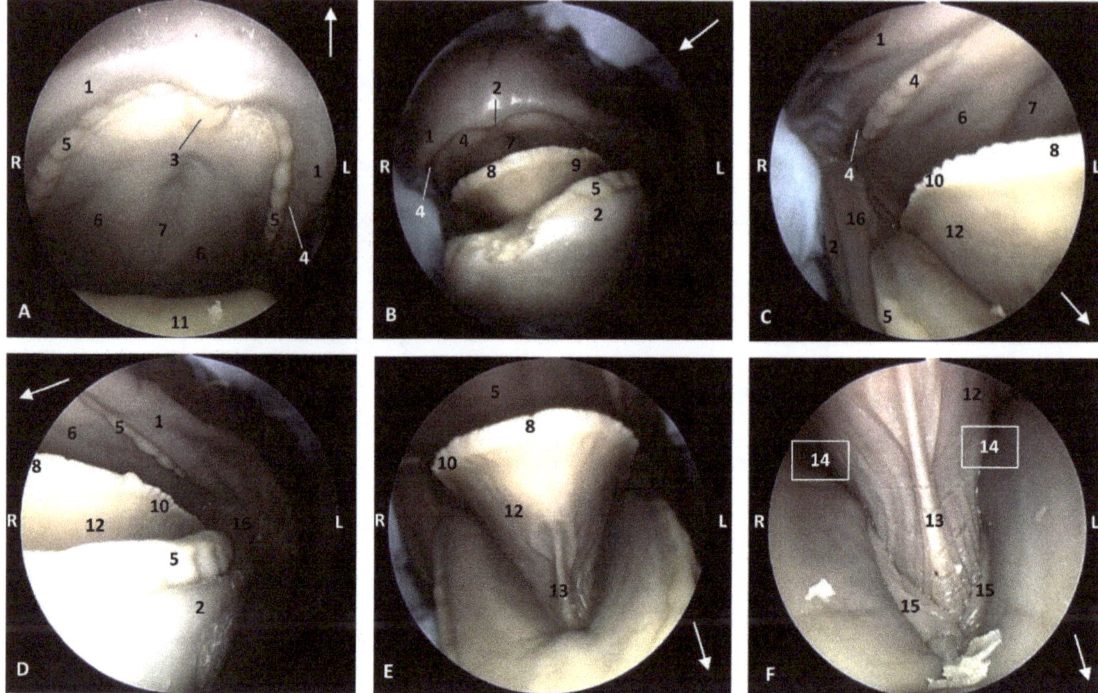

Figure 7. Endoscopic images of the oral cavity. In the different images, the arrow points to the tip of the mouth. **L** (Left) **R** (Right). (**A**,**B**) Oral cavity open. (**C**,**D**) Tongue: lateral part. (**E**,**F**) Oral cavity proper. 5 months, gma1. 1, Upper lip; 2, Lower lip; 3, Incisive papilla; 4, Oral vestibule; 5, Gums covering teeth; 6, Hard palate; 7, Palatine raphe; 8, Tongue: tip; 9, Tongue: border; 10, Marginal papilla; 11, Dorsum of the tongue; 12, Tongue: ventral part; 13, Lingual frenulum; 14, Lateral sublingual recess; 15, Lateral sublingual folds; 16, Angulus oris.

In a more developed fetus (dde8) *Delphinus delphis,* it was possible to observe marginal papillae extending from the tip to the middle of the tongue, and the lateral sublingual folds were thickening throughout their length (Figure 8).

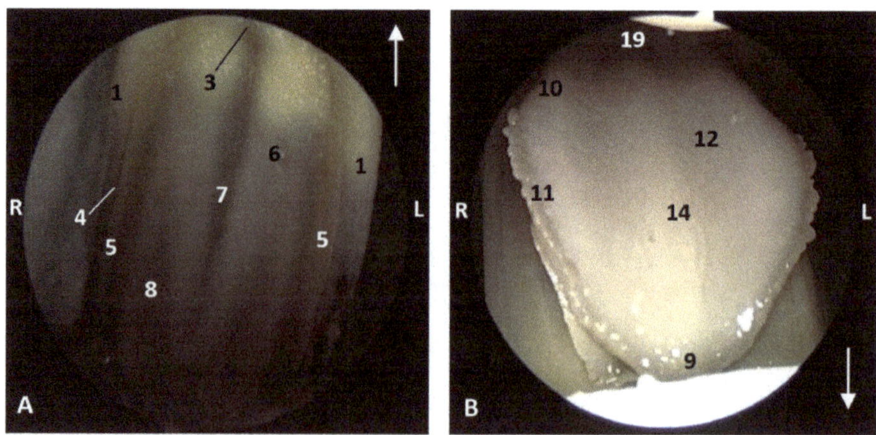

Figure 8. *Cont.*

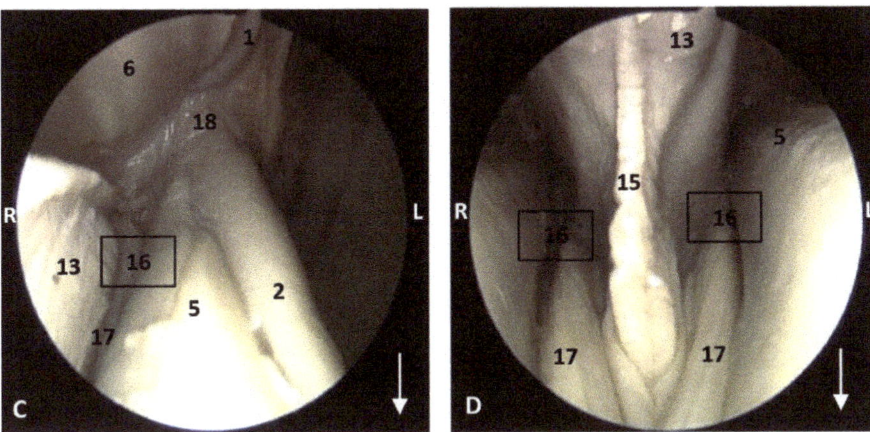

Figure 8. Endoscopic images of the oral cavity. In the different images, the arrow points to the tip of the mouth. **L** (Left) **R** (Right). (**A**) Hard palate. (**B**) Tongue. (**C**,**D**) Oral cavity proper. 6 months, dde8. 1, Upper lip; 2, Lower lip; 3, Incisive papilla; 4, Labial vestibule; 5, Gums; 6, Hard palate; 7, Palatine raphe; 8, Greater palatine groove; 9, Tip of the tongue; 10, Tongue: border; 11, Marginal papilla; 12, Tongue: dorsum; 13, Tongue: ventral part; 14, Tongue: longitudinal prominence; 15, Lingual frenulum; 16, lateral sublingual recesses; 17, Lateral sublingual folds; 18, Angulus oris; 19, Tongue: root.

In a *Delphinus delphis* (dde13) close to birth, the teeth of the superior arcade are well developed, unlike those of the inferior arcade. Additionally, the lateral sublingual folds are well dilated (Figure 9).

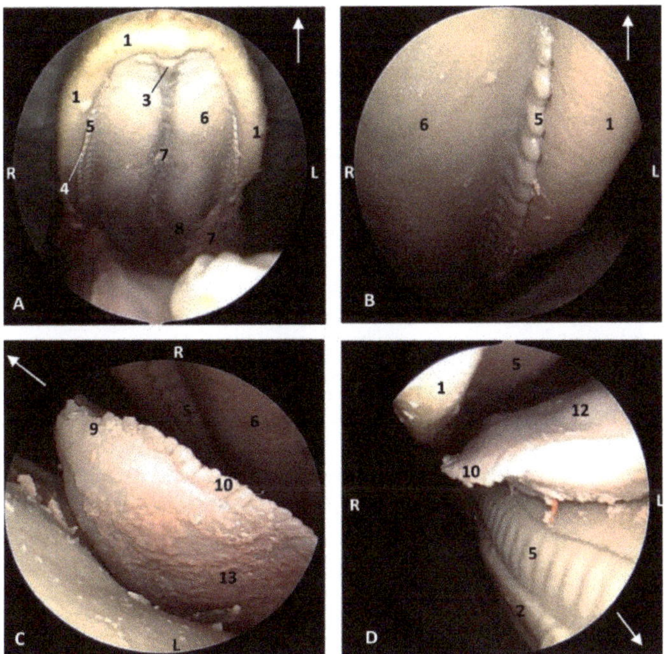

Figure 9. *Cont.*

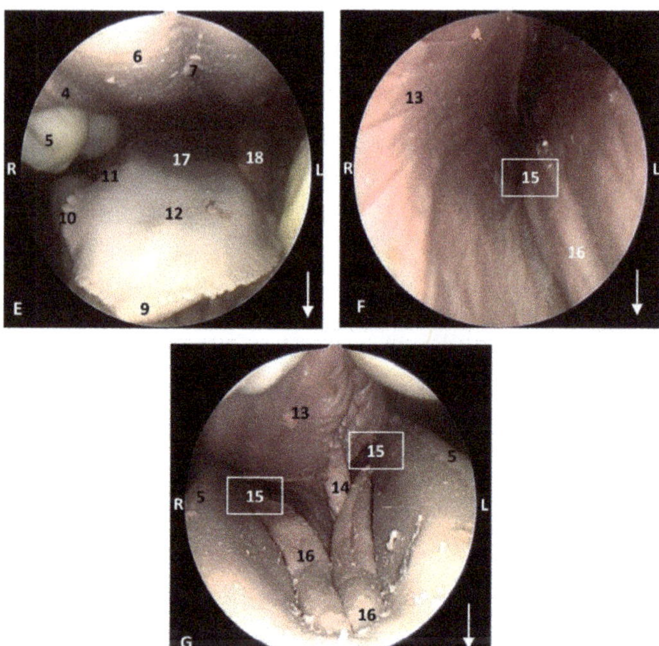

Figure 9. Endoscopic images of the oral cavity. In the different images, the arrow points to the tip of the mouth. **L** (Left) **R** (Right). (**A**) Hard palate. (**B**) Detail of gums. (**C**) Tip of the tongue and prefrenular space. (**D**) Right mandible and lateral part of the tongue. (**E**) Tongue. (**F**) Detail of left lateral sublingual recess. (**G**) Oral cavity proper: prefenular space. 9 months, dde13. 1, Upper lip; 2, Lower lip; 3, Incisive papilla; 4, Labial vestibule; 5, Gums; 6, Hard palate; 7, Palatine raphe; 8, Greater palatine groove; 9, Tongue: tip; 10, Tongue: marginal papilla; 11, Tongue: border; 12, Tongue: dorsum; 13, Tongue: ventral part; 14, Lingual frenulum; 15, lateral sublingual recesses; 16, Lateral sublingual folds; 17, Tongue: root; 18, Angulus oris.

In the youngest newborn *Stenella coeruleoalba* (scoce1), all teeth are covered by gums. Teeth eruption can be seen in the two older newborn specimens (scomu1 and scomu2) where the caudal teeth start to erupt, but the rostral teeth are covered by gums and the incisive teeth are not well developed. Also, the marginal papillae are well developed and decrease progressively towards the root. (Figure 10).

3.1.2. MRI Study

The T2 MRI image of the oral cavity in this *Delphinus delphis* (dde11) shows that the maxillary bone is medium hyperintense with respect to the hypointense hard palate. The superficial mucosa of tongue is slightly hyperintense with respect to the hypointense depressor, protractor and retractor muscles of the tongue. A hyperintense stratum under the tongue muscles is probably due to the high rate of irrigation of these muscles. We observed that non-erupted teeth can be seen under the gums (Figure 11).

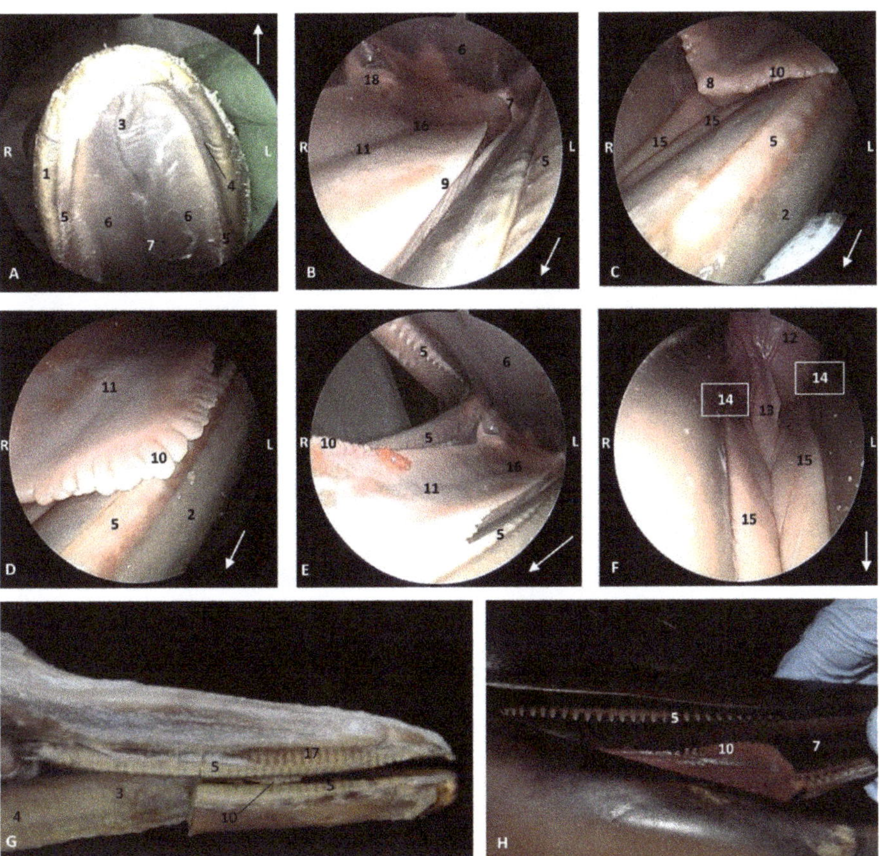

Figure 10. Endoscopic images of the oral cavity. In the different images, the arrow points to the tip of the mouth. **L** (Left) **R** (Right). (**A**) Hard palate. (**B**) Tongue. (**C–F**) Tongue, oral cavity proper and prefenular space. newborn, scumu2. (**G**) Deep dissection of the newborn dolphin head after removing skin and partial section of the right mandible. Observe that rostral teeth during lactation period do not erupt to protect mother's nipple. newborn, scocel. (**H**) Detailed image of the marginal papilla and teeth without gum covering the clinic crown. A palatal raphe in hard palate is present. newborn, scomul. 1, Upper lip; 2, Lower lip; 3, Incisive papilla; 4, Labial vestibule; 5, Gums; 6, Hard palate; 7, Palatine raphe; 8, Tongue: tip; 9, Tongue: border; 10, Tongue: marginal papilla; 11, Tongue: dorsum (middle tongue groove); 12, Tongue: ventral part; 13, Lingual frenulum; 14, lateral sublingual recess; 15, Lateral sublingual folds; 16, Tongue: root; 17, Teeth roots; 18, Angulus oris.

3.1.3. Histological Study

Vestigial incisive papillae appear with a keratinized pseudostratified epithelium. Conduits arriving at papilla were not seen, nor was the vomeronasal organ *Delphinus delphis* fetus (dde14). Abundant lymphatic vessels were observed in the dermis. Developing teeth were observed covered by gum tissue. The histological structure of teeth shows the inner dentin, covered by enamel and all parts (crown, neck, and root) covered by cementum. The sublingual lateral folds have a mucosa with abundant vessels and mucous glands while neither the sublingual caruncule nor the orobasal organ were observed (Figures 12 and 13).

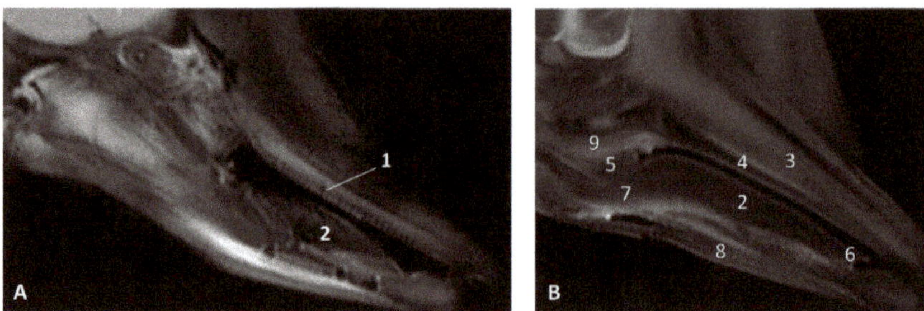

Figure 11. Images of the oral cavity. MR sagittal images is oriented so that the rostral is to the right. (**A**) T2 FrFSE sagittal plane. 6 months, dde8. Image of the oral cavity. (**B**) T2 FrFSE sagittal plane. Quadknee coil. 8 months, dde11. 1, Teeth (under gum); 2, Tongue: body; 3, Maxillary bone; 4, Hard palate; 5, Soft palate; 6, Tongue: apex; 7, Tongue: root; 8, Mandibles; 9, Pterygoid and palatine bones.

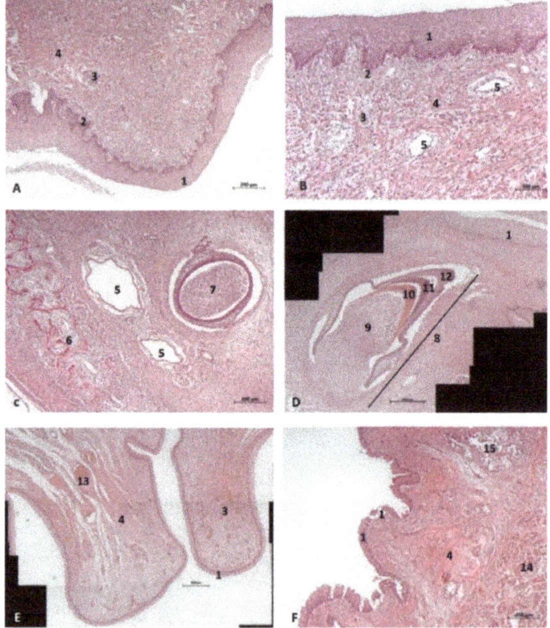

Figure 12. (**A**,**B**) Histological study of the oral cavity. (**A**–**C**) Hard palate: incisive papilla. (**D**) Tooth. (**E**) Oral cavity proper: sublingual lateral fold. (**F**) Tongue: root. 10 months, dde14. 1, Epidermis; 2, Papillary stratum; 3, Nervous tissue; 4, Connective tissue; 5, Lymphatic vessels; 6, Bony tissue; 7, Dental structure initial development transverse sectioned; 8, Dental structure development sagittal sectioned; 9, Dental papilla; 10, Dentin; 11, Enamel; 12, Cementum; 13, Venous vessels; 14, Striated muscle; 15, Mucous glands.

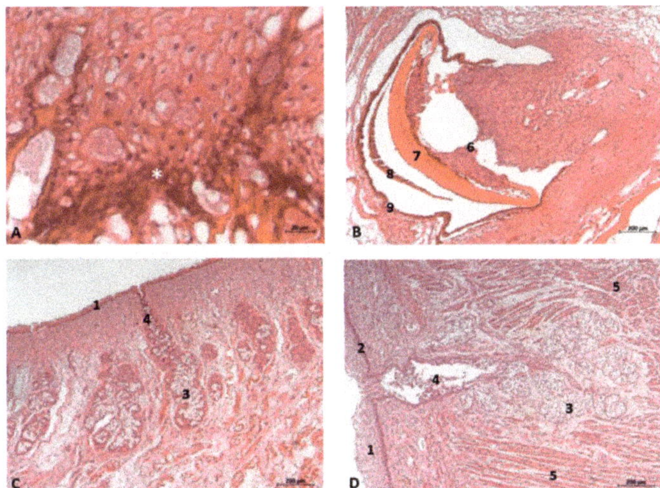

Figure 13. (**A**,**B**) Histological study of the oral cavity. (**A**) Sublingual lateral fold: pigmented epithelium in basal stratum (*). (**B**) Tooth. (**C**,**D**) Tongue: root. 7.5 months, dde10. 1, Epidermis; 2, Papillary stratum; 3, Mucous glands; 4, Secretor duct; 5, Striated muscle: proper muscle tongue; 6, Dental papilla 7, Dentin; 8, Enamel; 9, Cementum.

The sublingual lateral fold showed a pigmented epithelium at its basal stratum. The tongue shows a striated muscle base and abundant sub-epithelial mucous glands are visible, along with their secretory ducts (Figure 13).

In an adult *Stenella coeruleoalba* (scomu6), the incisive papilla showed a well-developed papillary stratum. The dermis contains abundant fat tissue and nests of epithelial ducts in regression (Figure 14A,B). The lateral sublingual recess has a keratinized epithelium with mucous glands and a well-developed papillary stratum (Figure 14).

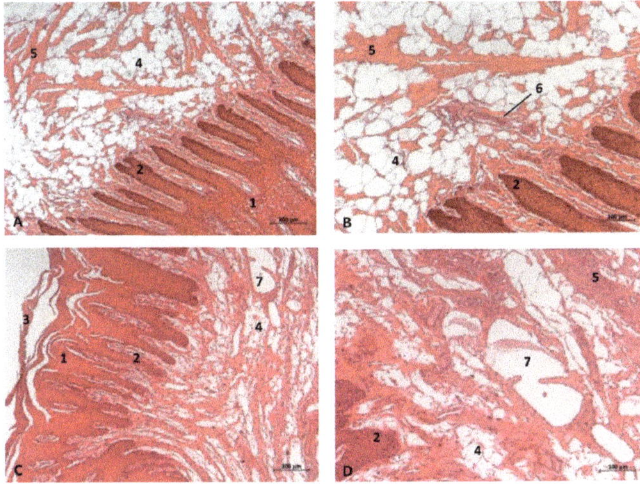

Figure 14. (**A**,**B**) Histological study of the oral cavity. (**A**,**B**) Hard palate: incisive papilla. (**C**,**D**) Oral cavity proper: sublingual lateral fold. Adult, scomu6. 1, Epidermis; 2, Papillary stratum; 3, Corneum stratum; 4, Fat tissue; 5, Connective tissue; 6, Remains of epithelial duct; 7, Lymphatic vessels.

3.2. *The Pharyngeal Cavity*

The pharynx is a musculo-membranous cavity divided into three parts: the oropharynx linking the oral cavity with the oesophagus, the nasopharynx connecting the nasal cavity with the larynx, and the laryngopharynx, which is an intermediate cavity caudal to the oropharynx and caudoventral to the nasopharynx. The laryngopharynx connects the oral cavity to the stomach and allows the aditus laryngis enter to the nasopharyngeal cavity crossing, only in cetaceans (Figure 15), the intrapharyngeal orifice.

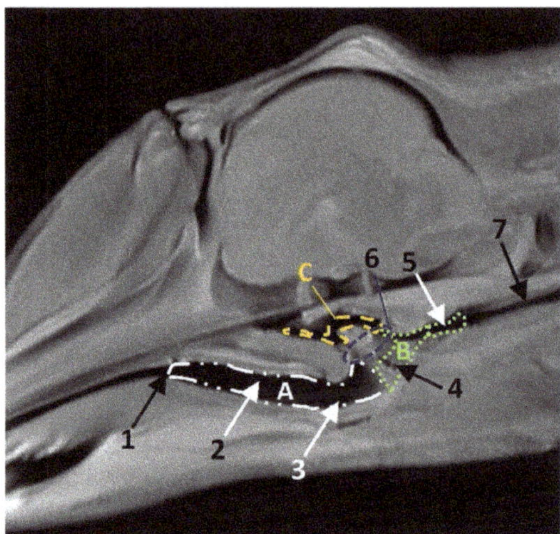

Figure 15. Head of a dolphin fetus, showing the pharyngeal cavity. (**A**) Oropharynx; (**B**) Laryngopharynx; (**C**) Nasopharynx. MRI sagittal T1 SE sequence. 10 months, dde14. 1, Isthmus of the fauces; 2, Fauces; 3, Epiglottic vallecula; 4, Piriform recess; 5, Oesophageal vestibule; 6, Intrapharyngeal orifice: 7, Oesophagus mucosa.

The bony roof of oropharynx is composed of the palatine bone and the pterygoid bone laminae (lateral and medial). Additionally, the nasopharynx is delimited dorsally by the vomer wings and the medial lamina of the pterygoid bone (Figure 2). Ventrally, the hyoid apparatus attaches to the root of the tongue and the larynx to the base of the cranium.

3.2.1. Study of Oropharynx, Nasopharynx and Laryngopharynx

The oropharynx begins at the isthmus of the fauces, continues with a conduit (fauces) and finishes at the lingual aspect of the epiglottic cartilage (Figure 15).

Endoscopic Study

The endoscopic study began at the **oropharynx**, showed a tightly closed isthmus of the fauces in a young fetus, a *Delphinus delphis* (dde2). The endoscope could not cross this gate (Figure 16).

The endoscope was passed into the fauces in an older fetus, a *Delphinus delphis* (dde3) showing a bright mucosa. No lymphoreticular tissue in the floor (tongue), walls (palatoglossus arches or folds) or roof (soft palate) of the fauces was observed. The soft palate is inserted into the ventral crest formed between the lateral and medial lamina of the pterygoid bones. The palatoglossal arches or folds connect to the soft palate through the tongue root. At the end of fauces a soft vallecula continues dorsally with the lingual aspect of the mucosa of epiglottic cartilage. At this level we dorsally observed the intrapharyngeal orifice to allow entry of the larynx into the **nasopharynx**. Additionally, the **laryngopharynx** begins with a piriform recess on either side of the larynx cartilages. The left piriform recess is wider than the right one. The dilated oesophageal vestibule is caudal to the recesses whose mucosa is arranged in longitudinal folds changing to small quadrangular folds where the oesophageal mucosa begins (Figure 17).

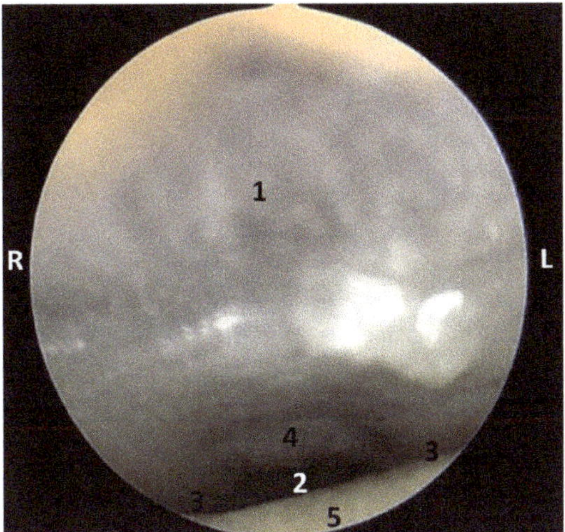

Figure 16. Endoscopic image of the pharyngeal cavity: oropharynx. **L** (Left) **R** (Right). Fauces: isthmus. 3.5 months, dde2. 1, Hard palate; 2, Isthmus of the fauces; 3, Arcus palatoglossus or palatoglossus folds; 4, Soft palate; 5, Tongue: root.

Both the *Stenella coeruleoalba* (scop1) and *Globicephala melas* (gma1) fetuses had a well-developed mucosa at the isthmus and only a narrow passage to the fauces which had a pale lingual mucosa and a grey/brown colour in its walls and roof (Figures 18 and 19).

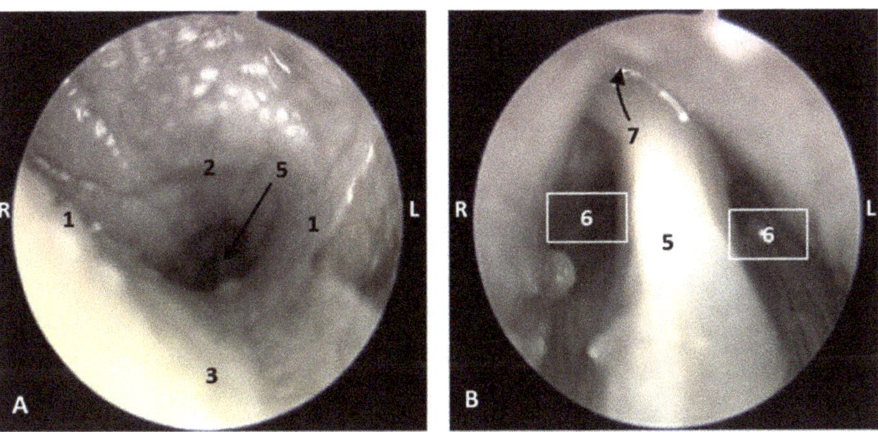

Figure 17. *Cont.*

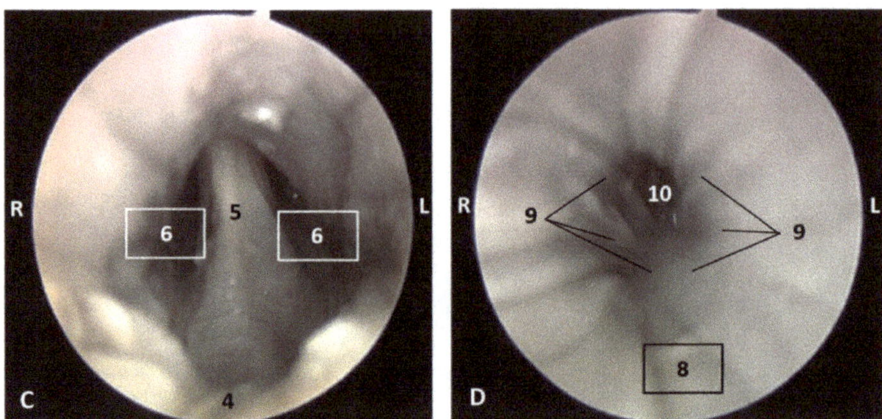

Figure 17. Endoscopic image of the pharyngeal cavity. **L** (Left) **R** (Right). (**A–C**) Oropharynx, (**A**) Fauces. (**B–D**) Laryngopharynx. 4 months, dde3. 1, Arcus palatoglossus or palatoglossus folds; 2, Soft palate; 3, Tongue: root; 4, Epiglottic vallecula; 5, Epiglottis: lingual surface (mucosa); 6, Piriform recess; 7, Intrapharyngeal orifice (nasopharynx); 8, Oesophageal vestibule; 9, Pharyngoesophageal limit; 10, Oesophageal mucosa.

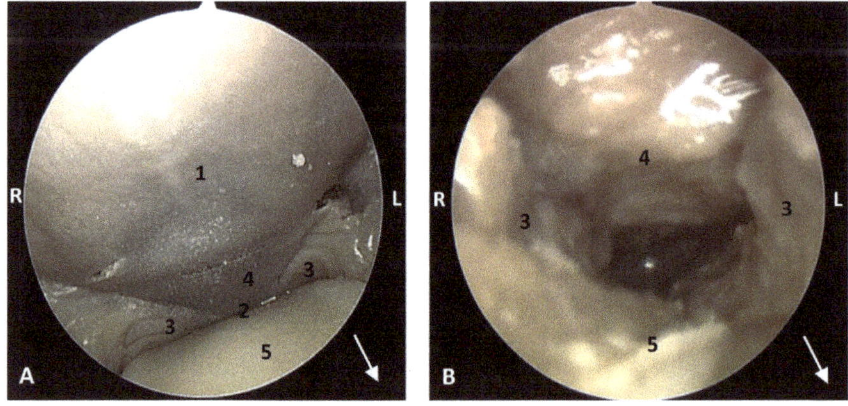

Figure 18. Endoscopic image of the pharyngeal cavity: oropharynx. The arrows show where is the tip of the mouth. **L** (Left) **R** (Right). (**A**) Fauces: isthmus. (**B**) Fauces: inside. 4.5 months, scop1. 1, Hard palate; 2, Isthmus of the fauces (closed); 3, Arcus palatoglossus or palatoglossus folds; 4, Soft palate; 5, Tongue: root.

A well-defined fauces was observed in an older *Delphinus delphis* (dde8) and also a broad left piriform recess, with longitudinal folds finishing at the oesophageal vestibule (Figure 20).

In this well-developed fetus (*Delphinus delphis*) (dde9), the endoscope could pass into the choanae to see the nasopharynx and the pharyngeal orifice of the auditory tube; alsothe longitudinal folds changing to small quadrangular folds where the oesophageal mucosa begins (Figure 21). The oropharyngeal mucosa is thickening, the longitudinal folds in the piriform recesses of the laryngopharynx are thin, and a clear difference between the mucosa of the oesophageal vestibule and oesophagus was seen.

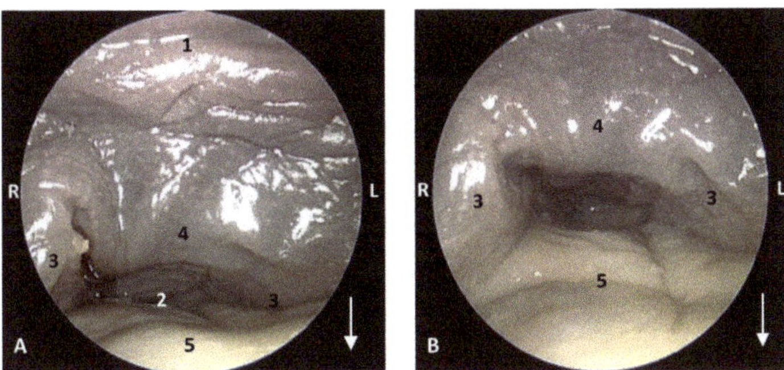

Figure 19. Endoscopic image of the pharyngeal cavity: oropharynx. The arrows show where is the tip of the mouth. **L** (Left) **R** (Right). (**A**) Fauces: isthmus. (**B**) Fauces: inside. 5 months, gma1. 1, Hard palate; 2, Isthmus of the fauces (closed); 3, Arcus palatoglossus or palatoglossus folds; 4, Soft palate; 5, Tongue: root.

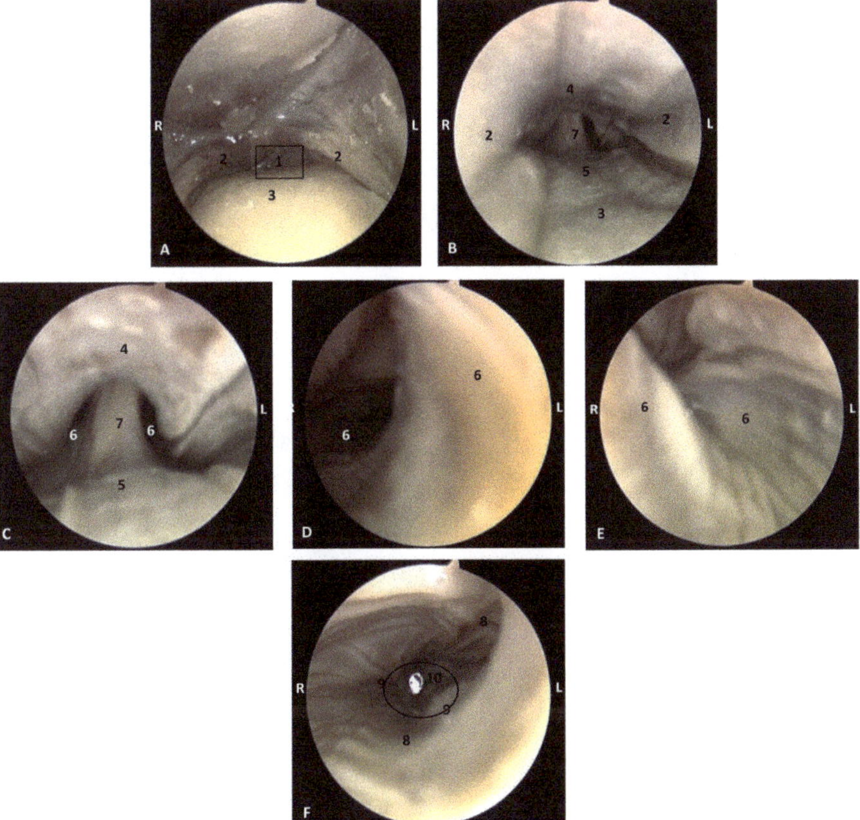

Figure 20. Endoscopic images of the pharyngeal cavity. **L** (Left) **R** (Right). (**A–C**) oropharynx. (**B**) Fauces. (**C–F**) Laryngopharynx. 6 months, dde8. 1, Isthmus of the fauces; 2, Arcus palatoglossus; 3, Tongue: root; 4, Soft palate; 5, Epiglottic vallecula; 6, Piriform recess; 7, Epiglottis: mucosa. 8, Oesophageal vestibule; 9, Pharyngoesophageal limit; 10, Oesophageal mucosa.

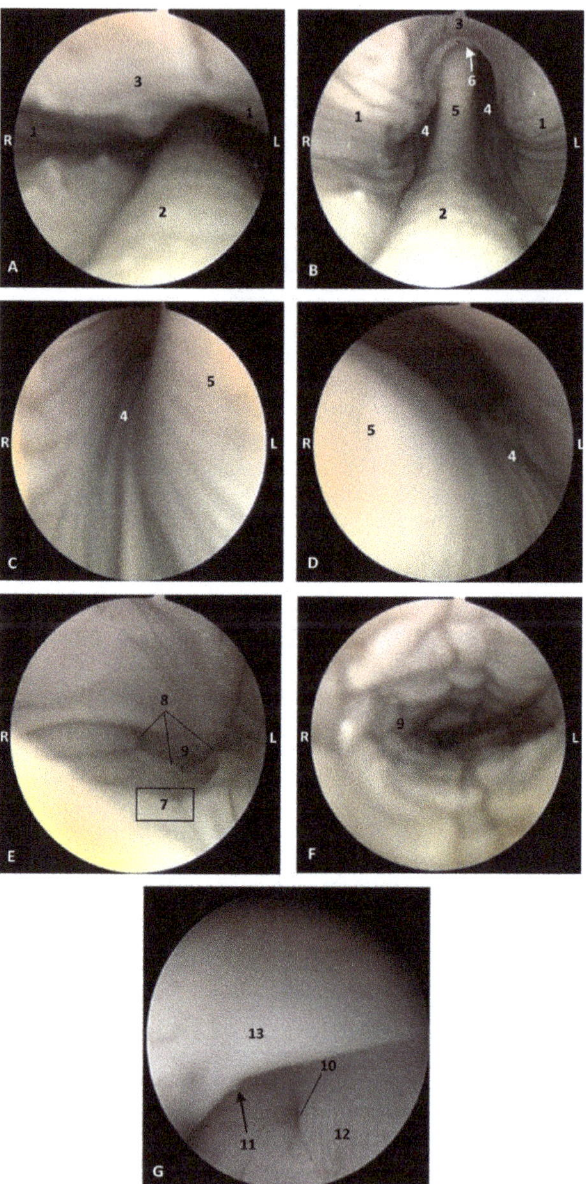

Figure 21. Endoscopic images of the pharyngeal cavity. **L** (Left) **R** (Right). (**A**,**B**) oropharynx: fauces. (**B**–**F**) Laryngopharynx. (**G**) Left nasopharynx. 7 months, dde9. 1, Arcus palatoglossus; 2, Tongue: root; 3, Soft palate; 4, Piriform recesses (laryngopharynx); 5, Epiglottis: mucosa. 6, Intrapharyngeal orifice (entrance to nasopharynx); 7, Oesophageal vestibule; 8, Pharyngoesophageal limit; 9, Oesophageal mucosa; 10, Pharyngeal orifice of the auditory tube; 11, Choanae; 12, Nasopharyngeal mucosa: longitudinal or striated folds; 13, Nasal septum: vomer bone.

The mucosa of the fauces continues to thicken and has a bright aspect in a *Delphinus delphis* fetus (dde11). Additionally, in the nasopharynx, the mucosa shows longitudinal folds and small openings surrounding the pharyngeal orifice of the auditory tube (Figure 22).

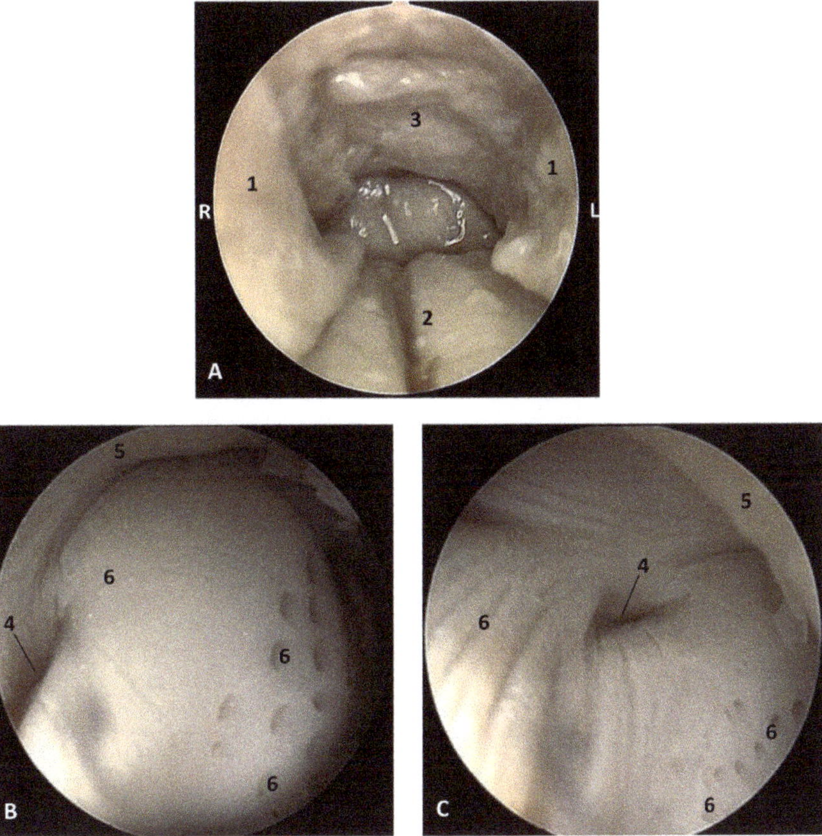

Figure 22. Endoscopic images of the pharyngeal cavity. **L** (Left) **R** (Right). (**A**) Oropharynx: fauces. (**B**) Left nasopharynx. (**C**) Right nasopharynx. 8 months, dde11. 1, Arcus palatoglossus or palatopharyngeal folds; 2, Tongue: root; 3, Soft palate; 4, Pharyngeal orifice of the auditory tube; 5, Nasal septum: vomer bone; 6, Nasopharyngeal mucosa: longitudinal or striated folds and small openings.

In a juvenile dolphin, we could observe the pinkish mucosa of the nasopharynx with longitudinal folds, but the small holes had less border definition (Figure 23).

Histological Study

The histological results show that the epidermis of the oropharynx at the soft palate level has a tightly papillary stratum with deep mucous glands and abundant mucous glands in its submucosa. Additionally, at the level of the isthmus of the fauces, histology shows a connective tissue stratum deep to the epidermis, containing many deep mucous glands (Figure 24A,B). The nasopharynx shows a respiratory mucosa with an anfractuous papillary stratum, below which is a wide connective stratum, and finally a deep serous gland close to striated muscle. Additionally, we have located Vater–Paccini corpuscles near the auditory duct between striated muscles (Figure 24C–E). No lymphoreticular tissue was observed.

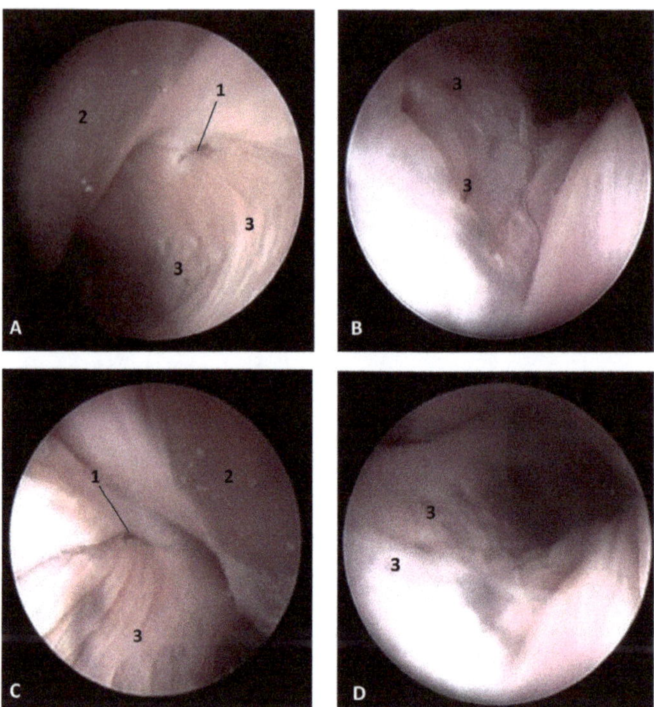

Figure 23. Endoscopic images of the pharyngeal cavity. **L** (Left) **R** (Right). (**A**,**B**) Left nasopharynx. (**C**,**D**) Right nasopharynx. Juvenile, scomu4. 1, Pharyngeal orifice of the auditory tube; 2, Nasal septum: vomer bone; 3, Nasopharyngeal mucosa: longitudinal or striated folds with small holes.

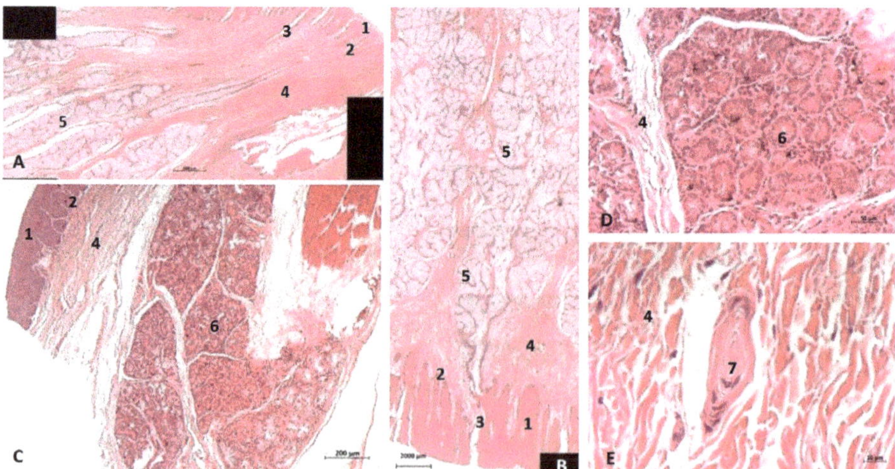

Figure 24. (**A**,**B**) Histological study of the pharyngeal cavity. (**A**) Fauces: soft palate. (**B**) Tongue: root. (**C**) Pharynx: mucosa (**D**,**E**). Detail of pharyngeal mucosa. Adult, scomu6. 1, Epidermis; 2, Papillary stratum; 3, Secretor ducts; 4, Connective tissue; 5, Deep mucous glands; 6, Deep serosa glands; 7, Vater-Paccini corpuscle.

MRI Study

The MRI sagittal images show a pharyngeal cavity in a *Globicephala melas* fetus (gma1) and we could appreciate the oropharynx (fauces), the **nasopharynx** and the oesophageal vestibule hypointense in both T1 and T2 sequences(Figure 25A,B). Coronal T1 and T2 sequences show the piriform recess alongside the larynx (Figure 25C,D).

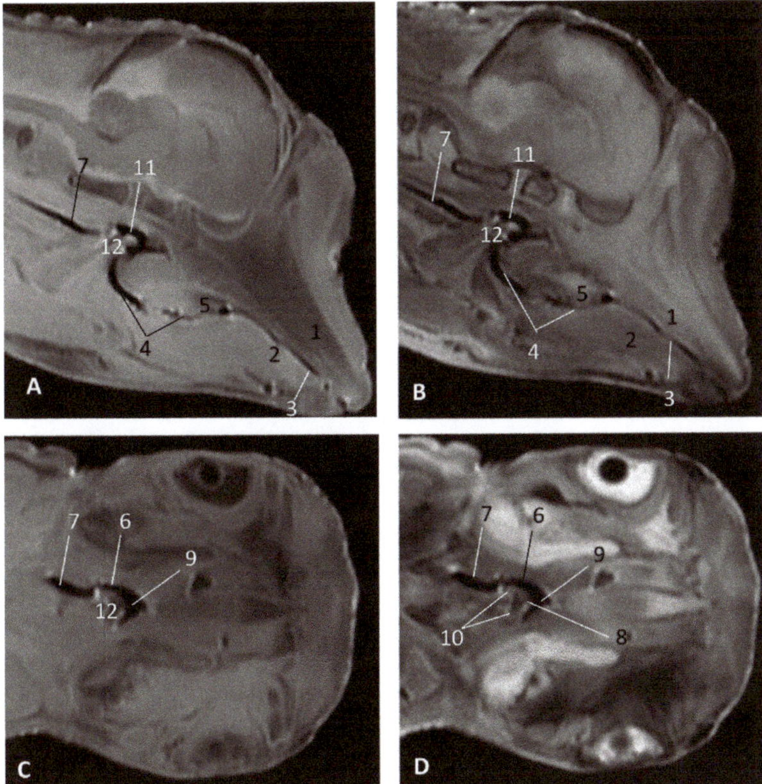

Figure 25. Images of the oral and pharyngeal cavity. MR sagittal and coronal images are oriented so that the rostral is to the right. (**A**) T1 SE sagittal, (**B**) T2 FrFSE sagittal, (**C**) T1 SE coronal and (**D**) T2 FrFSE coronal planes. 5 months, gma1. 1, Hard palate; 2, Tongue; 3, Oral cavity (closed); 4, Oropharynx: fauces; 5, Oropharynx: soft palate; 6, Laryngopharynx: left piriform recess; 7, Laryngopharynx: oesophageal vestibule; 8, Epiglottis cartilage; 9, Epiglottic vallecula; 10, Arytenoid cartilages; 11, Nasopharynx; 12, Larynx.

3.2.2. Special Study of Nasopharynx and Pharyngeal Diverticulum of the Auditory Tube (PDAT)

(a) MRI study

In MRI, we can appreciate, in early fetal stages, a bilateral structure within the laryngopharyngeal cavity, each named as a *pharyngeal diverticulum of the auditory tube* (PDAT). These are connected through the musculotubaric channel with the middle ear (temporal bone: petrous and tympanic part). In a young *Delphinus delphis* fetus (dde3), it appears in sagittal sections as a hyper/hypointense area seen caudal and rostrally, respectively (Figure 26A,B), and also in coronal sections (Figure 26C,D).

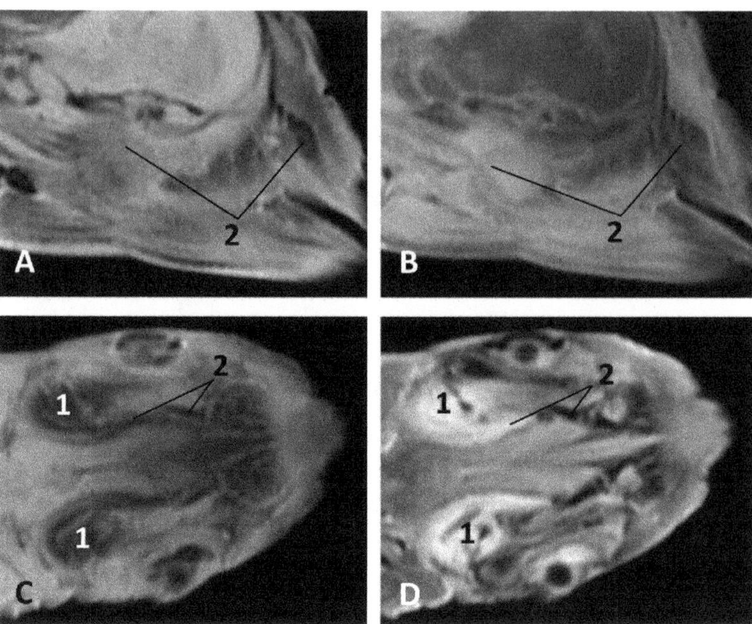

Figure 26. Images of the pharyngeal cavity. MR sagittal and coronal images are oriented so that the rostral is to the right. (**A**) T1 SE sagittal, (**B**) T2 FrFSE sagittal, (**C**) T1 SE coronal and (**D**) T2 FrFSE coronal planes. 4 months, dde3. 1, Inner and middle ear; 2, Pharyngeal diverticulum of the auditory tube.

In older *Delphinus delphis* fetuses (dde5, dde8, dde11) this double space at both sides of the laryngopharynx is more evident and shows the same intensity, but now we can distinguish the vascular area (hyperintense) and the air-filled area (hypointense) (Figures 27–31).

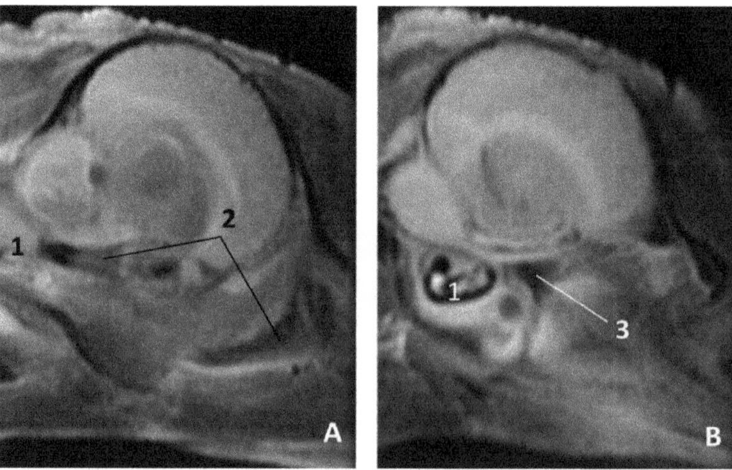

Figure 27. *Cont.*

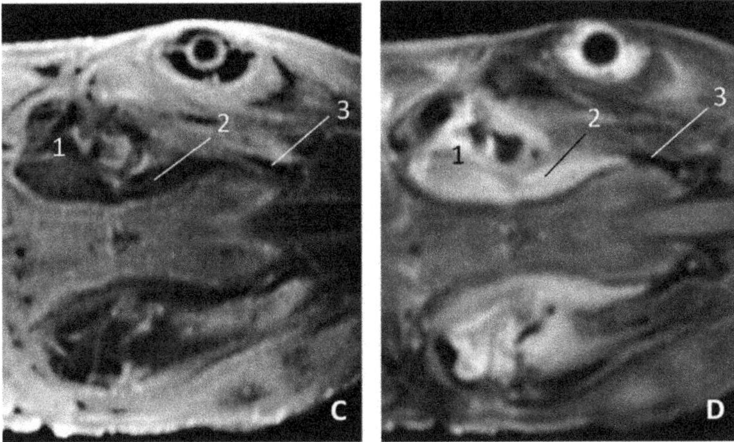

Figure 27. Images of the pharyngeal cavity. MR coronal and sagittal images are oriented so that the rostral is to the right. (**A**,**B**) T2 FrFSE sagittal, (**C**) T1 SE and (**D**) T2 FrFSE coronal planes. 5.5 months, dde5. 1, Inner ear; 2, Pharyngeal diverticulum of the auditory tube: moderate hyperintense area (vascular); 3, Pharyngeal diverticulum of the auditory tube: hypointense area (air).

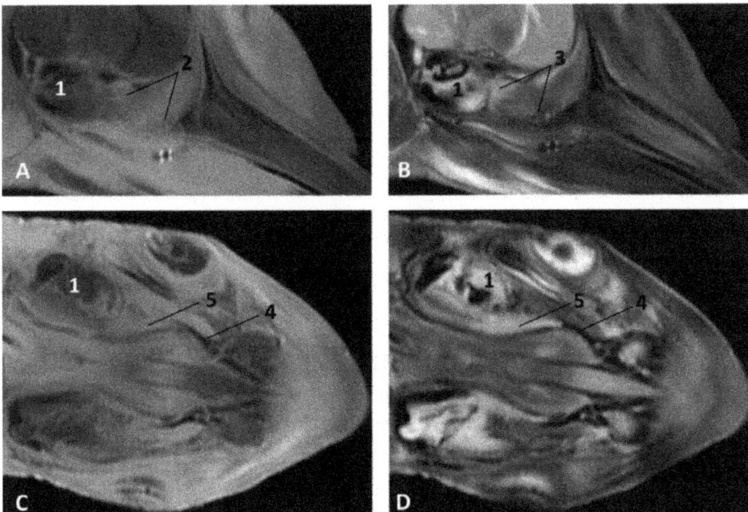

Figure 28. Images of the pharyngeal cavity. MR sagittal and coronal images are oriented so that the rostral is to the right. (**A**) T1 SE sagittal, (**B**) T2 FrFSE sagittal, (**C**) T1 SE coronal and (**D**) T2 FrFSE coronal planes. 6 months, dde8. 1, Inner ear; 2, Pharyngeal diverticulum of the auditory tube: moderate hyperintense area (vascular); 3, Pharyngeal diverticulum of the auditory tube: moderate hypointense area; 4, Pharyngeal diverticulum of the auditory tube: hypointense area (air); 5, Pharyngeal diverticulum of the auditory tube: hyperintense area (vascular).

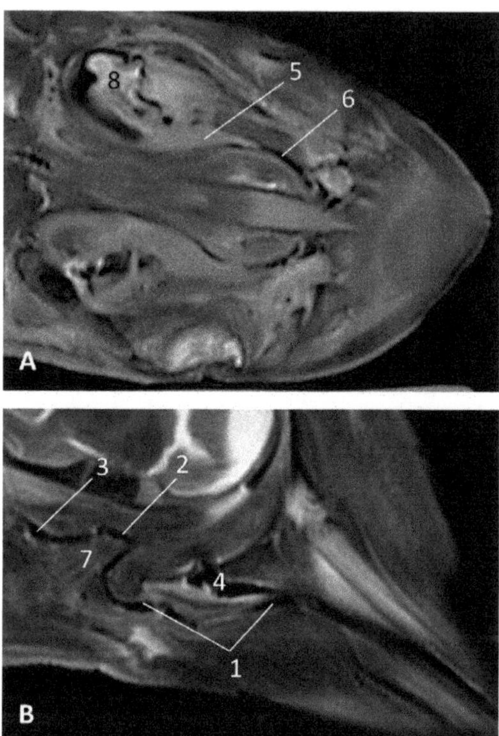

Figure 29. Images of the pharyngeal cavity. (**A**,**B**) MR coronal and sagittal images are oriented so that the rostral is to the right. (**A**,**B**) T2 FrFSE coronal and sagittal planes. 8 months, dde11. 1, Oropharynx: fauces; 2, Nasopharynx; 3, Laringopharynx: oesophageal vestibule; 4, Nasopharynx: pharyngeal diverticulum of the auditory tube; 5, Pharyngeal diverticulum of the auditory tube: hyperintense area (vascular); 6, Pharyngeal diverticulum of the auditory tube: hypointense area (air); 7, Larynx; 8, Middle and inner ear.

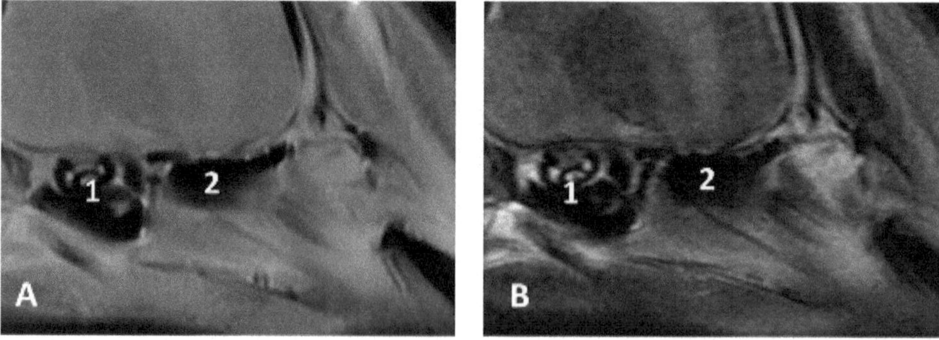

Figure 30. *Cont.*

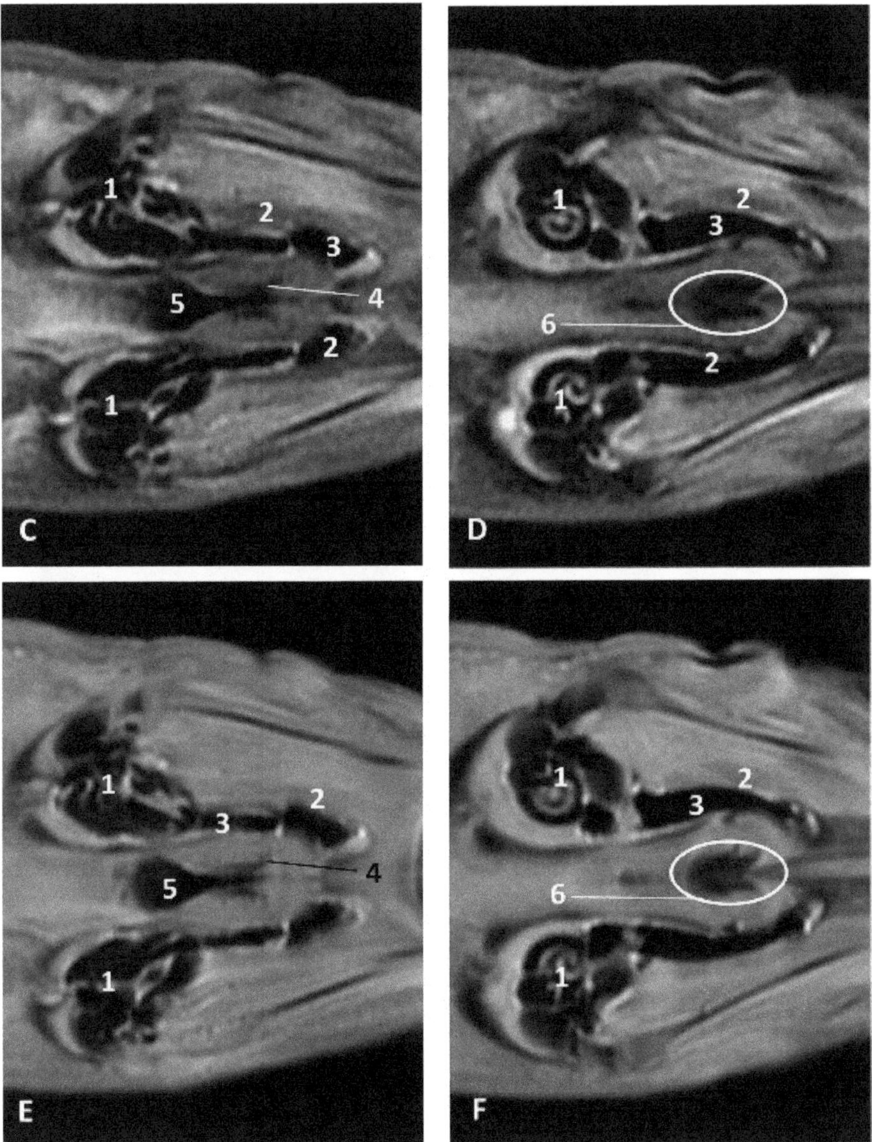

Figure 30. Images of the pharyngeal cavity. MR sagittal and coronal images are oriented so that the rostral is to the right. (**A**) T1 SE sagittal, (**B**) T2 FrFSE sagittal, (**C**,**E**) T1 SE coronal and (**D**,**F**) T2 FrFSE coronal planes. 4 months, dde14. 1, Inner and middle ear; 2, Pharyngeal diverticulum of the auditory tube (vascular); 3, Pharyngeal diverticulum of the auditory tube (air); 4, Auditory tube; 5, Nasopharinx; 6, Intrapharyngeal orifice.

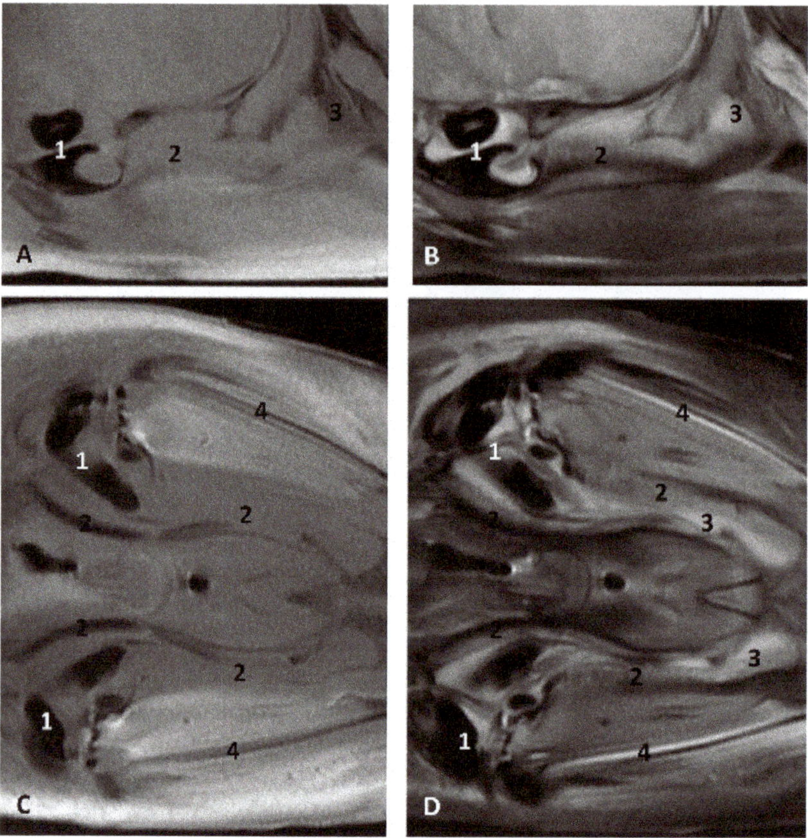

Figure 31. Images of the pharyngeal cavity. MR sagittal and coronal images are oriented so that the rostral is to the right. (**A**) T1 SE sagittal, (**B**) T2 FrFSE sagittal, (**C**) T1 SE coronal and (**D**) T2 FrFSE coronal planes. 9 months, grgr1. 1, Inner and middle ear; 2, Pharyngeal diverticulum of the auditory tube: vascular; 3, Pharyngeal diverticulum of the auditory tube: air; 4, Mandibles.

In more advanced fetal development, it is possible to observe air (hypointense) and vascular (moderate hyperintense) areas, and even the auditory tube (slightly hypointense) (Figure 30).

PDAT were clearly seen in sagittal and coronal sections in a *Grampus griseus* fetus (grgr1). The T2 sequences are clearer than T1 because they differentiate two areas: slightly hypointense (vascular) and hyperintense (air) (Figure 31).

(b) Histological study

The histological analysis of the PDAT shows two well-defined areas inside: a pharyngeal vascular plexus (Figure 32A) with abundant and dilated vascular endothelium and the respiratory epithelium area in contact with air (Figure 32 E). A detailed image reveals the luminal vessels filled with blood (Figure 32B,C). The wall of PDTA is filled with air, which is in contact with respiratory epithelium (Figure 32D,E).

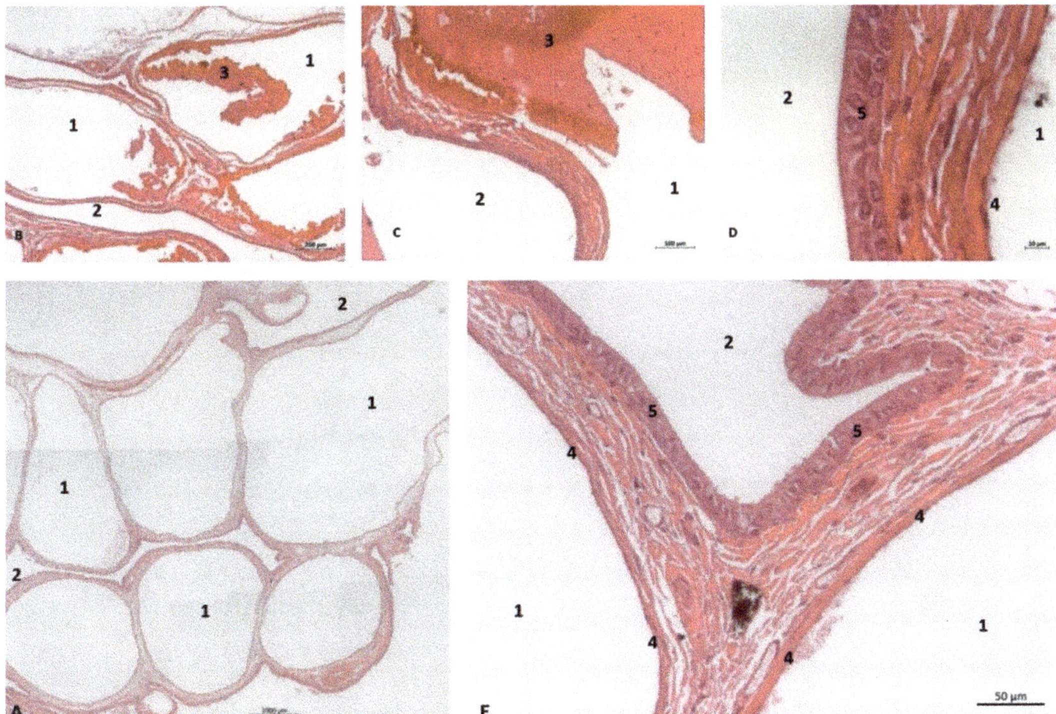

Figure 32. Histological study of the pharyngeal cavity: pharyngeal diverticulum of the auditory tube. (**A**) Pharyngeal vascular plexus wide. (**B**,**C**) Detail of plexus with blood in lumen. (**D**,**E**) Detail of plexus wall. Adult, scomu6. 1, Vascular lumen; 2, Respiratory lumen; 3, Blood vessel lumen; 4, Vascular endothelium; 5, Respiratory epithelium.

(c) Anatomical study

(c1) Osteology

The PDAT is a well delimited area, even in the early stages of fetal development (Figure 2). This area is medially extended to the bony choanae and extends dorsally to the maxillopalatine fossa, medially to the pterygopalatine recess (pterygoid sinus) and rostrally to the petrous and tympanic parts of the temporal bone (cochlea) (Figure 33).

(c2) Sectional anatomy

The three coronal sections, in a newborn *Stenella coeruleoalba* (scomu2), extend from the floor (Figure 34A) to the roof of the oral and pharyngeal cavities (Figure 34B,C). In these images it was possible to observe the proximity to the mandible channel tissue and the pharyngeal orifices of the auditory tubes crossing the pharyngeal muscles. It is easy to medially differentiate the air area (near the auditory tube and nasopharynx and laterally to the cribriform area) (Figure 34).

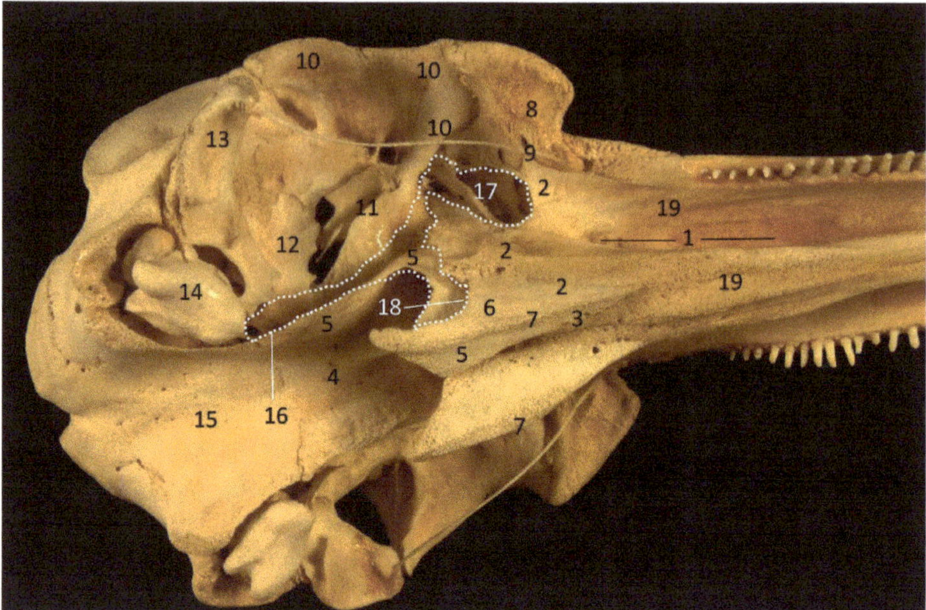

Figure 33. Common dolphin skull. Dots show the extension and form of the right pharyngeal diverticulum of the auditory tube. Photography Francisco Gil Cano. Courtesy from Ángel Tórtola. Spanish naturalist. Oblique view. dde15. 1, Greater palatine groove; 2, Palatine bone: perpendicular lamina; 3, Palatine bone: horizontal lamina; 4, Vomer bone; 5, Pterygoid bone: medial lamina; 6, Pterygoid bone: lateral lamina; 7, Pterygoid bone: crest; 8, Lacrimal and zygomatic bone; 9, Temporal process of the zygomatic bone; 10, Frontal bone; 11, Presphenoid bone: wings; 12; Basisphenoid bone: wings; 13, Temporal bone: squamous part; 14, Temporal bone: petrous and tympanic parts; 15, Occipital bone: basilar part; 16, PDAT area; 17, Maxilopalatine fossa (pterygopalatine fossa in mammals); 18, Pterygopalatyne recess (pterygoid sinus); 19, Maxillary bone: palatine process.

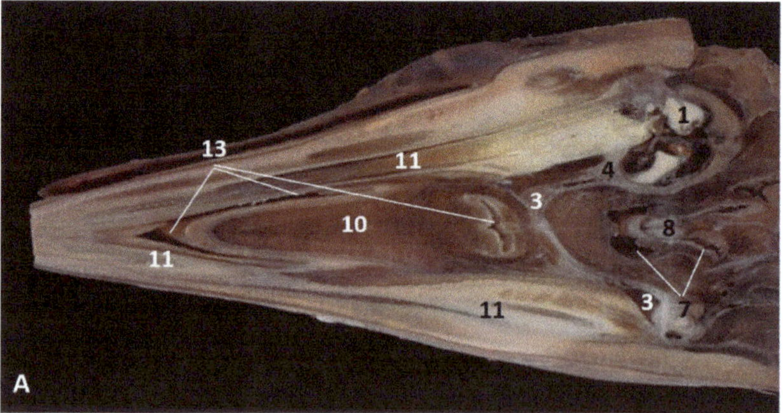

Figure 34. *Cont.*

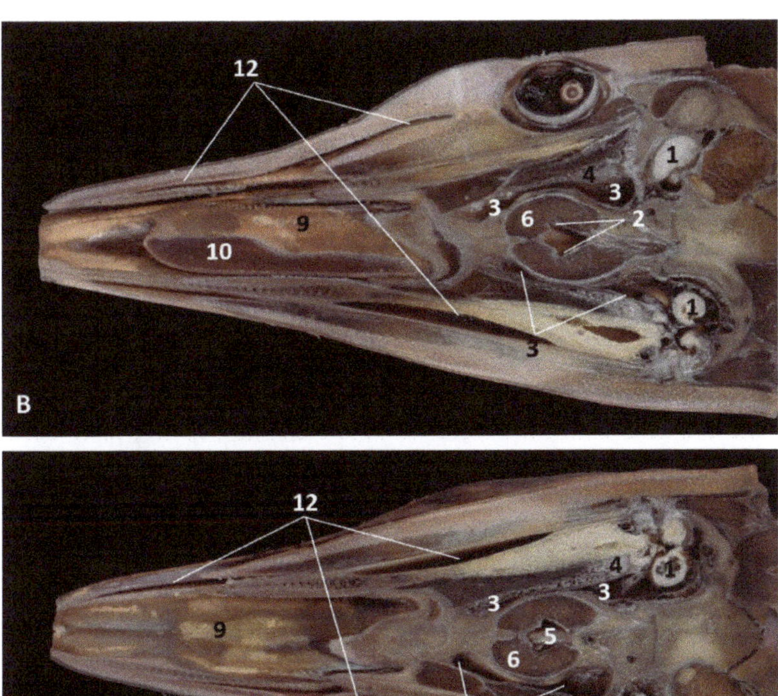

Figure 34. (**A–C**) Coronal sections of head at level of eyes, ear, pharyngeal and oral cavity. These three sections show the extension and connection between the pterygopalatine recess (pterygoid sinus) and the PDAT and between the nasopharynx and PDAT. (**A**,**B**) Dorsal view (**C**) Ventral view. scomu2. 1, Middle and inner ear; 2, Pharyngeal orifices of the auditory tube; 3, Pharyngeal diverticulum of the auditory tube: air area; 4, Pharyngeal diverticulum of the auditory tube: vascular area; 5, Vomer and choanas; 6, Pharyngeal muscles; 7, Piriform recess; 8, Laryngeal cartilages: aditus laryngis; 9, Hard palate (maxillary bones); 10, Tongue (sectioned) 11, Mandibles; 12, Labial vestibule; 13, Oral cavity.

The two sagittal sections in a juvenile *Stenella coeruleoalba* (scomu3) were made parasagittally at the level of the ear. It shows that this area (PDAT) extends rostrally to the inner and middle ear crossing below the basal bones of the cranium to arrive to the pterygopalatine recess (pterygoid sinus) and finish dorsally at the maxillopalatine fossa (Figure 35).

In the adult *Stenella coeruleoalba* (scomu6), the pharyngeal orifice of the auditory tube is canalized by a trocar (Figure 36A) and the PDAT area is located ventrally. The enlarged image shows, after removing the pharyngeal muscles, the trajectory of the auditory tube towards the pharyngeal diverticulum (Figure 36B).

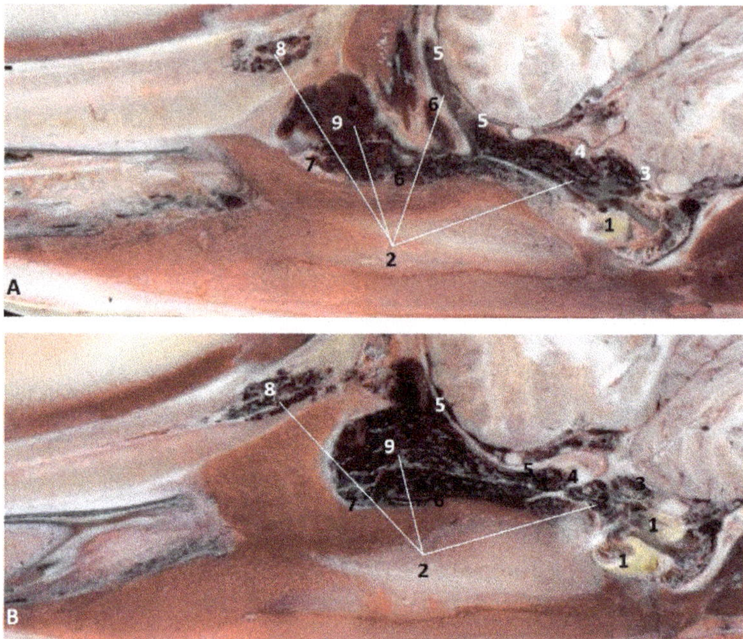

Figure 35. (**A**,**B**) Detailed serial sagittal sections at level of the pharyngeal diverticulum of the auditory tube with an anfractuous mucosa filled with a heterogeneous content. It extends up to the maxillopalatine fossa rostral to the eyeball. scomu3. 1, Middle and inner ear; 2, Pharyngeal diverticulum of the auditory tube; 3, Occipital bone: basilar part; 4, Basisphenoid bone; 5, Presphenoid and ethmoid bones; 6, Pterygoid bone; 7, Palatine bone; 8, Maxilopalatine fossa (pterygopalatine fossa in domestic mammals); 9, Pterygopalatine recess (pterygoid sinus).

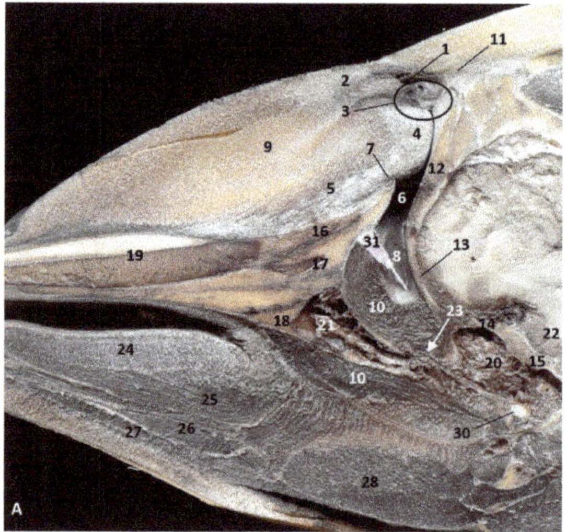

Figure 36. *Cont.*

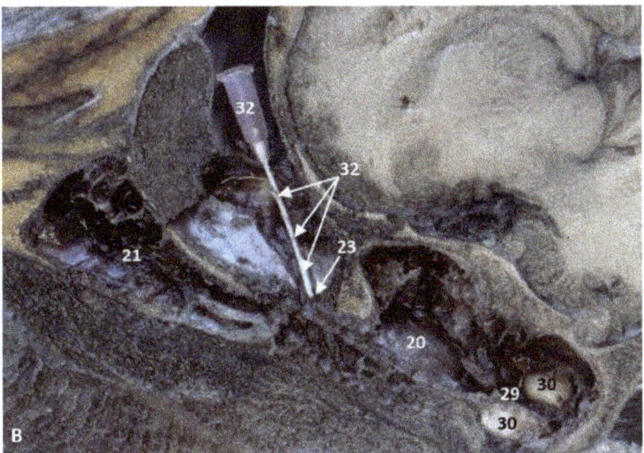

Figure 36. (**A**) Sagittal section of head at level of nasal, pharyngeal and oral cavity. (**B**) Detail of the trajectory of the trocar towards the pharyngeal diverticulum after removing the pharyngeal muscles around the auditory tube. Adult, scomu6. 1, Nasal cavity: vestibule; 2, External nares muscles; 3, Phonic lips; 4, Nasal plug; 5, Nasal plug muscles; 6, Nasal cavity: respiratory part; 7, Nasal cavity: incisive recess; 8, Choanae; 9, Melon; 10, Pharyngeal muscles; 11, Nasal bone; 12, Frontal bone; 13, Ethmoid bone; 14, Presphenoid bone; 15, Basisphenoid bone; 16, Incisive bone; 17, Maxillary bone; 18, Pterygoid bone; 19, Mesethmoid cartilage; 20, Pharyngeal diverticulum of the auditory tube (rostral part is pterygoid sinus); 21, Pterygopalatine recess (pterigoyd sinus); 22, Hypophysis; 23, Connection orifice; 24, Tongue: proper lingual muscle; 25, Hyoglossus muscle; 26, Geniohyoid muscle; 27, Mylohyoid muscle; 28, Digastricus muscle; 29, Musculotubaric channel; 30, Middle ear (petrotympanic bone); 31, Trocar inserted in the pharyngeal orifice of the auditory tube; 32, Trocar (showing duct trajectory).

(c3) Dissection

A deep dissection in a newborn *Stenella coeruleoalba* (scoce1) shows the extent of the PDAT area (Figure 37).

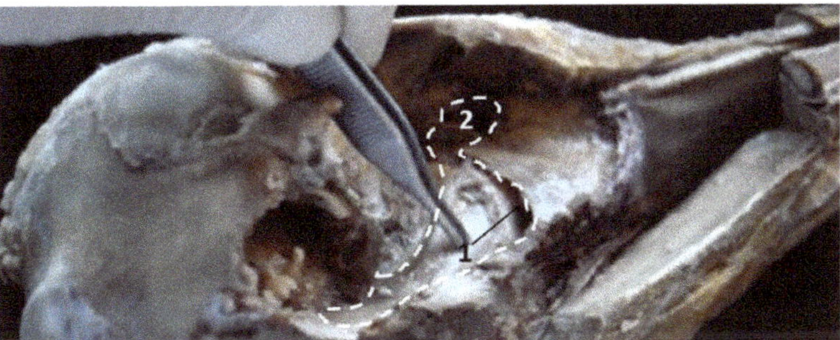

Figure 37. Deep dissection of the dolphin head after removing petrous and tympanic part of the temporal bone. Discontinuous line shows the extension and form of the right pharyngeal diverticulum of the auditory tube. Newborn, scoce1. 1, Pterygopharyngeal recess (pterygoid sinus); 2, Maxilopalatine fossa (pterygopalatine fossa in domestic mammals).

4. Discussion

4.1. Oral Cavity

4.1.1. Vestibule

The oral vestibule in terrestrial mammals could be sub-divided in oral and labial vestibules, due to the presence of a space between lips and teeth (labial vestibule), and between cheeks and teeth (buccal vestibule). In cetaceans, which lack cheeks and, therefore, masseter muscles during development [3], only the labial one is present.

Additionally, in terrestrial mammals, there are superior and inferior labial frenula to capture air and grab food. In cetaceans, the lips are tightly sealed, so the mouth is hermetically closed as it happens with the blowhole. Therefore its lips lack the mobility of the terrestrial mammal. Therefore, the labial vestibule is unique in cetaceans because it lacks a buccal vestibule and it is different from terrestrial mammals.

4.1.2. Oral Cavity Proper

(a) Roof

The rostral part of hard palate has only a vestigial incisive papilla. Since the ontogeny reflects the phylogeny, the ducts are more formed in early stages of development and degenerate as gestation progresses. The innervation observed in this incisive papilla is probably related to the tactile sensitivity to test the mucosa of some prey. This sensation is collected by the major palatine nerve from the maxillary branch of trigeminal nerve.

(b) Tongue

The lingual papillae in the *Globicephala melas*, a teuthophagous (consuming cephalopods) suction feeder [2], may enable its specific feeding strategy, especially when combined with the modifications to its hyoid apparatus, and often to the skull and jaws [28]. On the other hand, *Stenella coeruleoalba* and *Delphinus delphis* dolphin's lingual papillae do not extend caudally, and these disappear as the animal matures (as in terrestrial mammals).

The regressing papillae may be due to the adult type of feeding and with the prolonged lactation period, 19–20 months in *Stenella coeruleoalba* and *Delphinus delphis* and 24 months or even longer in *Globicephala melas* [26]. Newborn carnivores and suids also have marginal papillae [24].

In birds, there are blade-like lamellae in the inner and outer margin of the bill. For instance, anatids generally have filtering papilla, while ducks have a double row of overlapping bristles in the tongue that interdigitate with a double row of lamellae on the bill [29].

4.1.3. Histological Considerations

Refs. [10,26] show the lingual seromucous and mucous glands in *Delphinus delphis*. We also confirm the presence of mucous glands in an adult *Stenella coeruleoalba* (scomu6) at these levels as well as in the oesophagus, confirmed by the endoscopic images. We also observed mucous glands in *Delphinus delphis* fetus (dde10). The large number of mucous glands has a double function, first to serve as mechanical protection against the abrasive diet, i.e., the fins and scales of fishes and second, to favour lubrication to enable a proper food movement during swallowing. The same conclusions apply to the oropharynx. In addition, the absence of salivary glands explains the abundant mucous glands as a substitute.

In birds, there is a similar compensation process [29], which is of great utility in the digestive process of certain avian species such as galliforms, which ingest seeds, grains and even stones and mud to help trituration.

As for the immune system, the large number of lymphatic vessels (Figure 13) explains the absence of nodular formations (tonsils and nodules) which, in cetaceans, could jeopardize deglutition. The absence of lymphoreticular tissue does not indicate lack of immune function.

The frenulum in most terrestrial animals is wider and more mobile. In dogs and horses, it is sharper and more angled like *Delphinus delphis*, *Stenella coeruleoalba* and *Globicephala melas*. In cetaceans the tongue has limited mobility, serving first during lactation and then during the rest of life, as it is useful to grab food and expel extra water.

The small grooves may, in fact, be the location of the hitherto unveiled taste receptors [10]. We did not locate receptors for taste or olfaction in the oral or nasal cavity [3].

(c) Floor of oral cavity

The sublingual recess and labial vestibule are quite shallow, perhaps due to the absence of mastication, which nullifies the need for a functional space outside the vestibule (outer) surface of the teeth [26].

We see this narrowness of the labial vestibules and sublingual recess during fetal development. Within the lateral sublingual folds, we did not observe the presence of any duct transporting saliva from the major salivary glands, as would happen in terrestrial mammals. Neither did we observe sublingual papillae, as in domestic mammal species.

The monostomatic salivary glands are absent in cetaceans [3]. Stomas opened along the lateral sublingual folds (polystomatic gland) and major sublingual and mandibular ducts, with the latter atrophied in cetaceans, as we have observed in histology (Figure 14).

Even though the mouth is a tight opening, when the dolphin feeds under the water, the entrance of some salt water is unavoidable, but this will be ejected later with the help of the tongue. In a humid environment, the saliva would be unnecessary.

(d) Teeth

Tooth formation occurs below the gums [3]. The final eruption takes place during the perinatal period after lactation (this, along with the few development of the incisive teeth, protects the mother's nipple), probably by the scratching of the gums against the rough surfaces of food.

In *Delphinus delphis,* teeth are developed just in the molar region; the mandibular teeth develop later. *Grampus griseus* (Figure 31) and *Globicephala melas* (Figure 7) have fewer teeth due to their shorter and rounder jaw. In this last species, the caudal teeth do not develop, since the mandible grows and the teeth remain rostral in position. In *Phocoena phocoena* (phog1) the conic teeth (Figure 2) with which they are born will flatten and round with age by the process of wear.

For the first time, we discovered that the teeth, during development, are covered by cementum under the gums. We can state that odontocete teeth are phylogenetically closer to the hypsodont dentition of ruminants and equids, in which teeth are composed from the crown to root as follows: a core of dentine covered internally and externally by enamel which is, in turn, covered by cementum (Figures 12d and 13d).

Cetacean teeth are designed to grab and swallow but not to chew, so molars are not necessary.

4.2. Pharyngeal Cavity

During fetal development, we observed that the soft palate of odontocetes stays attached caudally along the crest of the pterygoid bone (Figure 13), while in terrestrial mammals, it remains attached to the horizontal lamina of the palatine bone (palatine aponeurosis).

The isthmus of the fauces is tightly closed, as happens with the blowhole [3], because it is formed by musculomembranous walls, like those of the oesophagus.

The tightly closed oropharynx (Figure 12) can be opened (Figure 15) during gestation to allow the entrance of amniotic fluid into the fetal alimentary canal to stimulate the glands producing gastric enzymes [30,31]. Once born, neither the oropharynx nor the oesophagus open until food is swallowed.

The endoscope overcomes the resistance of the isthmus of the fauces by exerting pressure similar to the effect produced by the ingestion of prey.

(a) Histological considerations

The lateral asymmetry in dolphins exists to the right externally, and to the left internally. [28] This defends a trophic explanation since the internal asymmetry allows the ingestion of larger prey, though we believe the asymmetry described is the opposite, since it is not the right piriform recess but the left one which is the larger (Figure 21). The voluntary dislocation of the larynx to ingest larger prey has already been described [32].

Nasopharynx

The pharyngeal orifice of the auditory tube is an open membranous duct (Figure 36a) close to the pharyngeal muscles. The PDAT connects to the musculotubaric channel and finishes in the tympanic orifice of the auditory tube.

The pterygopalatine recess (pterygoid sinus) is not isolated but connected to the pharyngeal diverticulum of the auditory tube at the level of the auditory tube notch [33]). We do not consider the pterygopalatine recess as an isolated space of the auditory tube diverticulum, but instead as part of the structure.

In the horse, the auditory tube diverticulum extends from the skull base to the atlas dorsally and ventrally to the pharyngeal cavity roof [34,35].

The fact that the auditory tube is not a closed duct but dilating and folding according to pressure, has its application in decompression. The auditory tube protects from threats (i.e., strong sounds which are potentially dangerous to the small bones) forcing the dolphin to emerge quickly releasing high-pressure air (or liquid) through the orifice. This theory can be reinforced by the fact that the Vater–Paccini corpuscles (Figure 24) are pressure receptors observed in the nasopharyngeal mucosa close to the pharyngeal orifice of the auditory tube (Figure 36).

This would also prevent the collapse of the tracheal and respiratory walls while simultaneously ensuring that the air at great depths is distributed to essential areas.

The blood present can have a double function of refrigeration and a reservoir of extra oxygen, available for exchange with blood vessels in part of the epithelium and within the sinuses. It could be very interesting to further examine this process in Mysticetes. A similar process happens in other parts of the lower respiratory system [36].

(b) Comparative Anatomy

In the horse, the function of these diverticula is to cool the blood going to the brain via the internal carotid and the external carotid artery. In equids, the diverticula are divided by a membranous septum extending to the ventral face of the atlas. In odontocetes, it extends only to the pterygopalatine recess (pterygoid sinus) and medially to the pterygoid bone. The opposite happens in terrestrial mammals (where the diverticulum of the auditory tube is divided by the basilar portion ridge, the basihyoid, and the medial lamina of the pterygoid bone [34] and [35]. Both diverticula are divided by bone, so the entrance to the larynx (aditus laryngis) is not interrupted at this level.

(c) Histological considerations

The deep serosal glands observed in the nasopharynx secrete towards the small openings observed in the deepest area of the nasal cavity [3] and to the small holes observed in the nasopharynx area close to the pharyngeal orifice of the auditory tube (Figures 21–23). Close to these orifices, a pressure corpuscle (Figure 24) was observed. This possibly indicates that, when the vascular plexus observed in the pharyngeal diverticulum of the auditory tube is filled with blood (Figures 27–32 and 34–36), this process promotes the expulsion of the air stored to the auditory tube into the nasopharynx (Figure 36). The air pressure within the nasopharynx is detected by the Vater–Paccini corpuscles (Figure 24).

5. Conclusions

Inside the oral cavity proper, incisive papilla are present in all specimens to test food texture and hardness and, during lactation, this is also reinforced by the lack of incisive teeth to prevent nipple harming. Comparatively, we can state that *Delphinus delphis* palate is morphologically different from the other species due to an additional groove.

We have also seen the form and function of the tongue showing the marginal papillae disappearing gradually after lactation, lasting more in some species like *Globicephala melas*. The teeth in formation present three layers under the gum, developed during fetal life but erupting through the gums after lactation.

We found mucous glands in the oral and pharyngeal cavities, with a double function of lubrication and mechanical protection. In the nasopharynx, we found serous glands to humidify the area as well as pressure corpuscles.

In the laryngopharynx, we realized that the left piriform recess was larger than the right one, probably to allow the ingestion of larger prey.

In the nasopharynx, one of the achievements of this study was finding and visualizing, for the first time, the diverticula of the auditory tube from early stages of the development, using MRI and the dissection of fetus skulls. This space extends rostrodorsally from the ear to the nasopharynx and contains air and a vascular plexus, similar to those mammals which possess such a structure.

Buccopharyngeal cavity endoscopies would also be useful to highlight alterations in the anatomical structures at these levels, creating situations incompatible with life.

Supplementary Materials: The following are available online at https://www.mdpi.com/article/10.3390/ani11061507/s1, Table S1: Other parameters observed in this study. Table S2: MRI parameters used in this study.

Author Contributions: Conceptualization, Á.G.d.l.R.y.L. and G.R.Z.; Formal analysis, A.A.E. and F.G.C.; Investigation, Á.G.d.l.R.y.L.; Methodology, A.A.E., M.S.L., F.M.G., A.L.F., J.S.A. and C.S.C.; Resources, F.M.G., A.L.F., F.G.C., J.S.A. and C.S.C.; Supervision, Á.G.d.l.R.y.L., A.A.E. and G.R.Z.; Writing—Original draft, Á.G.d.l.R.y.L. and G.R.Z.; Writing—Review and editing, Á.G.d.l.R.y.L., A.A.E., M.S.L. and G.R.Z. All authors have read and agreed to the published version of the manuscript.

Funding: MRI acquisitions were financed by Departamento de Anatomía y Anatomía Patológica Comparadas. Facultad de Veterinaria. Universidad de Murcia. Spain. The Galician stranding network is supported by the regional government Xunta de Galicia—Dirección Xeral de Patrimonio Natural. CESAM/FCT: thanks are due to FCT/MCTES for the financial support to CESAM (UIDP/50017/2020+UIDB/50017/2020), through national funds. Norma transitoria. Alfredo López is funded by national funds (OE), through FCT—Fundaçâo para a Ciência e a Tecnologia, I.P. in the scope of the framenetwork contract foreseen in numbers 4, 5 and 6 of the article 23, of the Decree-Law 57/2016, of August 29, changed by Law 57/2017, of July 19.

Institutional Review Board Statement: Not applicable.

Informed Consent Statement: Not applicable.

Data Availability Statement: Not applicable.

Acknowledgments: We are grateful to Consejería de Sanidad, Ceuta, Spain. Many thanks to Mariano Orenes Hernández for bony preparation. Spanish technician. Anatomía Veterinaria. Murcia. Thanks to Ángel Tórtola, a Spanish Naturalist, for the adult dolphin skull picture. Special thanks to Xunta de Galicia and CESAM. We thank all the CEMMA volunteers that helped with necropsies and sample collection. Thanks are due to the Valencia Oceanographic and the University of Valencia for allowing us to perform image diagnostic analysis on the Risso's dolphin fetus, and especially to their Veterinarian Daniel García-Párraga and José Luis Crespo and the biologist Patricia Gozalbes. We are also thankful to image technician Oscar Blázquez Pérez for the MRI scan performed at Centro Veterinario de Diagnostico por Imagen del Levante, Ciudad Quesada, Rojales Alicante, Spain. We give special thanks to M. José Gens Abujas (Oficina de impulso Socioeconómico del Medio Ambiente, Dirección General de Medio Natural, Consejería de Empleo, Universidades, Empresas y Medio Ambiente, Región de Murcia, Spain). Thank you very much to the CRFS Veterinary Team, in a special way, Fernando Escribano Cánovas, Luisa Lara Rosales and Alicia Gómez de Ramón Ballesta, El Valle, Murcia, Spain, for allowing us to have access to the carcasses stranded in their regional area.

Conflicts of Interest: The authors of this manuscript have no conflict of interest to declare.

References

1. Thewisen, J.G.M.; Cooper, L.N.; George, J.C.; Bajpai, S. From land to water: The origin of whales, dolphins, and porpoises. *Evo. Edu. Outreach.* **2009**, *2*, 272–288. [CrossRef]
2. Werth, A.J. Adaptations of the Cetacean Hyolingual Apparatus for Aquatic Feeding and Thermoregulation. *Anat. Rec.* **2007**, *290*, 546–568. [CrossRef] [PubMed]
3. García de los Ríos y Loshuertos, A.; Soler Laguía, M.S.; Arencibia Espinosa, A.A.; López Fernández, A.L.; Covelo Figueiredo, P.C.; Martínez Gomariz, F.M.; Sánchez Collado, C.S.; García Carrillo, N.G.; Ramírez Zarzosa, G.R. Comparative Anatomy of the Nasal Cavity in the Common Dolph min *Delphinus delphis* L., Striped Dolphin *Stenella coeruleoalba* M. and Pilot Whale *Globicephala melas* T: A Developmental Study. *Animals* **2021**, *11*, 441. [CrossRef] [PubMed]
4. Raven, H.C.; Gregory, W.K. The spermaceti organ and nasal passages of the Sperm Whale (*Physeter catodon*) and other odontocets. *Amer. Mus. Novitatis.* **1993**, *677*, 1–18.
5. Mead, J.G. Anatomy of the external nasal passages and facial complex in the delphinidae (Mammalia: Cetacea). *Smithson. Contr. Zool.* **1975**, *207*, 1–72. [CrossRef]
6. Heyning, J.E.; Mead, J. Evolution of the nasal anatomy of cetaceans. In *Sensory Abilities of Cetaceans*; Thomas, J., Kastelain, R., Eds.; Plenum Press: New York, NY, USA, 1990; pp. 67–79.
7. Schenkkan, E.J. On the comparative anatomy and function of the nasal tract in odontocetes (Mammalia, Cetacea). *Bijdr. Dierkd.* **1973**, *43*, 127–159. [CrossRef]
8. Schenkkan, E.J.; Purves, P.E. The comparative anatomy of the nasal tract and the function of the spermaceti organ in the physeteridae (Mammalia, Odontoceti). *Bijdr. Dierkd.* **1973**, *43*, 93–112. [CrossRef]
9. Lawrence, B.; Schevill, W.E. Gular musculature in delphinids. *Bull. Mus. Comp. Zool.* **1965**, *133*, 1–65.
10. Sokolov, V.E.; Volkova, O.V. Structure of the dolphin's tongue. In *Morphology and Ecology of Marine Mammals*; Chapskii, K.K., Sokolov, V.E., Eds.; Halsted Press (John Wiley & Sons): New York, NY, USA, 1973; pp. 119–127.
11. Yoshimura, K.; Kobayashi, K. A comparative morphological study on the tongue and the lingual papillae of some marine mammals —Particularly of four species of odontoceti and zalophus—. *Shigaku Odontol.* **1997**, *85*, 385–407. [CrossRef]
12. Hohn, A.A. Age estimation. In *Encyclopedia of Marine Mammals*, 2nd ed.; Perrin, W.F., Würsig, B., Thewissen, J.G.M., Eds.; Academic Press: San Diego, CA, USA, 2009; pp. 11–17.
13. Berta, A.; Ekdale, E.G.; Zellmer, N.T.; Deméré, T.A.; Kienle, S.S.; Smallcomb, M. Eye, Nose, Hair, and Throat: External Anatomy of the Head of a Neonate Gray Whale (Cetacea, Mysticeti, Eschrichtiidae). *Anat. Rec.* **2015**, *298*, 648–659. [CrossRef]
14. Kienle, S.S.; Ekdale, E.G.; Reidenberg, J.S.; Demere, T.A. Tongue and Hyoid Musculature and Functional Morphology of a Neonate Gray Whale (Cetacea, Mysticeti, Eschrichtius robustus). *Anat. Rec.* **2015**, *298*, 660–674. [CrossRef] [PubMed]
15. Yamasaki, F.; Komatsu, S.; Kamiya, T. An observation on the papillary projections at the lingual margin in the striped dolphin. *Sci. Rep. Whales Res. Inst.* **1976**, *28*, 137–140.
16. Yamasaki, F.; Komatsu, S.; Kamiya, T. Taste buds in the pits at the posterior dorsum of the tongue of *Stenella coeruleoalba*. *Sci. Rep. Whales Res. Inst.* **1978**, *30*, 285–290.
17. Reidenberg, J.S.; Laitman, J.T. Prenatal development in cetaceans. In *Encyclopedia of Marine Mammals*, 2nd ed.; Perrin, W.F., Würsig, B., Thewissen, J.G.M., Eds.; Academic Press: San Diego, CA, USA, 2009; pp. 220–230.
18. Chaplin, M.; Kamolnick, T.; Van Bonn, W.; Carder, D.; Ridgway, S.; Cranford, T. Conditioning Tursiops truncatus for nasal passage endoscopy. In Proceedings of the IAAAM International Association for Aquatic Animal Medicine, 27th Conference Tennessee Aquarium, Chatanooga, TN, USA, 11–15 May 1996.
19. Dover, S.R.; Van Bonn, W. *Flexible and rigid endoscopy in marine mammals In CRC Handbook of Marine Mammals Medicine*, 2nd ed.; Dierauf, L., Gulland, M.D., Eds.; CRC Press L.L.C.: Boca Raton, FL, USA, 2001; pp. 621–642.
20. Harrel, J.H.; Reiderson, T.H.; McBain, J.; Sheetz, H. Bronchoscopy of the bottlenose dolphin. In Proceedings of the IAAAM International Association for Aquatic Animal Medicine, 27th Conference Tennessee Aquarium, Chatanooga, TN, USA, 11–15 May 1996.
21. Tsang, W.K.; Kinoshita, R.; Rouke, N.; Yuen, Q.; Hu, W.; Lam, W.K. Bronchoscopy of Cetaceans. *J. Wildl. Dis.* **2002**, *38*, 224–227. [CrossRef] [PubMed]
22. Reiderson, T.H.; McBain, J.; Harrel., J.H. The use of bronchoscopy and fungal serology to diagnose Aspergillus fumigatus lung infection in a bottlenose dolphin. In Proceedings of the IAAAM International Association for Aquatic Animal Medicine, 27th Conference Tennessee Aquarium, Chatanooga, TN, USA, 11–15 May 1996.
23. Van Bonn, W.; Cranford, T.; Chaplin, M.; Carder, D.; Ridgway, S. Clinical Observations During Dynamic Endoscopy of the Cetacean Upper Respiratory Tract. IAAAM 1997. Available online: https://www.vin.com/apputil/content/defaultadv1.aspx?id=3864239&pid=11257& (accessed on 16 March 2021).
24. Schaller, O. *Illustrated Veterinary Anatomical Nomenclature*; Ferdinand Enke Verlag: Stuttgart, Germany, 1992; pp. 1–614.
25. *Whales of the World*; Brill, E.J. (Ed.) Bess Pr Inc.: Leiden, The Netherlands, 1988; pp. 1–310.
26. Cozzi, B.; Huggenberger, S.; Oelschläger, H. *Anatomy of Dolphins. Insights into Body Structure and Function*; Academic Press: Cambridge, MA, USA, 2017.
27. García de los Ríos y Loshuertos, A.; Ocaña, O. *Cetáceos de Ceuta y Aguas Próximas: Estudio Faunístico, Ecológico y Veterinario de los Cetáceos de Ceuta, Estrecho de Gibraltar y Mar de Alborán*; Septem Nostra: Ceuta, Spain, 2006; pp. 1–150.

28. Macleod, C.C.; Reidenberg, J.S.; Weller, M.; Santos, M.B.; Herman, J.; Goold, J.; Piercel, G.J. Breaking symmetry: The marine environment, prey size, and the evolution of asymmetry in cetacean skulls. *Anat. Rec.* **2007**, *290*, 539–545. [CrossRef] [PubMed]
29. McLelland, J. *A Colour Atlas of Avian Anatomy*; Wolfe Publishing Ltd.: Aylesbury, UK, 1990; p. 113.
30. Sadler, T.W. *Langman Embriología Médica con Orientación Clínica*, 9th ed.; Panamericana: Buenos Aires, Argentina, 2004; pp. 145–146.
31. McGeady, T.A.; Quinn, P.J.; Fitzpatrick, E.S.; Ryan, M.T.; Kilroy, D.; Lonergan, P. *Veterinary Embryology*; Wiley-Blackwell: Hoboken, NJ, USA, 2017; p. 88.
32. Dold, C.; Ridgway, S. Cetaceans. In *Zoo Animal and Wildlife Immobilization and Anesthesia*; West, G., Heard, D., Caulkett, N., Eds.; Wiley-Blackwell: Ames, IA, USA, 2007; pp. 485–496.
33. Mead, J.G.; Fordyce, R.E. The therian skull. A lexicon with emphasis on the odontocetes. *Smithson. Contr. Zool.* **2009**, *627*, 1–248. [CrossRef]
34. Sandoval, J. *Tratado de Anatomía Veterinaria. Tomo III: Cabeza y Sistemas Viscerales*; Imprenta Sorles: León, Spain, 2000; pp. 1–457.
35. Nickel, R.; Schummer, A.; Seiferle, E. *The Anatomy of the Domestic Animals. The Locomotor System of the Domestic Mammals*; Verlag Paul Parey: Berlin/Hamburg, Germany, 1986; Volume 1, pp. 1–499.
36. Garcia Parraga, D.; Moore, M.; Fahlman, A. Pulmonary ventilation–perfusion mismatch: A novel hypothesis for how diving vertebrates may avoid the bends. *Proc. R. Soc. B Biol. Sci.* **2018**, *285*, 20180482. [CrossRef]

Article

Cranial Structure of *Varanus komodoensis* as Revealed by Computed-Tomographic Imaging

Sara Pérez [1], Mario Encinoso [2], Juan Alberto Corbera [1], Manuel Morales [1], Alberto Arencibia [3], Eligia González-Rodríguez [1], Soraya Déniz [2], Carlos Melián [2], Alejandro Suárez-Bonnet [3] and José Raduan Jaber [3,*]

1 Instituto Universitario de Investigaciones Biomedicas y Sanitarias (IUIBS), Facultad de Veterinaria, Universidad de Las Palmas de Gran Canaria, Trasmontaña, Arucas, 35413 Las Palmas, Spain; saraperezalberto@gmail.com (S.P.); juan.corbera@ulpgc.es (J.A.C.); manuel.morales@ulpgc.es (M.M.); eligia.gonzalez102@alu.ulpgc.es (E.G.-R.)

2 Hospital Clínico Veterinario, Facultad de Veterinaria, Universidad de Las Palmas de Gran Canaria, Trasmontaña, Arucas, 35413 Las Palmas, Spain; mencinoso@gmail.com (M.E.); soraya.deniz@ulpgc.es (S.D.); carlos.melian@ulpgc.es (C.M.)

3 Departamento de Morfologia, Facultad de Veterinaria, Universidad de Las Palmas de Gran Canaria, Trasmontaña, Arucas, 35413 Las Palmas, Spain; alberto.arencibia@ulpgc.es (A.A.); alejandro.suarez@ulpgc.es (A.S.-B.)

* Correspondence: joseraduan.jaber@ulpgc.es

Simple Summary: We investigated the head of Komodo dragons using CT imaging. Cross-sections show that all cranial bones can be delineated, while soft tissue structures are evident but not clearly identifiable without an anatomical atlas. Additional three-dimensional reconstructed and maximum intensity projection images of the head were presented to depict bony structures. The anatomical structures identified on the CT images could help further assess the head of the Komodo dragon.

Abstract: This study aimed to describe the anatomic features of the normal head of the Komodo dragon (*Varanus komodoensis*) identified by computed tomography. CT images were obtained in two dragons using a helical CT scanner. All sections were displayed with a bone and soft tissue windows setting. Head reconstructed, and maximum intensity projection images were obtained to enhance bony structures. After CT imaging, the images were compared with other studies and reptile anatomy textbooks to facilitate the interpretation of the CT images. Anatomic details of the head of the Komodo dragon were identified according to the CT density characteristics of the different organic tissues. This information is intended to be a useful initial anatomic reference in interpreting clinical CT imaging studies of the head and associated structures in live Komodo dragons.

Keywords: computed tomography; head; Komodo dragon

Citation: Pérez, S.; Encinoso, M.; Corbera, J.A.; Morales, M.; Arencibia, A.; González-Rodríguez, E.; Déniz, S.; Melián, C.; Suárez-Bonnet, A.; Jaber, J.R. Cranial Structure of *Varanus komodoensis* as Revealed by Computed-Tomographic Imaging. *Animals* **2021**, *11*, 1078. https://doi.org/10.3390/ani11041078

Academic Editors: Matilde Lombardero and Mar Yllera Fernández

Received: 1 March 2021
Accepted: 7 April 2021
Published: 9 April 2021

1. Introduction

The introduction of imaging diagnostic techniques has revolutionized the knowledge in reptile medicine. The radiographic evaluation has been traditionally used by clinicians [1]. Nonetheless, the progressive increase in modern imaging modalities such as computed tomography (CT) and magnetic resonance imaging has improved diagnostic abilities in reptile medicine and research [2]. Therefore, these techniques represent an enormous resource that allows for fast, non-invasive anatomy visualizations of internal structures that are challenging to interpret [2].

In recent years, the contributions of zoo veterinarians, researchers, and specialized technicians (anatomists, radiologists, and wildlife and exotic specialists) working with captive and free-ranging animals to prevent and treat diseases that threaten the survival of species in wildlife conservation have increased [3]. Since 1996, the Komodo dragon (*Varanus komodoensis*) is listed as vulnerable by the Red List of the World Conservation Union [4]. To our knowledge, the anatomy of different species of reptiles has already been thoroughly described by radiology and CT [1,5–10], but only sparse numbers of these studies reported comprehensive descriptions of computed tomographic features of the head [1,5,6,9–12]. To date, not one of these reports investigates to what extent structures of the varanid head could be visualized and identified in low-resolution clinical CT-image data. In the Komodo dragon and other reptiles, the head conforms to a complex structure, which is challenging to interpret. The purpose of this study was to describe the normal anatomy of the head of the *Varanus komodoensis* by computed tomography, and three-dimensional head reconstructed images to assist in the understanding of the head and its associated structures.

2. Materials and Methods

2.1. Animals

Two 17-year-old female specimens born in captivity at Reptilandia Park (Las Palmas, Spain) were imaged at the Veterinary Clinic Hospital of Las Palmas de Gran Canaria University. One female had a length of 225 cm (snout-vent length) and weighed 36 kg, whereas the other had a length of 190 cm and weighed 24 kg. No physical abnormalities were detected before the study. The Ethical Committee of Las Palmas de Gran Canaria University, College of Veterinary Medicine Section authorized the research protocol (MV-2019/04). The owner of the animals was informed of the study and signed consent for participation in the study.

2.2. CT Technique

Sequential transverse CT slices were obtained using a 16-slice helical CT scanner (Toshiba Astelion, Toshiba Medical System, Madrid, Spain). The animals were positioned symmetrically in ventral recumbency on the CT couch, and a standard clinical protocol (120 kVp, 80 mA, 512 × 512 acquisition matrix, 1809 × 858 field of view, a spiral pitch factor of 0.94, and a gantry rotation of 1.5 s) was used to acquire sequential transverse CT images of 1 mm thickness slice. The original transverse data were stored and transferred to the CT workstation. No CT density or anatomic variations were detected in the head of the dragons used in the study. In the CT technique, tissue density can be assessed directly on the image using the Hounsfield Unit scale, The range of values assumes 2000 shades of gray (between −1000 and +1000 HU). However, as the human eye cannot distinguish more than 30 shades, representing the entire range of values in an image implies not being able to visualize a large amount of information. Therefore, only a partial sector of the TC values previously selected by the operator (window selection) is represented by grayscale. Ultimately, the use of windows allows extracting the information that the computer has, showing only a part of it, which is of interest in each anatomical region. Therefore, bone window, soft tissue windows, brain window, and pulmonary window can be applied, delivering alternate streams of information. Thus, two CT windows were applied by adjusting the window widths (WW) and window levels (WL): a bone window setting (WW = 1500; WL = 300), and a soft tissue window setting (WW = 350; WL = 40). The original data were used to generate head volume-rendered reconstructed images after manual editing of the transverse CT images to remove soft tissues using a standard dicom 3D format (OsiriX MD, Geneva, Switzerland). In addition, maximum intensity projection (MIP) images were obtained to better display the outlines between bones and other lower-attenuation structures using an image viewer (OsiriX MD, Apple, Cupertino, CA, USA). MIP is a specific type of rendering in which the brightest voxel is projected into the 3D image. There tends to be much less variability in MIP image reconstruction than in volume rendering because fewer parameters are factored into the MIP algorithm [13].

3. Results

3.1. Transverse Computed Tomography Images

Transverse sections are provided that demonstrate critical anatomical features of the varanid cranium (Figures 1–6). Figure 1 consists of three images: (C) Sagittal image of the head, where each line and number (I–VI) represents the approximate level of the following transverse CT images, (A) CT bone window, (B) CT soft tissue window. Figures 2–6 represent transverse CT images where (A) CT bone window, and (B) CT soft tissue window. The CT images are presented in a cranial to caudal progression from the septomaxilla level (Figure 1) to the brain stem level (Figure 6). The comparison between available literature and CT images enabled us to identify most of the clinically relevant anatomic structures of the head. These features were identified according to location and the degree of attenuation of the different tissues.

With regards to hard tissues, the CT images acquired using the bone window setting (Figures 1–6A) provided good differentiation between the bones and the soft tissues of the head. Thus, the bones of the cranium (prefrontal, frontal, postorbital, parietal, squamosal, quadrate, jugal, pterygoid, basioccipital, parabasisphenoid, and maxilla), the mandible (dentary, angular, surangular, and articular bones) and hyoid bones were easily recognizable because of the high CT density in cortical bone and the low CT density in their medullary cavities. Most of these structures were also visualized with the soft tissue window setting (Figures 1–6B).

Air-filled structures, such as the nasal cavity, larynx, trachea, and the oral cavity gave negligible CT-tissue density and appeared black with both window settings.

Soft-tissue structures—such as the jaw muscles, the labial and nasal glands, the eyes, and the Harderian glands—gave an intermediate CT density and appeared grey. The nervous structures (brain, cerebellum, lateral ventricles, brain stem, and spinal cord) were appreciated in both CT window modalities (Figures 3–6).

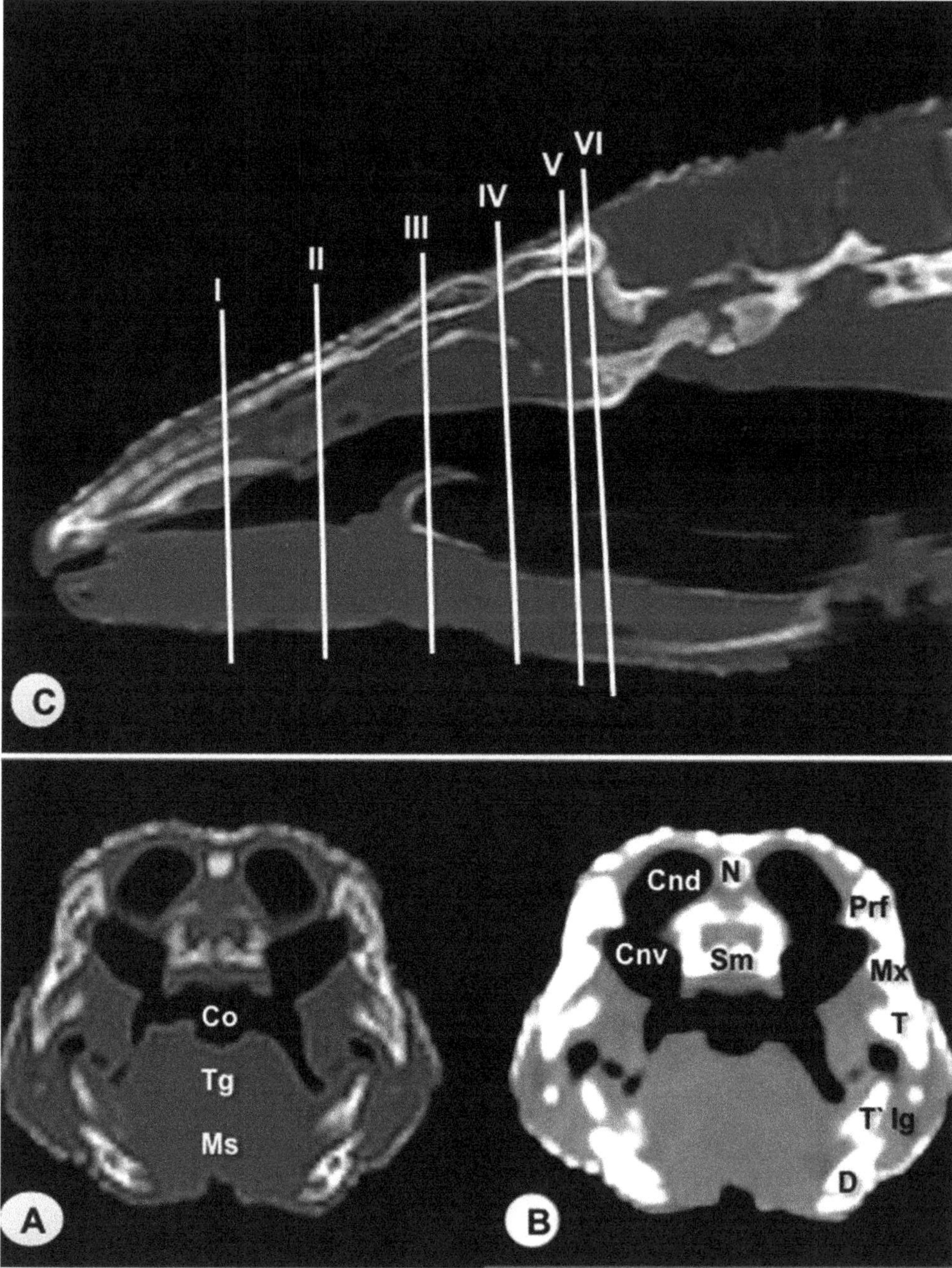

Figure 1. Sagittal image of the head of *Varanus komodoensis*. The lines and numbers (I–VI) represent the approximate level of the following transverse CT images (**C**). Transverse CT image of the head at the level of the nasal cavity corresponding to line I. (**A**) CT bone window. (**B**) CT soft tissue window. These images are displayed so that the right side of the head is to the viewer's left and the dorsal view is at the top. N: Nasal bone. Sm: Septomaxilla. Cnd: Dorsal nasal conchae. Cnv: Ventral nasal conchae. Prf: Prefrontal bone. 6. Maxillary bone. T: Tooth. T': Tooth. Co: Cavum oris. Tg: Tongue. Ms: Musculus intermandibularis + Musculus geniohyoideus + Musculus genioglossus. D: Dentary bone. Ig: Infralabial glands.

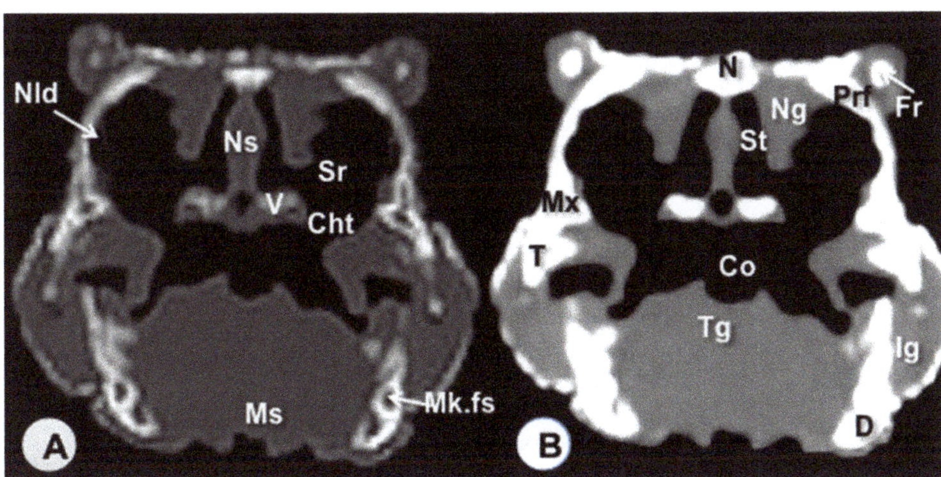

Figure 2. Transverse CT image of the head of *Varanus komodoensis* at the level of the nasal bone corresponding to line II. (**A**) CT bone window. (**B**) CT soft tissue window. These images are displayed so that the right side of the head is to the viewer's left and the dorsal view is at the top. N: Nasal bone. Prf: Prefrontal bone. Ns: Nasal septum. Ng: Nasal glands. St: Stammteil, V: Vomer. Sr: Subconchal recess. Cht: Choanal tube. Nld: Nasolacrimal duct. Mx: Maxillary bone. T: Tooth. Co: Cavum oris. Tg: Tongue. Ms: Musculus intermandibularis + Musculus geniohyoideus + Musculus pterygoideus + Musculus hyoglossus. D: Dentary bone. Mk.fs: Meckelian fossa. Ig: Infralabial glands. Fr: Frontal bone.

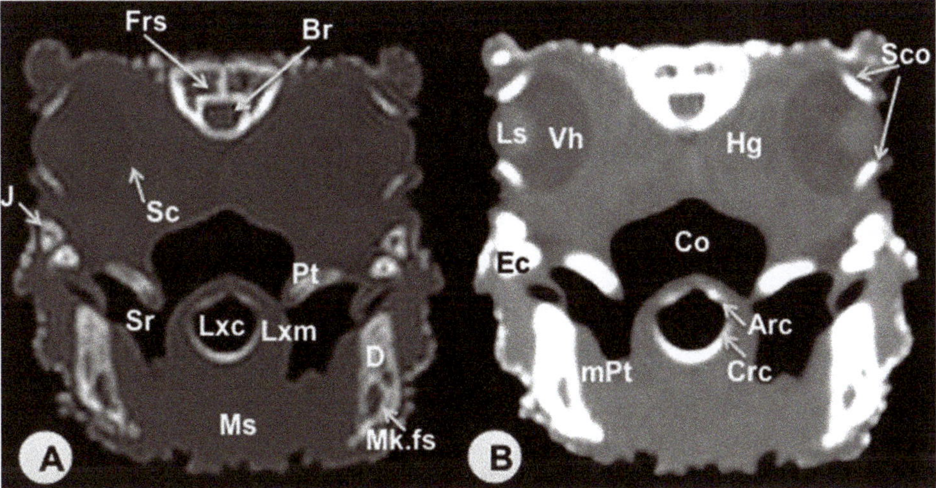

Figure 3. Transverse CT image of the head of *Varanus komodoensis* at the level of the eyes corresponding to line III. (**A**) CT bone window. (**B**) CT soft tissue window. These images are displayed so that the right side of the head is to the viewer's left and the dorsal view is at the top. Frs: Frontal sinus. Br: Brain. Sco: Scleral ossicles. Ls: Lens. Vh: Vitreous humor. Hg: Harderian gland. Sc: Sclera. J: Jugal bone. Ec: Ectopterygoid bone. Pt: Pterygoid bone. D: Dentary bone. Mk.fs: Meckelian fossa. mPt: Musculus pterygoideus. Lxc: Laryngeal cavity. Lxm: Laryngeal muscles. Ms: Musculus intermandibularis + Musculus geniohyoideus + Musculus genioglossus + Musculus hyoglossus. Crc: Cricoid cartilage. Arc: Arytenoid cartilage. Co: Cavum oris. Sr: Sublingual recess.

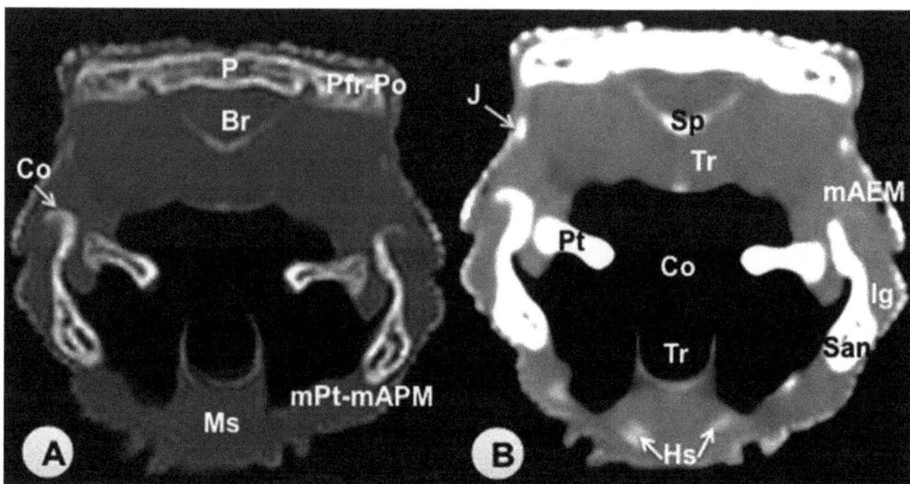

Figure 4. Transverse CT image of the head of *Varanus komodoensis* at the level of the parietal bone corresponding to line IV. (**A**) CT bone window. (**B**) CT soft tissue window. These images are displayed so that the right side of the head is to the viewer's left and the dorsal view is at the top. P: Parietal bone (frontoparietal suture). Pfr-Po: Postfrontal + Postorbital bone. Br: Brain. Sp: Sphenoid bone. Pt: Pterygoid bone. San: Surangular bone. Co: Coronoid bone. Co: Cavum oris. Tr: Trachea. Hs: Hyobranchial skeleton. Ms: Musculus intermandibularis + Musculus geniohyoideus + Musculus hyoglossus. J: Jugal bone. Ig: Infralabial glands. mAEM: Musculus adductor mandibulae externus. mPt-mAPM: Musculus pterygoideus + Musculus adductor mandibularis posterior.

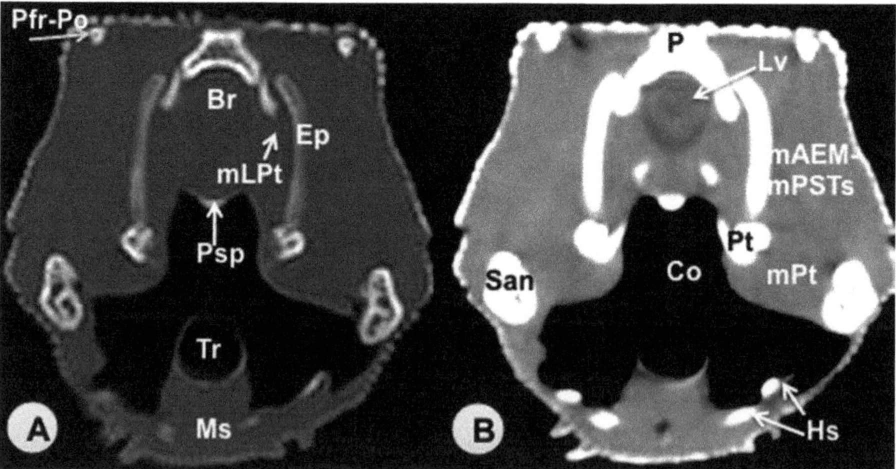

Figure 5. Transverse CT image of the head of *Varanus komodoensis* at the level of the postorbital + postfrontal bone corresponding to line V. (**A**) CT bone window. (**B**) CT soft tissue window. These images are displayed so that the right side of the head is to the viewer's left and the dorsal view is at the top. P: Parietal bone. Lv: Lateral ventricle. Br: Brain. Ep: Epipterygoid bone. Pt: Pterygoid bone. Psp: Parabasisphenoid bone. Pfr-Po: postfrontal + postorbital bone. San: Surangular bone. Co: Cavum oris. Tr: Trachea. Hs: Hyobranchial skeleton. Ms: Musculus intermandibularis + Musculus geniohyoideus + Musculus hyoglossus. mAEM-mPSTs: Musculus adductor mandibulae externus + Musculus pseudotemporalis superficialis. mLPt: Musculus levator pterygoideus. mPt: Musculus pterygoideus.

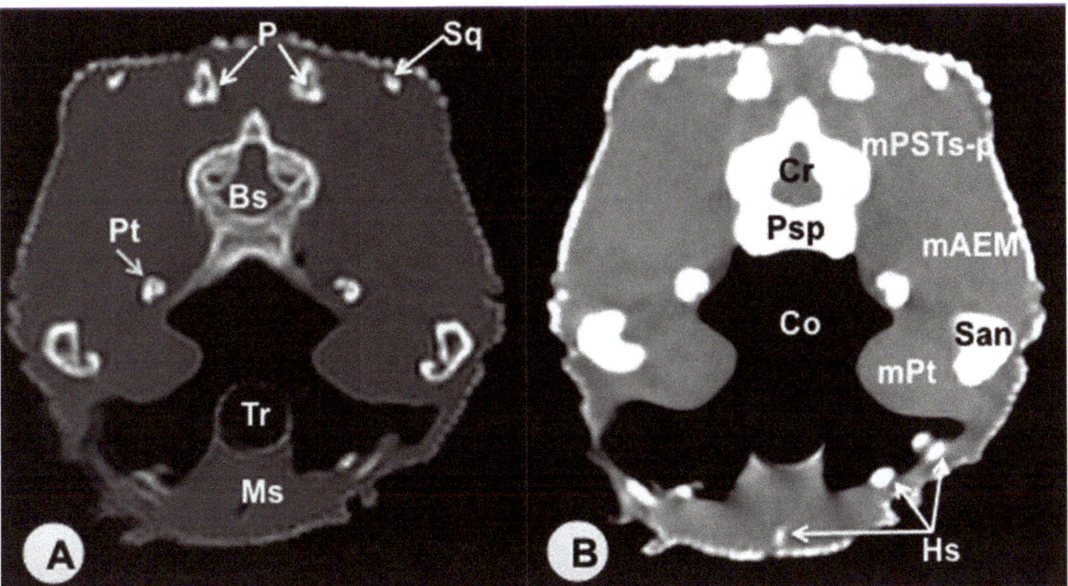

Figure 6. Transverse CT image of the head of *Varanus komodoensis* at the level of the squamosal bone corrsponding to line VI. (**A**) CT bone window. (**B**) CT soft tissue window. These images are displayed so that the right side of the head is to the viewer's left and the dorsal view is at the top. Sq: Squamosal bone. P: Parietal bone. Cr: Cerebellum (vermis). Bs: Brain stem. Psp: Parabasisphenoid bone. Pt: Pterygoid bone. San: Surangular bone. Co: Cavum oris. Tr: Trachea. Hs: Hyobranchial skeleton. Ms: Musculus intermandibularis + Musculus geniohyoideus + Musculus hyoglossus. mAEM: Musculus adductor externus mandibularis. mPSTs-p: Musculus pseudotemporalis superficialis and profundus. mPt: Musculus pterygoideus.

3.2. Head Volume-Rendered Reconstructed Images

We provide images of the three-dimensional structure of the Varanus cranium in dorsal and ventral view (Figures 7 and 8, respectively) and the left lateral view (Figure 9). Volume-rendered reconstructed CT images provided good visualization of the different bones that compose the skull. Thus, the orbital border was circumscribed by the lacrimal, the prefrontal, and the jugal bones (Figures 7 and 9). Moreover, the jugal bone was distinguishable from the ectopterygoid (Figure 9). At the posterodorsal border of the orbit, the fusion of postorbital and postfrontal bones could be seen in the lateral and dorsal reconstructed CT images (Figures 7 and 9). In ventral view, the following bones of the neurocranium were clearly delineated: the parabasisphenoid, the basioccipital, and the prootic (Figure 8). The junction between premaxilla and maxilla with the tooth arranged in a curved row was identified in the lateral and ventral view (Figures 8 and 9). In the lateral view, this tooth row curved with the margin of the mandible and maxillary. Besides, the primary curvature of the maxilla was convex, whereas that of the mandible was concave. In addition, the coronoid process was quite prominent, and the surangular and articular bones were observed extending caudally (Figure 9).

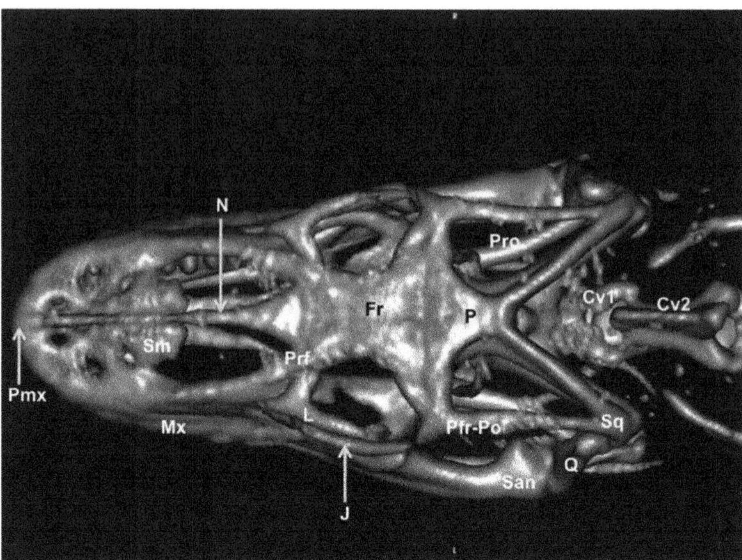

Figure 7. Three-dimensional volume-rendered reconstruction image of the cranium of *Varanus komodoensis*. Dorsal aspect. Pmx: Premaxillary bone. Mx: Maxillary bone. Sm: Septomaxilla. N: Nasal bone. Prf: Prefrontal bone. Fr: Frontal bone. L: Lacrimal bone. J: Jugal bone. Q: Quadrate bone. Sq: Squamosal. Pfr-Po: Postfrontal + postorbital. P: Parietal. Pro: Prootic. San: Surangular bone. Cv1: First cervical vertebra. Cv2: Second cervical vertebra.

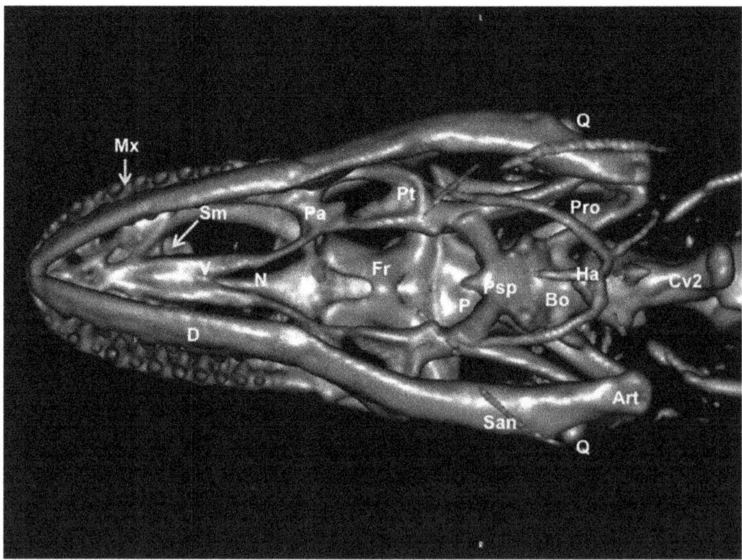

Figure 8. Three-dimensional volume-rendered reconstruction image of the cranium of *Varanus komodoensis*. Ventral aspect. Mx: Maxillary bone. D: Dentary bone. San: Surangular bone. Art: Articular bone. Q: Quadrate bone. Bo. Basioccipital bone. Psp: Parabasisphenoid bone. Pro: Prootic bone. Pt: Pterygoid bone. Pa: Palatine bone. V: Vomer. Sm: Septomaxilla. N: Nasal bone. Fr: Frontal bone. P: Parietal bone. Ha: Hyoid apparatus. Cv2: Second cervical vertebra.

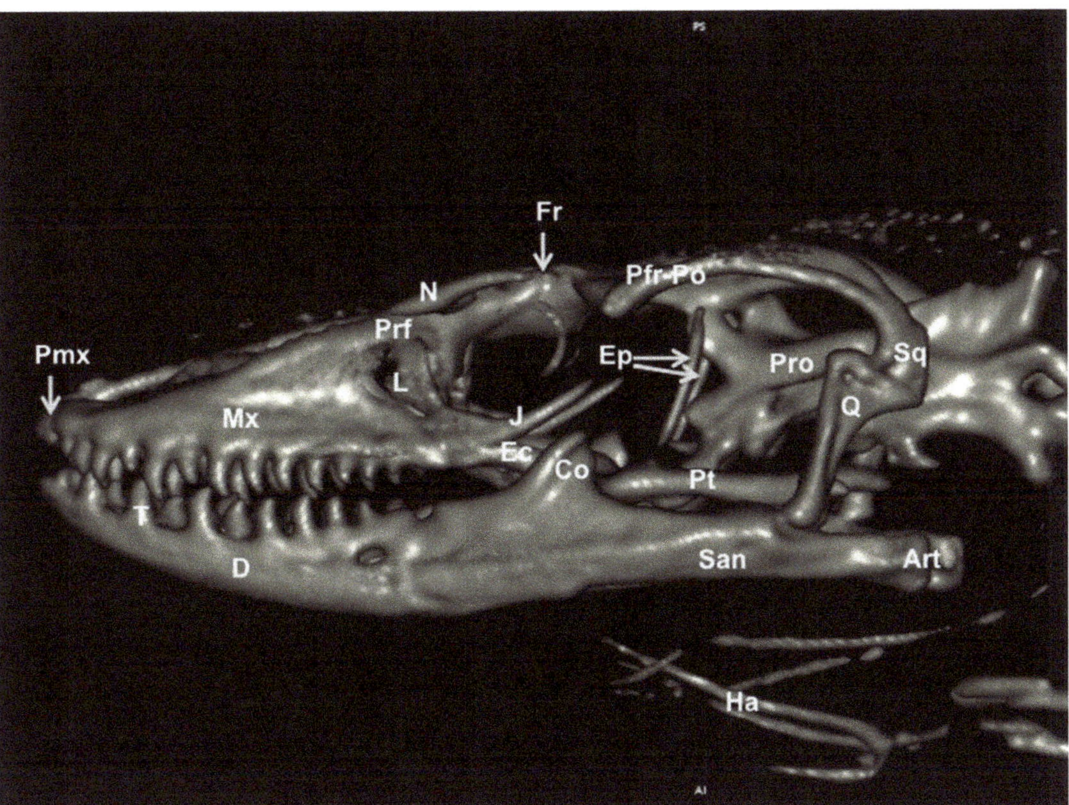

Figure 9. Three-dimensional volume-rendered reconstruction image of the cranium of *Varanus komodoensis*. Lateral aspect. Pmx: Premaxillary bone. Mx: Maxillary bone. Prf: Prefrontal bone. N: Nasal bone. Fr: Frontal bone. L: Lacrimal bone. J: Jugal bone. Ec: Ectopterygoid bone. Pt: Pterygoid bone. Ep: Epipterygoid. Q: Quadrate bone. Sq: Squamosal. Pfr-Po: Postfrontal + postorbital. Pro: Prootic. D: Dentary bone. T: Tooth. San: Surangular bone. Art: Articular bone. Co: Coronoid bone. Ha: Hyoid apparatus.

3.3. *Maximum Intensity Projection (MIP) Images*

Two MIP images corresponding to dorsal (Figure 10) and ventral (Figure 11) views of the varanid skull were selected. These images were able to resolve the relation between the bones that comprise the head. The dorsal MIP image showed the junction between the premaxilla and the maxilla. We were also able to show how the laminar disposition of the vomer supports the septomaxilla (Figure 10). This last finding could be better distinguished in the ventral MIP image (Figure 11). The relation between the lacrimal, the prefrontal, and the frontal bones was seen in the dorsal view (Figure 10). At the posterodorsal border of the orbit, the fusion of postorbital and postfrontal bones could be easily seen. In addition to these observations, the junction of the frontal and the parietal bones were identified in dorsal (Figure 10) and ventral (Figure 11) MIP images. The ventral MIP image showed excellent visualization of the pterygoid, a flat, Y-shaped bone. This bone provides a rounded process that contacts the caudal border of the palatine. Moreover, this view displayed the junction between the parabasisphenoid, the prootic, and the basioccipital bones. This last bone forms the ventral portion of the occipital condyle.

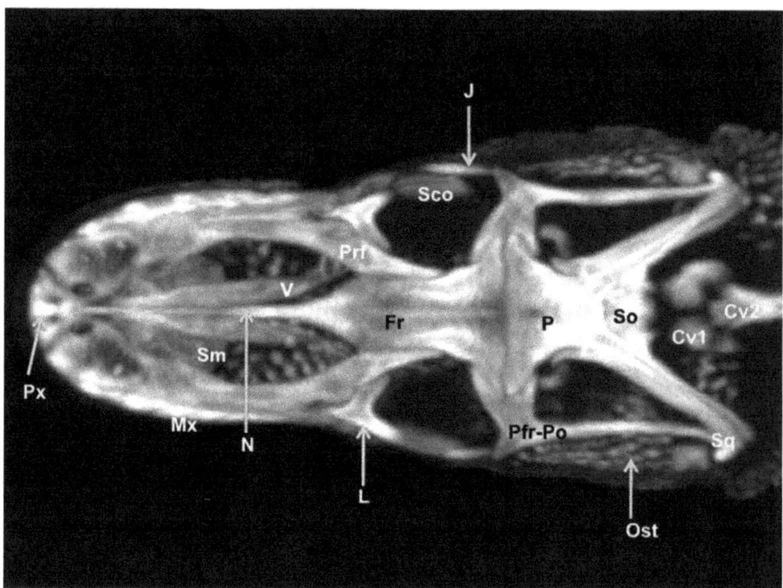

Figure 10. Dorsal MIP image of the cranium of *Varanus komodoensis*. Px: Premaxillary bone. Mx: Maxillary bone. Sm: Septomaxilla. V: Vomer. N: Nasal bone. Prf: Prefrontal bone. Fr: Frontal bone. L: Lacrimal bone. Sco: Scleral ossicles. J: Jugal bone. Pfr-Po: Postfrontal + postorbital. P: Parietal. So: Supraoccipital. Sq: Squamosal. Ost: Osteoderms. Cv1: First cervical vertebra. Cv2: Second cervical vertebra.

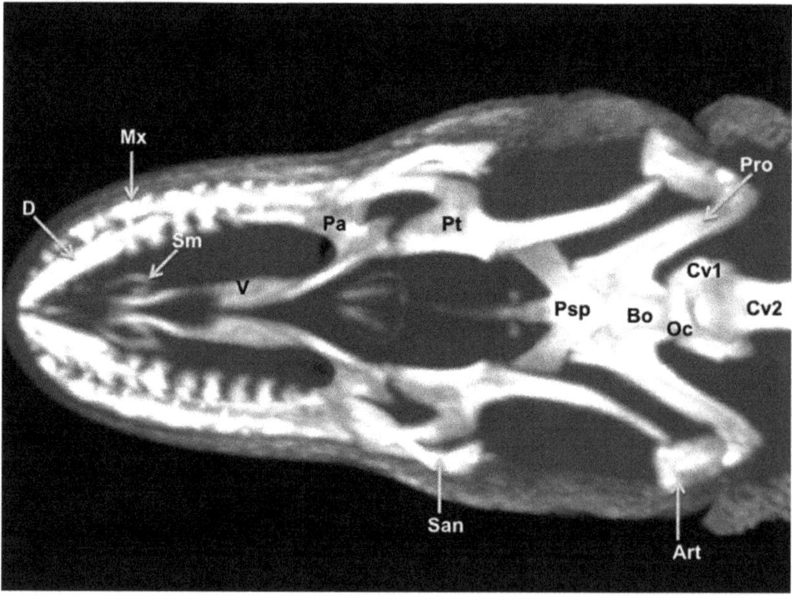

Figure 11. Ventral MIP image of the cranium of *Varanus komodoensis*. D: Dentary bone. Mx: Maxillary bone. San: Surangular bone. Art: Articular bone. Bo. Basioccipital bone. Psp: Parabasisphenoid bone. Oc: Occipital condyle. Pro: Prootic bone. Sm: Septomaxilla. V: Vomer. Pa: Palatine bone. Pt: Pterygoid bone. Cv1: First cervical vertebra. Cv2: Second cervical vertebra.

4. Discussion

In recent years, the contribution of imaging techniques to reptile medicine has increased the knowledge in veterinary practice and research [4–6,12]. Traditionally, radiography and ultrasonography have been used to obtain information on the bony and the main soft-tissue structures of different reptile regions [14,15]. More advanced imaging techniques, such as computed tomography, have become increasingly common in veterinary clinical practice [15]. Third and fourth generation CT scanners give considerable advantages over traditional radiography: body sections from different tomographic planes, fair anatomic resolution without superimposition of the tissues, and a higher differentiation of tissue densities allow better detection of several diseases [6,15,16]. Nonetheless, its use in reptile medicine is still limited because of the cost of the equipment, availability, and logistic problems of acquiring CT images in these animals [6,15].

The cranium of the genus *Varanus* is a complex structure that has received some attention in morphofunctional studies [12,17], perhaps due to the enormous disparity in the form that evolved among varanid lizards [18]. The head of this iconic varanid represents a complex structure, composed of various tissues with varying degrees of attenuation in radiographic images, making it a challenging object to assess. The two window settings used in our CT study facilitated the identification of the main head structures such as the bones of the skull, mandible, muscles, air-filled structures of the respiratory and digestive system, the nervousranium and other associated structures. Visualizing images through use of the "bone window" provided good resolution for skeletal structures, whereas the "soft tissue window" allowed us to distinguish the eyes and the nervous structures from the remaining soft tissues. Similar results were described in other studies conducted in reptiles [1,5,6,15]. Several causes have been reported to explain the low resolution of soft tissue structures showed in our study [1,15], such as the small volume of these species, the impossibility of reducing the field of view of the CT scanner, and the presence of bones embedded within the skin. These bones, called cephalic osteoderms, vary in shape and complexity and serve primarily as a defensive anatomical system to protect individuals during aggressive confrontations with other specimens [11]. To avoid this low resolution, some investigations reported the use of micro-CT scanners [17], although this equipment is not usually available in veterinary clinics [15].

Employing computed tomography, we were able to fully visualize the cranium in virtual reconstructions and MIP images. Thus, the reconstructed images showed a broad, dorsoventrally compressed cranium. The mandible was curved, and teeth were laterally compressed. This morphology contrasts with most other varanids, which feature a relatively narrow rostrum, a dorsoventrally tall cranium, and a straight ventral margin of the maxilla [18]. Additionally, an enormous variation of the orbit was observed, especially along the posterior margin of the orbit, which is closed in lizards, or semiclosed in these varanids. This fact is determined by variation in the shape, size, and presence of the jugal bone and variations in the postorbital and postfrontal bones [19]. MIP images proved a helpful tool in delineating bones in volume-rendered images.

In conclusion, the CT images obtained in this study provided an adequate anatomical interpretation of the head of *Varanus komodoensis*. This information could be used to diagnose disorders involving the head of lizards, such as abscesses, metabolic bone diseases, fractures, and neoplasia.

Author Contributions: Conceptualization, A.A., S.P. and J.R.J.; Methodology, S.P., J.A.C., E.G.-R., A.S.-B. and M.E.; Formal analysis, J.R.J., M.E. and A.A.; Investigation, S.D. and M.M.; Writing—original draft, A.A. and J.R.J.; Writing—review and editing, C.M.; Supervision, J.R.J. All authors have read and agreed to the published version of the manuscript.

Funding: This research received no external funding and the University of Las Palmas de Gran Canaria funded the study.

Institutional Review Board Statement: The study was conducted according to the guidelines of the Declaration of Helsinki, and approved by the Ethical Committee of Las Palmas de Gran Canaria University, College of Veterinary Medicine Section (protocol code MV-2019/04).

Data Availability Statement: Not applicable.

Acknowledgments: In loving memory of Alvaro Domingo Rodriguez Garcia. We also thank Marisa Mohamad for her support and constructive comments.

Conflicts of Interest: The authors declare no conflict of interest.

References

1. Banzato, T.; Russo, E.; Di Toma, A.; Palmisano, G.; Zotti, A. Anatomic imaging of the Boa constrictor head: A comparison between radiography, computed tomography and cadaver anatomy. *Am. J. Vet. Res.* **2011**, *72*, 1592–1599. [CrossRef] [PubMed]
2. Lauridsen, H.; Hansen, K.; Wang, T.; Agger, P.; Andersen, J.L.; Knudsen, P.S.; Rasmussen, A.S.; Uhrenholt, L.; Pedersen, M. Inside Out: Modern Imaging Techniques to Reveal Animal Anatomy. *PLoS ONE* **2011**, *6*, e17879. [CrossRef] [PubMed]
3. Deem, S.L. Role of the zoo veterinarian in the conservation of captive and free ranging wildlife. *Int. Zoo Yearb.* **2007**, *41*, 3–11. [CrossRef]
4. IUCN/SSC. IUCN Red List of Threatened Species. 2002. Available online: www.redlist.org (accessed on 1 December 2020).
5. Banzato, T.; Selleri, P.; Veladiano, I.A.; Martin, A.; Zanetti, E.; Zotti, A. Comparative evaluation of the cadaveric, radiographic and computed tomographic anatomy of the heads of green iguana (*Iguana iguana*), common tegu (*Tupinambis merianae*) and bearded dragon (*Pogona vitticeps*). *BMC Vet. Res.* **2012**, *11*, 53. [CrossRef] [PubMed]
6. Arencibia, A.; Rivero, M.; De Miguel, I.; Contreras, S.; Cabrero, A.; Oros, J. Computed tomographic anatomy of the head of the loggerhead sea turtle (*Caretta caretta*). *Res. Vet. Sci.* **2006**, *81*, 165–169. [CrossRef] [PubMed]
7. Banzato, T.; Selleri, P.; Veladiano, I.A.; Zotti, A. Comparative Evaluation of the Cadaveric and Computed Tomographic Features of the Coelomic Cavity in the Green Iguana (*Iguana iguana*), Black and White Tegu (*Tupinambis merianae*) and Bearded Dragon (*Pogona vitticeps*). *Anat. Histol. Embryol.* **2013**, *42*, 453–460. [CrossRef] [PubMed]
8. Pees, M.; Kiefer, I.; Thielebein, J.; Oechtering, G.; Krautwald-Junghanns, M.E. Computed tomography of the lung of healthy snakes of the species *Python regius*, *Boa constrictor*, *Python reticulatus*, *Morelia viridis*, *Epicrates cenchria* and *Morelia spilota*. *Vet. Radiol. Ultrasound* **2009**, *50*, 487–491. [CrossRef] [PubMed]
9. Hernández, C.; Peloso, P.L.; Bolívar, W.; Daza, J.D. Skull Morphology of the Lizard *Ptychoglossus vallensis* (Squamata: Alopoglossidae) with Comments on the Variation within Gymnophthalmoidea. *Anat. Rec.* **2019**, *302*, 1074–1092. [CrossRef] [PubMed]
10. Maisano, J.A.; Rieppel, O. The skull of the Round Island boa, *Casarea dussumieri schlegel*, based on high-resolution X-ray computed tomography. *J. Morphol.* **2007**, *268*, 371–384. [CrossRef] [PubMed]
11. Maisano, J.A.; Laduc, T.J.; Bell, C.J.; Barber, D. The Cephalic Osteoderms of *Varanus komodoensis* Revealed by High-Resolution X-ray Computed Tomography. *Anat. Rec.* **2019**, *302*, 1675–1680. [CrossRef] [PubMed]
12. Moreno, K.; Wroe, S.; Clausen, P.; McHenry, C.; D' Amore, D.C.; Rayfield, E.J.; Cunningham, E. Cranial performance in the Komodo dragon (*Varanus komodoensis*) as revealed by high-resolution 3-D finite element analysis. *J. Anat.* **2008**, *212*, 736–746. [CrossRef] [PubMed]
13. Fishman, E.K.; Ney, D.R.; Heath, D.G.; Corl, F.M.; Horton, K.M.; Johnson, P.T. Volume rendering versus maxi mum intensity projection in CT angiography: What works best, when, and why. *Radiographics* **2006**, *26*, 905–922. [CrossRef] [PubMed]
14. Knipe, F.M. Principles of neurological imaging of exotic animal species. *Vet. Clin. N. Am. Exot. Anim. Pract.* **2007**, *10*, 893–907. [CrossRef] [PubMed]
15. Banzato, T.; Hellebuyck, T.; Van Caelenberg, A.; Saunders, J.H.; Zotti, A. A review of diagnostic imaging of snakes and lizards. *Vet. Rec.* **2013**, *173*, 43–49. [CrossRef] [PubMed]
16. George, T.F., II; Smallwood, J.E. Anatomic atlas for computed tomography in the mesaticephalic dog: Head and neck. *Vet. Radiol. Ultrasound* **1992**, *33*, 217–240. [CrossRef]
17. Wilken, A.T.; Middleton, K.M.; Sellers, K.C.; Cost, I.N.; Holliday, C.M. The roles of joint tissues and jaw muscles in palatal biomechanics of the savannah monitor (*Varanus exanthematicus*) and their significance for cranial kinesis. *J. Exp. Biol.* **2019**, *222*, jeb201459. [CrossRef] [PubMed]
18. Mertens, R. Die Familie der Warane (Varanidae). 2, Der Schädel. *Senck. Naturf. Ges.* **1942**, *465*, 1–118.
19. Evans, S. The skull of lizards and Tuatara. *Biol. Reptil.* **2008**, *20*, 1–347.

Article

Sex Determination in Two Species of Anuran Amphibians by Magnetic Resonance Imaging and Ultrasound Techniques

María José Ruiz-Fernández [1], Sara Jiménez [2], Encarnación Fernández-Valle [3], M. Isabel García-Real [1], David Castejón [3], Nerea Moreno [2], María Ardiaca [4], Andrés Montesinos [4], Salvador Ariza [4] and Juncal González-Soriano [5,*]

1 Departamento de Medicina y Cirugía, Facultad de Veterinaria, Universidad Complutense, Avenida Puerta de Hierro s/n, 28040 Madrid, Spain; mariajoseruizfernandez@gmail.com (M.J.R.-F.); isagreal@vet.ucm.es (M.I.G.-R.)
2 Departamento de Biología Celular, Facultad de Biología, Universidad Complutense, Avenida José Antonio Novais 12, 28040 Madrid, Spain; sajime01@ucm.es (S.J.); nerea@bio.ucm.es (N.M.)
3 Unidad de RMN—CAI Bioimagen Complutense, Universidad Complutense, Paseo de Juan XXIII 1, 28040 Madrid, Spain; evalle@ucm.es (E.F.-V.); dcastejon@ucm.es (D.C.)
4 Centro Veterinario Los Sauces, Calle de Santa Engracia 63, 28010 Madrid, Spain; ardiaca.m@outlook.com (M.A.); amontesinosbarcelo@gmail.com (A.M.); salvadorariza@msn.com (S.A.)
5 Departamento de Anatomía y Embriología, Sección Departamental de Anatomía y Embriología (Veterinaria), Facultad de Veterinaria, Universidad Complutense, Avenida Puerta de Hierro s/n, 28040 Madrid, Spain
* Correspondence: juncalgs@vet.ucm.es

Received: 18 October 2020; Accepted: 16 November 2020; Published: 18 November 2020

Simple Summary: Amphibians are of critical importance among vertebrates. They play a critical role in their ecosystems, are commonly used as environmental health indicators and are also considered as exotic pets throughout the world. Among amphibians, many anuran species are included in active conservation programs as they are listed as endangered species. Thus, it is important for veterinarians and biologists to examine their sanitary status and to find a non-invasive tool to evaluate the health status of these individuals, particularly the state of their reproductive system and to be able to carry out a sex determination in case of no sexual dimorphism. For the first time, we demonstrate that benchtop magnetic resonance imaging and high-resolution ultrasound are suitable non-invasive imaging techniques for an accurate sex determination of two anuran species. Both techniques allowed the identification of ovaries and testes. Therefore, our data constitute an important contribution for clinical diagnostic and conservation purposes in amphibians, as it is possible to distinguish males and females in a quick, safe and relatively easy way.

Abstract: The objective of the present study was to evaluate whether gender determination in two amphibian species (*Kaloula pulchra* and *Xenopus laevis*) can be reliably carried out by means of magnetic resonance imaging (benchtop magnetic resonance imaging; BT-MRI) or ultrasound (high-resolution ultrasound; HR-US) techniques. Two species of healthy, sexually mature anurans have been used in the present study. Eight *Kaloula* (blind study) and six *Xenopus* were used as controls. Magnetic resonance imaging experiments were carried out on a low-field (1 Tesla) benchtop-MRI (BT-MRI) system. HR-US examination was performed with high-resolution equipment. Low-field BT-MRI images provided a clear and quantifiable identification of all the sexual organs present in both genders and species. The HR-US also allowed the identification of testes and ovaries in both species. Results indicate that BT-MRI allowed a very precise sex identification in both anuran species, although its use is limited by the cost of the equipment and the need for anesthesia. HR-US allowed an accurate identification of ovaries of both species whereas a precise identification of testes is limited by the ultrasonographer experience. The main advantages of this technique are the possibility of performing it without anesthesia and the higher availability of equipment in veterinary and zoo institutions.

Keywords: amphibians; anurans; sex determination; magnetic resonance; ultrasonography

1. Introduction

Amphibians are one of the most interesting groups among vertebrates in terms of the diversity of their reproductive modes. Globally, they include over 7000 species of frogs (*Anura*), 700 species of salamanders (*Caudata*) and 200 species of caecilians (*Gymnophiona*) [1]. More than 2000 amphibian species are listed as endangered [2–4]. Amphibians play a critical role in their ecosystems and are commonly used as environmental health indicators in their habitats worldwide [4–6]. They are also frequently kept as exotic pets throughout the world. To evaluate these species properly, veterinarians need to be able to perform a clinical evaluation and a proper diagnosis in case of pathological problems, particularly those of the reproductive system [7].

Many anurans, such as *Xenopus laevis* (commonly known as the clawed frog), display several forms of sexual dimorphism, including differences in size, skin coloration, texture, secondary sex characteristics (nuptial pads, vowel sac color spines, glands, etc.) and/or behaviors, allowing for an easy differentiation between males and females [8,9]. In African populations, the average size of adult *Xenopus laevis* females is a 110 mm snout to vent length (SVL) (maximum 130 mm SVL) and males are three-fourths as large; however, under laboratory conditions the range varied from 50 to over 140 mm. Other species, such as *Kaloula pulchra* (Banded bullfrog), show very little dimorphism, with males presenting calling behavior during the mating season and being slightly smaller than females, ranging in size from 54 to 70 mm SVL for males and 57–75 mm SVL for females. These differences, however, overlap and are difficult to assess, making this species almost monomorphic [10]. In this sense, the use of non-invasive approaches such as diagnostic imaging methods is of particularly great value for sex identification in the weakly dimorphic or monomorphic species. The startup of non-invasive methods for identification of biological sex in wildlife and experimental animals is important, among other things, for the demographic or genetic management of captive breeding colonies [4–6]. It is especially important for many endangered species that need attention. In this context of the current global decline and extinction of many species, one of the main strategies in the case of amphibians has been to establish captive breeding populations for reintroduction programs. To this end, the first step is the identification of the gender of the individuals present within the colony and this is not always easily accomplished, particularly when working with species that have monomorphic or weakly dimorphic secondary sexual characteristics [2,4–6].

Several articles describe the use of ultrasonography in amphibians for medical purposes [11–13], assessment of reproductive status and sex identification in frogs with better specificity for female identification [14–17]. Magnetic resonance imaging (MRI) has been previously used in frogs focusing on different functional aspects of the Central Nervous System (CNS) [18,19]. However, there are no previous reports on MRI studies for clinical evaluation or sex determination, or for the use of the benchtop MRI (BT-MRI) in these species.

The objective of the present study was to evaluate the accuracy of BT-MRI and high-resolution ultrasound (HR-US) and assess their applicability for sex differentiation in two anuran species, one with "well defined sexual dimorphism, and another with very limited differences" (*Xenopus laevis* and *Kaloula pulchra*, respectively). Particularly, we tried to determine if the reproductive organs of both sexes, male and female, can be reliably visualized by these two techniques and determine the technical parameters that would allow the best imaging of the different organs of interest.

2. Material and Methods

2.1. Animals

A total of 14 healthy anurans (good overall appearance, shiny skin without lacerations, damage or changes in the color and normal motility and standard reflexes), 8 sexually mature *Kaloula pulchra*, without previous identification of males and females (blind study) and 6 *Xenopus laevis* (three males and three females) (Figure 1) were used. *Xenopus laevis* were purchased from the European Xenopus Resource Centre (EXRC; EXRC@xenopusresource.org). *Kaloula pulchra* specimens were obtained from commercial pet suppliers. Animals were handled in accordance with the guidelines for animal research set out in the European Community Directive 2010/63/EU and following the recommendations of the European Commission for the protection of animals used for scientific purposes [20]. All procedures were approved by the local ethics committee.

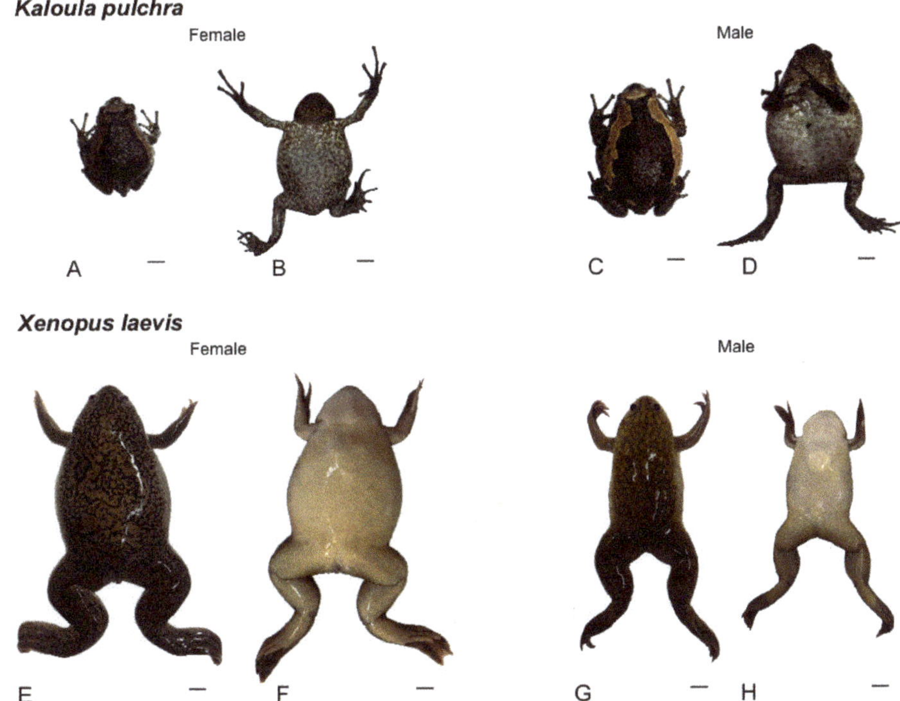

Figure 1. Photos of specimens in dorsal (left) and ventral (right) views of *Kaloula pulchra* (**A–D**) and *Xenopus laevis* (**E–H**) showing the differences between both species and females and males. Scale bars = 1 cm.

2.2. Procedures—Imaging Techniques

2.2.1. Magnetic Resonance Imaging

BT-MRI was performed at the Bioimaging Center of the Complutense, University of Madrid (UCM) using a 1-Tesla benchtop-MRI scanner (Icon (1T-MRI); Bruker BioSpin GmbH, Ettlingen, Germany). The BT-MRI spectrometer consists of a 1 T permanent magnet (without extra cooling required for the magnet) with a gradient coil that provides a gradient strength of 450 mT/m. The MRI system bed integrates both the oval-cylinder solenoid radiofrecuency (RF) coil (59 × 50 mm^2) and the animal monitoring system, which allows an accurate animal positioning and a full control of vital signs. A previous physical evaluation was carried out to ensure that the animal was healthy and well hydrated. Hydration was assured during the procedure, as it is a critical point for amphibians. For the performance of BT-MRI studies of *Kaloula pulchra,* animals were anesthetized by intramuscular injection of alfalxalone (5 mg/kg, Alfaxan 10 mg/mL). In the case of *Xenopus laevis*, specimens were deeply anesthetized by immersion in 0.1% tricaine methanesulfonate solution (MS222, pH 7.4; Sigma-Aldrich, Steinheim, Germany). After anesthesia, animals were positioned inside the RF coil in sternal recumbence.

The BT-MRI experiment consisted of two-dimensional gradient-echo T1-weighted coronal images. Two-dimensional images were acquired using a Fast Low Angle Shot (FLASH) MRI-sequence, the most commonly used gradient-spoiled gradient-echo imaging sequence. The main selected parameters were: repetition time = 302 ms, echo time = 3.05 ms, pulse flip angle = 80°, number of averaged experiments = 6, field of view = 74 × 46 mm^2, slice thickness = 1 mm and number of slices = 15. The acquired matrix size was 172 × 209, the reconstructed matrix size 370 × 230 (resolution 0.20 × 0.20 × 1.00 mm) and the total acquisition time ~5 min.

2.2.2. Ultrasonography

HR-US examinations were performed with an Aplio i800 (Canon Medical Systems, Otawara, Japan) equipped with 2 linear transducers (PLT 1202 BT with 4.5–17 MHz frequency range and PLI 2002 BT with 8.8–22 MHz frequency range). In order to optimize the image quality of the smallest anatomical structures, the highest frequency was selected in all cases. The animals were handled and examined without anesthesia. Frogs were positioned first in dorsal recumbence to obtain HR-US images from a ventral access ("ventral acoustic window"), and later in ventral recumbence, to obtain images from a dorsal access ("dorsal acoustic window"). The transducer was placed directly over the animal's skin, previously wetted with acoustic gel to improve ultrasound conduction. The coelomic cavity was examined in transverse and longitudinal planes, from cranial to caudal.

To avoid any subjective interpretation, all our images were analyzed by the same persons, which included personnel with long experience in diagnostic imaging and researchers with no previous training. Imaging interpretation was performed according to classical anuran anatomy descriptions [21].

3. Results

3.1. BT-MRI

A low-field (1T) BT-MRI system allowed to obtain non-invasive magnetic resonance (MR) images of the anuran coelomic cavity in a semi-automatic way, without manual optimization. Most of the structures (heart, lungs, kidneys, stomach, pancreas, spleen, small and large intestine) could be identified on the MR images in one viewing (see supplementary information, Figure S2, for further details). The BT-MRI protocol developed in this study allowed us to obtain a suitable balance between a good signal-to-noise ratio, acceptable time per animal (~10 min), absence of motion artifacts (due to the use of fast imaging techniques) and good contrast of the main reproductive organs. In both species, the contrast obtained between tissues allowed an easy segmentation of several structures as testes and kidneys and therefore, an easy volumetric quantification (see Table S1).

Figure 2A–F show the sexual organs of *Kaloula pulchra*. In both males and females, the gonads were identified in the dorsal part of the cavity, just cranial to the corresponding kidney. The female reproductive system mainly consists of ovaries and oviducts (Figure 2A–D). The pair of ovaries showed numerous follicles, which were visible on the ovary surface (Figure 2A,B). In the case of sexually matured females, it was possible to observe a great number of dark-colored follicles (Figure 2C,D). The oviducts were observed as two tubes, relatively long and coiled, that flow into the cloaca separately. The male reproductive apparatus consists of a pair of testes, *vasa efferentia* and two urinogenital ducts that open into the cloaca. Testes were easily visible and identified as two ovoid-elongated shaped structures, with a sufficient contrast in comparison to the surrounding tissues to be easily segmented (Figure 2E,F). See Figure S3 for more detailed information on the reproductive system.

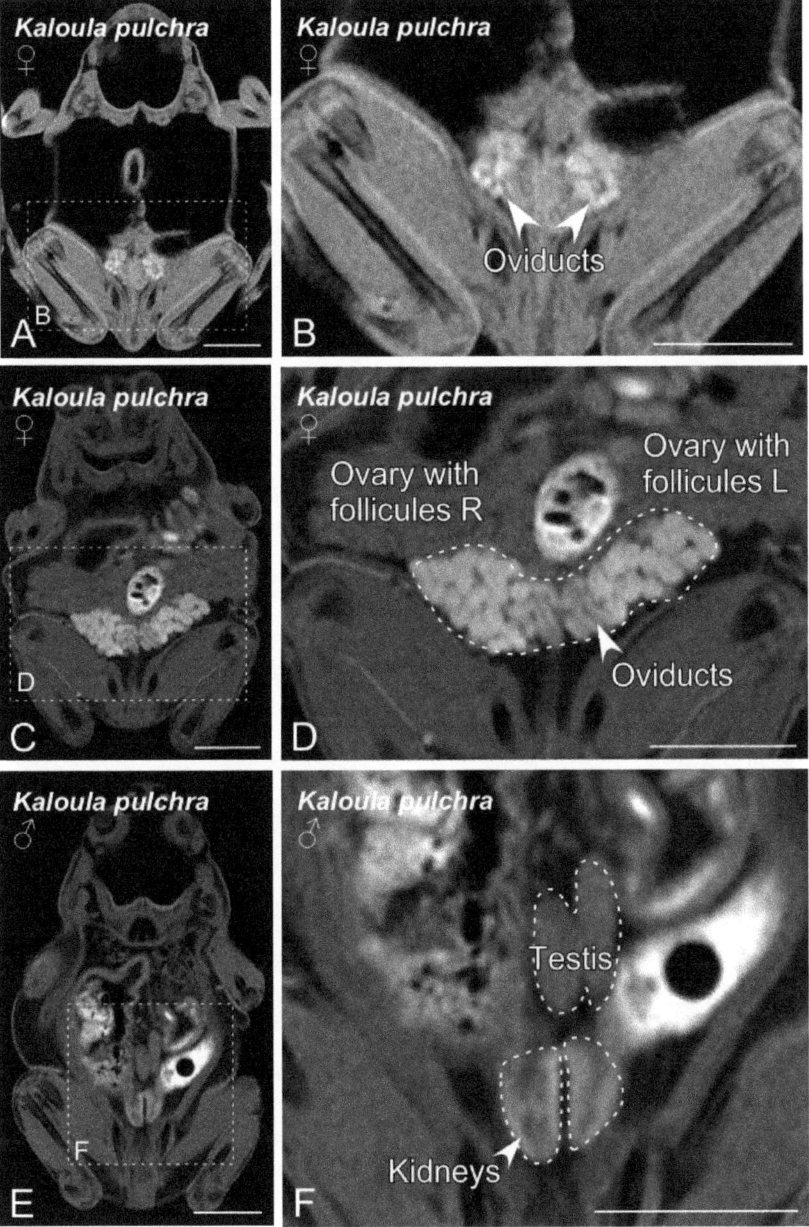

Figure 2. Magnetic resonance imaging (MRI) slices selected to classify *Kaloula pulchra* anurans by sex. Images from (**A**) to (**D**) show the main sexual structures identified in a non-gravid (**A**,**B**) and gravid (**C**,**D**) *Kaloula pulchra* female. Images (**E**) and (**F**) show testis and kidneys in a *Kaloula pulchra* male. The dashed lines in (**A**,**C**,**E**) indicate the magnification photos showed in (**B**,**D**,**F**). Scale bars = 1 cm.

The *Xenopus* reproductive system identified by means of BT-MRI is shown in Figure 3A–F. In both males and females, the gonads were placed adjacent to the kidneys and were surrounded by consistent fat bodies (Figure 3). In the case of females, the identification was particularly easy, because of the recognizable image of the ovaries with well-developed follicles (Figure 3A–D). The oviducts were also easily identifiable. In males, two oval-shaped testes were placed on the dorsal aspect of the coelomic cavity (Figure 3E,F).

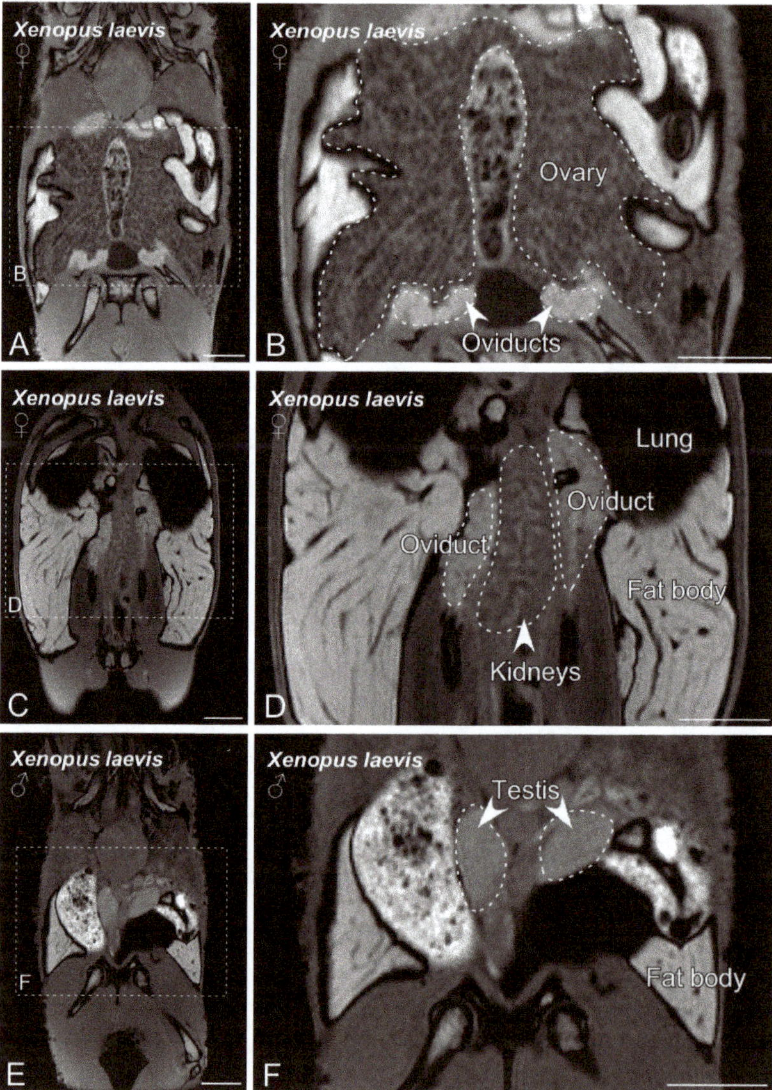

Figure 3. MRI slices selected to illustrate the main sex organs of the *Xenopus laevis* anurans. Images from (**A**) to (**D**) show the structures identified in a gravid *Xenopus laevis* female. Images (**E**,**F**) show testis and fat body in a *Xenopus laevis* male. The dashed lines in (**A**,**C**,**E**) indicate the magnification photos showed in (**B**,**D**,**F**). Scale bars = 1 cm.

3.2. Ultrasonography

As pointed out before, all females included in our experiment had the ovaries full of follicles at the time of examination. Follicles appeared as a complex of anechoic (black) or hypoechoic (dark gray) rounded or oval areas separated by hyperechoic (bright) lines (Figure 4). Both structures were easily recognized not only by an experienced ultrasonographer, but also by researchers without previous training in the use of this imaging technique. Although the follicles were clearly identified in ventral and dorsal acoustic windows (Figure 5A,B), the ventral acoustic window was considered more adequate for female gonad examinations. On the contrary, the kidneys were better identified using the dorsal acoustic window (Figure 5B). The ultrasound examination was repeated in one female *Kaloula* just after oviposition (Figure 6). In this case, both ovaries with follicles were clearly identified as a bilateral hyperechoic structure with multiple hypoechoic foci.

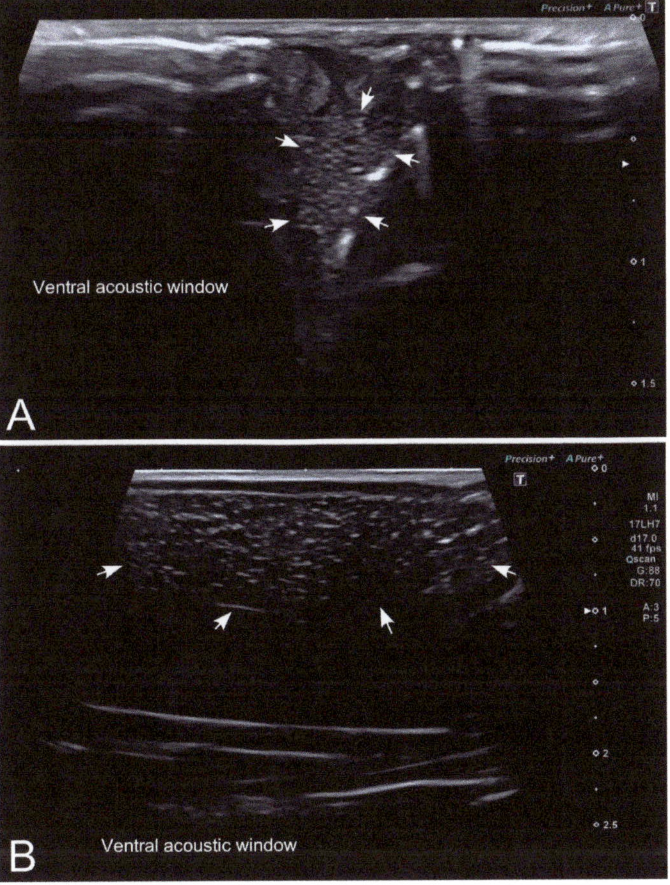

Figure 4. Representative high-resolution ultrasound (HR-US) images of a female *Kaloula pulchra* obtained in the transverse plane (**A**) and a female *Xenopus laevis* (**B**) obtained in the longitudinal plane using, in both cases, a ventral acoustic window. Follicles appeared as a complex of anechoic or hypoechoic rounded or oval areas separated by hyperechoic lines.

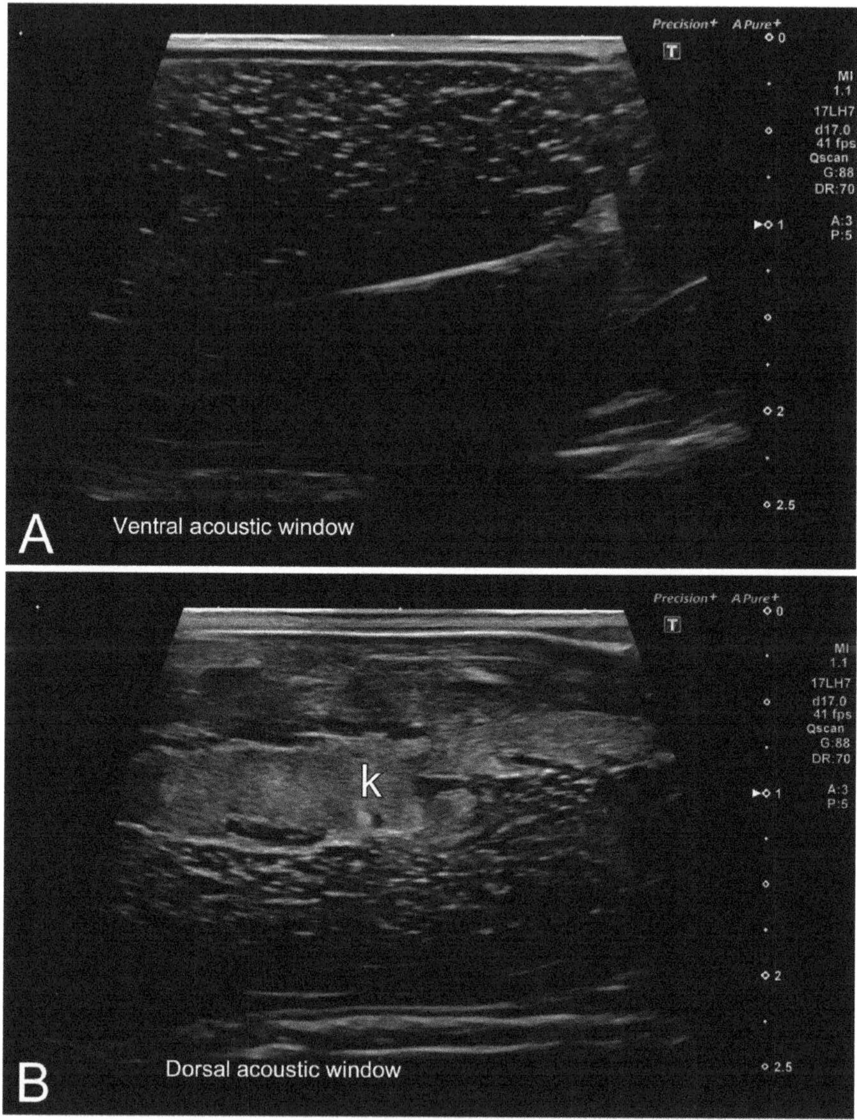

Figure 5. HR-US images of a female *Xenopus laevis* in the longitudinal plane using ventral (**A**) and dorsal (**B**) acoustic windows. Although the complex of follicles is clearly identified in both, the ventral acoustic window was considered more adequate for the examination of female gonads. Nevertheless, the kidney (k) was better identified using the dorsal acoustic window.

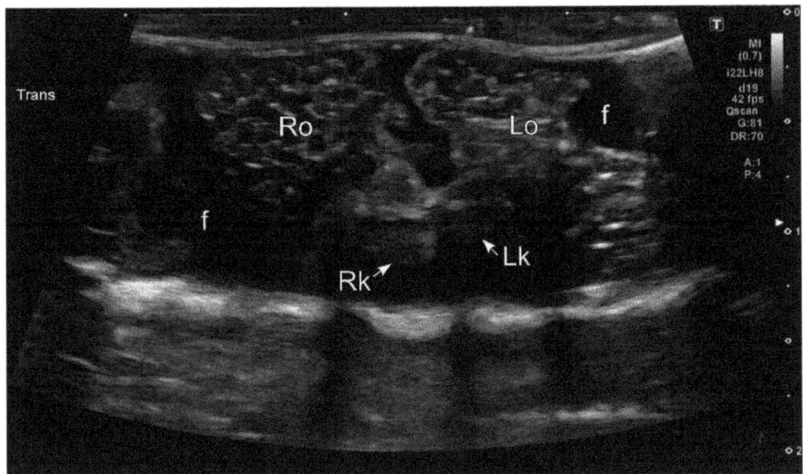

Figure 6. HR-US image of a female *Kaloula pulchra* obtained in the transverse plane using a ventral acoustic window. The image was taken just after oviposition. Both ovaries (right ovary: Ro; left ovary: Lo) appeared as hyperechoic structures with multiple hypoechoic foci which represent follicles. The kidneys are also visible (right kidney: Rk; left kidney: Lk). This animal had free fluid (f), which is considered normal in amphibians. The fluid around the ovaries improves its margin definition.

Testes were identified in all males of both species (Figure 7A,B). Testes appeared as structures with an oval morphology, mid echogenicity (medium gray) and homogenous echotexture. Testes were located just ventral or ventrolateral with respect to the ipsilateral kidney. Kidneys had a similar appearance to testes, but their size was bigger, and the margins were slightly irregular, while testes had a smooth contour. A dorsal window offered an easier identification and differentiation of testes and kidneys than a ventral acoustic window. However, even when using very high-resolution equipment, as is the case of this study, the differentiation between both organs is difficult for non-experienced ultrasonographers.

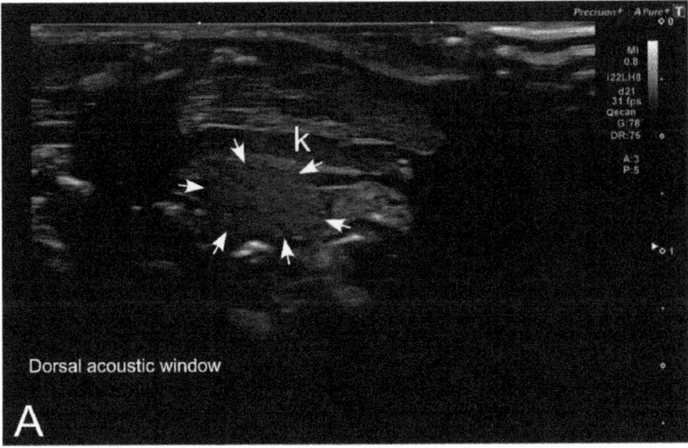

Figure 7. *Cont.*

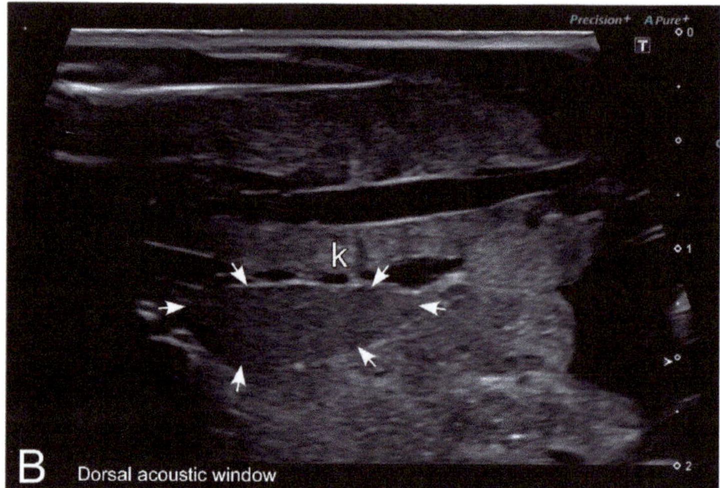

Figure 7. Representative HR-US images of *Kaloula pulchra* (**A**) and *Xenopus laevis* (**B**) males obtained in the longitudinal plane using a dorsal acoustic window. The testes (delimited by arrows) appeared as structures with oval morphology, mid echogenicity and homogenous echotexture. Testes were located just ventral or ventrolateral of the ipsilateral kidney (k).

4. Discussion

Due to the important role of amphibians in their ecosystems, as well as to their increasing importance as exotic pets, problems related to conservation and reproduction represent a big concern for veterinarians and biologists. Amphibian sex identification through different non-invasive techniques has been a subject of interest for many researchers. Genetic analysis is likely to be a difficult and expensive method due to the fact that most amphibians do not have distinct sex chromosomes and as oddities such as aneuploidy and polyploidy can occur [22]. Other sex identification methods in anurans, such as hormone measurement in blood, fecal or urine samples, present variable degrees of effectiveness, cost, technical availability and invasiveness [14,23–29]. In some cases, they are expensive, technically difficult, lightly discriminative, or even risky for the animal's life or require sacrifice, making them unsuitable for conservation purposes. However, the use of imaging techniques for sex determination has been poorly investigated. In fact, there is only one previous study, focused on the use of Near Infrared Reflectance spectroscopic (NIRS) for sex identification in the Mississippi gopher frog (*Lithobates sevosa*). Results were successful in females with developing eggs, although males and undeveloped females remained indistinguishable [30]. For the first time, the present study investigated the applicability of two routine imaging techniques (BT-MRI and HR-US) for sex identification in two anuran species with a well-defined and a very limited sexual dimorphism (*Xenopus laevis* and *Kaloula pulchra*).

Magnetic resonance imaging (MRI) techniques have become more accessible and popular in recent years, for both veterinarians and conservation experts. In comparison with other imaging techniques, MRI constitutes a valuable tool that allows the easy visualization and differentiation of soft tissue structures. In addition, the possibility of an automatic self-adjustment of the equipment's main technical parameters and the lower cost of the new low-field benchtop-MRI equipment has contributed to its more generalized use in preclinical research and in a daily veterinary clinic routine [31–35]. The present study is the first report of the use of BT-MRI for the purpose of sex determination. This technique allowed a fast and precise differentiation of the genital organs in the two species studied (Figures 2 and 3), as well as the identification of most of the structures in the coelomic cavity. Because of its characteristics compared to other imaging techniques, BT-MRI would be ideal for monitoring reproductive organs in a wide variety of animals that do not have sexual dimorphism. We think that this approach could be a useful possibility in a high variety of species, including those with a low SVL range—20 mm or even lower. This opens an important opportunity for the captive breeding populations for reintroduction programs, since this will, for example, maintain adequate sex ratios in captive breeding programs that are sexing metamorphic individuals (which is not possible or very difficult using other techniques) and thus increase the efficiency of conservation programs. The BT-MRI dataset obtained allowed us to not only determine the sex of the studied specimens, but it also provided valuable additional information, such as the detection and identification of anomalies and pathologies if present, and volumetric information from sexual and non-sexual internal organs (see supplementary information Figure S4). For example, it could be a very interesting and useful application in track cases of sex reversal in amphibians in response to pollution, something that it is threatening amphibian populations throughout the world. The use of BT-MRI required a minimal dose of anesthesia to keep the animal unconscious for 10–15 min. The need for anesthesia could be a limiting factor, especially when working with sensible species, but in the present study all the frogs woke up normally after a mild anesthesia. Additionally, anesthesia will be likely substituted in the future by simple sedation or restraint techniques such as those used with mice [36]. In any case, the advantages of BT-MRI clearly outweigh the possible limitations, as the results are clear, acute, of very good quality and require minimal training for their interpretation.

Ultrasonographic visualization of reproductive organs in anurans has been considered difficult, especially in males [13–17]. However, in our study, ultrasound examination allowed an easy identification of testes and ovaries in both anuran species (Figures 5–7). Yet, the identification of testes required trained ultrasonographers to make a correct interpretation of the images. In our study, we used state-of-the-art HR-US equipment and two transducers with a frequency range up to 17 and 22 MHz, respectively, which provided excellent quality images from the skin to 3–5 cm deep, while the transducers previously used in similar studies had lower frequency ranges [14,16,17,37]. The high technical characteristics of the HR-US equipment and transducers used in this study can explain the discrepancies with previous studies on the feasibility of ultrasonography to visualize testes and ovaries in anurans, as pointed out in studies pertaining to reptiles [38]. In comparison with other imaging techniques, HR-US has clear advantages, such as the good differentiation of soft tissue structures, its relatively low cost and wide availability in veterinary and zoo institutions. In our experience, correct gentle animal handling was sufficient to carry out a complete HR-US examination of the coelomic cavity in a few minutes (between 1 and 5). Thus, the possibility of handling animals without anesthesia is another important advantage of the HR-US technique for sex determination in amphibians.

Another interesting issue to consider for ultrasound examination is the animal's position, that can be in dorsal recumbence, using a ventral acoustic window [13,14,16] or in ventral recumbence, using a dorsal acoustic window [13]. We found that the ventral acoustic window was the most suitable to evaluate ovaries in females, while kidneys and testes were better examined using the dorsal acoustic window.

Limitations of ultrasonography are the presence of artifacts caused by gas, the effect of bone and other mineralized structures, which avoid the conduction of HR-US waves. This drawback was particularly evident in *Kaloula pulchra* as these frogs greatly inflate the lungs as a defensive mechanism, complicating the visualization of organs especially when using the ventral acoustic window. Gentle handling and being patient until the frog deflated after 10–30 s allowed the ultrasonographic examination to continue with minimal delay.

5. Conclusions

In summary, we show that both BT-MRI and HR-US are suitable non-invasive imaging techniques for the accurate sex determination in both *Kaloula pulchra* and *Xenopus laevis* species. BT-MRI required minimal training for image interpretation and could be applied to low SVL range animals, but compared to HR-US, has the drawbacks of the higher cost of the equipment and the need for anesthesia. HR-US also allowed an accurate identification of ovaries of both species, even with minimal training, but the precise imaging and identification of testes in males required experienced ultrasonographers. The main advantages of this technique, compared to BT-MRI, are the shorter time required to perform the study (a few minutes per animal), the possibility of not using anesthesia and the availability of appropriate equipment in many veterinary and zoo institutions. In addition, the interpretation of data depends on experienced technicians, and anesthesia is not mandatory. Thus, the use of one or the other depends on the objective in each case.

This study represents a promising initial step that could promote the widespread use of these non-invasive imaging techniques for sex identification in amphibians or the clinical diagnosis of reproductive problems of these species. Our data constitute an important contribution for clinical diagnosis and conservation purposes in amphibians. However, further studies in other anuran species are needed to confirm the suitability of these imaging techniques for sex determination in these individuals.

Supplementary Materials: The following are available online at http://www.mdpi.com/2076-2615/10/11/2142/s1, Figure S1: MRI identification of structures contained in the coelomic cavity, Graph S1: MRI-Volumetric quantification of testes and kidneys, Figure S2: Additional sex-gender system structures identified by MRI, Figure S3: MRI identification of reproductive system disorders.

Author Contributions: Conceptualization, M.J.R.-F., E.F.-V., M.I.G.-R. and J.G.-S.; methodology, E.F.-V., M.I.G.-R. and D.C.; data collection and validation, M.J.R.-F. and S.J.; formal analysis, A.M., M.A. and S.A.; writing—original draft preparation, N.M., D.C., E.F.-V. and M.I.G.-R.; writing—review and editing, M.J.R.-F., S.J., N.M. and J.G.-S.; supervision, J.G.-S.; project administration, E.F.-V., D.C. and M.I.G.-R.; funding acquisition, E.F.-V. and M.I.G.-R. All authors have read and agreed to the published version of the manuscript.

Funding: This research received no external funding.

Acknowledgments: This research did not receive any significant grant from funding agencies in the public, commercial, or not-for-profit sectors.

Conflicts of Interest: The authors declare no conflict of interest.

References

1. Frost, D. Amphibian Species of the World: An Online Reference. Version 6.1. Available online: https://amphibiansoftheworld.amnh.org/ (accessed on 12 July 2020).
2. Stuart, S.N.; Chanson, J.S.; Cox, N.A.; Young, B.E.; Rodrigues, A.S.L.; Fischman, D.L.; Waller, R.W. Status and Trends of Amphibian Declines and Extinctions Worldwide. *Science* **2004**, *306*, 1783–1786. [CrossRef]
3. IUCN Red List of Threatened Species. Available online: https://www.iucnredlist.org/search?query=amphibian&searchType=species (accessed on 12 July 2020).
4. Nori, J.; Lemes, P.; Urbina-Cardona, N.; Baldo, D.; Lescano, J.N.; Loyola, R.D. Amphibian conservation, land-use changes and protected areas: A global overview. *Biol. Conserv.* **2015**, *191*, 367–374. [CrossRef]
5. Amaral, C.; Chaves, A.C.S.; Júnior, V.N.T.B.; Pereira, F.; Silva, B.M.; Silva, D.A.; Amorim, A.; Carvalho, E.F.; Rocha, C.F.D. Amphibians on the hotspot: Molecular biology and conservation in the South American Atlantic Rainforest. *PLoS ONE* **2019**, *14*, e0224320. [CrossRef]
6. Kouba, A.J.; Vance, C.K. Applied reproductive technologies and genetic resource banking for amphibian conservation. *Reprod. Fertil. Dev.* **2009**, *21*, 719–737. [CrossRef]
7. De La Navarre, B.J. Common Procedures in Reptiles and Amphibians. *Veter. Clin. N. Am. Exot. Anim. Pract.* **2006**, *9*, 237–267. [CrossRef]
8. Duellman, E.; Trueb, L. *Biology of Amphibians*; John Hopkins University Press: Baltimore, MD, USA, 1994.
9. Chai, N. *Veterinary Clinics of North America—Exotic Animal Practice*; Saunders, W.B., Ed.; Elsevier: Amsterdam, The Netherlands, 2017; pp. 307–325.
10. Lalremsanga, H. *Studies on the Ecology, Breeding Behavoir and Development of Ranid and Microhylid Anurans Prevalent in Mizoram*; North Eastern Hill University: Shillong, India, 2012.
11. Stettler, M. Imaging amphibians: Techniques, normal anatomy and pathological structures. In Proceedings of the TNVAC, Orlando, FL, USA, 10–14 January 1998; pp. 818–819.

12. Stetter, M.D. Noninfectious medical disorders of amphibians. *Semin. Avian Exot. Pet Med.* **1995**, *4*, 49–55. [CrossRef]
13. Schildger, B.; Triet, H. Ultrasonography in amphibians. *Semin. Avian Exot. Pet Med.* **2001**, *10*, 169–173. [CrossRef]
14. Graham, K.M.; Kouba, A.J.; Langhorne, C.J.; Marcec, R.M.; Willard, S.T. Biological sex identification in the endangered dusky gopher frog (*Lithobates sevosa*): A comparison of body size measurements, secondary sex characteristics, ultrasound imaging, and urinary hormone analysis methods. *Reprod. Biol. Endocrinol.* **2016**, *14*, 1–14. [CrossRef]
15. Calatayud, N.; Stoops, M.; Durrant, B. Ovarian control and monitoring in amphibians. *Theriogenology* **2018**, *109*, 70–81. [CrossRef] [PubMed]
16. Graham, K.M.; Langhorne, C.J.; Vance, C.K.; Willard, S.T.; Kouba, A.J. Ultrasound imaging improves hormone therapy strategies for induction of ovulation and in vitro fertilization in the endangered dusky gopher frog (*Lithobates sevosa*). *Conserv. Physiol.* **2018**, *6*, coy020. [CrossRef] [PubMed]
17. Calatayud, N.E.; Chai, N.; Gardner, N.R.; Curtis, M.J.; Stoops, M.A. Reproductive Techniques for Ovarian Monitoring and Control in Amphibians. *J. Vis. Exp.* **2019**, e58675. [CrossRef] [PubMed]
18. Ringler, E.; Coates, M.; Cobo-Cuan, A.; Harris, N.G.; Narins, P.M. MEMRI for Visualizing Brain Activity after Auditory Stimulation in Frogs. In Proceedings of the Behavioral Neuroscience, Brisbane, Australia, 15–20 July 2018; American Psychological Association Inc.: Washington, DC, USA, 2019; Volume 133, pp. 329–340.
19. Sandvig, A.; Sandvig, I.; Berry, M.; Olsen, Ø.; Pedersen, T.B.; Brekken, C.; Thuen, M. Axonal tracing of the normal and regenerating visual pathway of mouse, rat, frog, and fish using manganese-enhanced MRI (MEMRI). *J. Magn. Reson. Imaging* **2011**, *34*, 670–675. [CrossRef] [PubMed]
20. Directive 2010/63/eu of the European Parliament and of the Council of 22 September 2010 on the Protection of Animals Used for Scientific Purposes (Text with EEA Relevance). Available online: https://eur-lex.europa.eu/LexUriServ/LexUriServ.do?uri=OJ:L:2010:276:0033:0079:en:PDF (accessed on 12 July 2020).
21. Vitt, L.J.; Caldwell, J.P. Anatomy of amphibians and reptiles. In *Herpetology: An Introductory Biology of Amphibians and Reptiles*; Academic Press: London, UK, 2008; pp. 35–83. ISBN 9780123743466.
22. Arregui, L.; Diaz-Diaz, S.; Alonso-López, E.; Kouba, A.J. Hormonal induction of spermiation in a Eurasian bufonid (*Epidalea calamita*). *Reprod. Biol. Endocrinol.* **2019**, *17*, 1–10. [CrossRef] [PubMed]
23. Narayan, E.J. Non-invasive reproductive and stress endocrinology in amphibian conservation physiology. *Conserv. Physiol.* **2013**, *1*, cot011. [CrossRef]
24. Lynch, K.S.; Wilczynski, W. Gonadal steroids vary with reproductive stage in a tropically breeding female anuran. *Gen. Comp. Endocrinol.* **2005**, *143*, 51–56. [CrossRef]
25. Szymanski, D.; Gist, D.; Roth, T. Anuran gender identification by fecal steroid analysis. *Zoo Biol.* **2006**, *25*, 35–46. [CrossRef]
26. Hogan, L.A.; Lisle, A.T.; Johnston, S.D.; Goad, T.; Robertston, H. Adult and Juvenile Sex Identification in Threatened MonomorphicGeocriniaFrogs Using Fecal Steroid Analysis. *South Am. J. Herpetol.* **2013**, *47*, 112–118. [CrossRef]
27. Germano, J.M.; Molinia, F.; Bishop, P.; Cree, A. Urinary hormone analysis assists reproductive monitoring and sex identification of bell frogs (*Litoria raniformis*). *Theriogenology* **2009**, *72*, 663–671. [CrossRef]
28. Germano, J.; Molinia, F.; Bishop, P.; Bell, B.; Cree, A. Urinary hormone metabolites identify sex and imply unexpected winter breeding in an endangered, subterranean-nesting frog. *Gen. Comp. Endocrinol.* **2012**, *175*, 464–472. [CrossRef]
29. Narayan, E.J.; Molinia, F.C.; Christi, K.S.; Morley, C.G.; Cockrem, J. Annual cycles of urinary reproductive steroid concentrations in wild and captive endangered Fijian ground frogs (*Platymantis vitiana*). *Gen. Comp. Endocrinol.* **2010**, *166*, 172–179. [CrossRef]
30. Vance, C.; Graham, K.; Kouba, A.; Willard, S. In vivo sex identification of the endangered Mississippi Gopher Frog (*Lithobates sevosa*) using Near Infrared Reflectance spectroscopy. In Proceedings of the 17th International Conference on Near Infrared Spectroscopy, Foz do Iguassu, Brazil, 18–23 October 2015.
31. Inderbitzin, D.; Stoupis, C.; Sidler, D.; Gass, M.; Candinas, D. Abdominal magnetic resonance imaging in small rodents using a clinical 1.5T MR scanner. *Methods* **2007**, *43*, 46–53. [CrossRef] [PubMed]

32. Nardini, G.; Di Girolamo, N.; Leopardi, S.; Paganelli, I.; Zaghini, A.; Origgi, F.C.; Vignoli, M. Evaluation of liver parenchyma and perfusion using dynamic contrast-enhanced computed tomography and contrast-enhanced ultrasonography in captive green iguanas (Iguana iguana) under general anesthesia. *BMC Vet. Res.* **2014**, *10*, 112. [CrossRef] [PubMed]
33. Di Ianni, F.; Volta, A.; Pelizzone, I.; Manfredi, S.; Gnudi, G.; Parmigiani, E. Diagnostic Sensitivity of Ultrasound, Radiography and Computed Tomography for Gender Determination in Four Species of Lizards. *Vet. Radiol. Ultrasound* **2014**, *56*, 40–45. [CrossRef] [PubMed]
34. Banzato, T.; Russo, E.; Finotti, L.; Milan, M.C.; Gianesella, M.; Zotti, A. Ultrasonographic anatomy of the coelomic organs of boid snakes (Boa constrictor imperator, Python regius, Python molurus molurus, and Python curtus). *Am. J. Vet. Res.* **2012**, *73*, 634–645. [CrossRef]
35. Bucy, D.S.; Guzman, D.S.-M.; Zwingenberger, A.L. Ultrasonographic anatomy of bearded dragons (*Pogona vitticeps*). *J. Am. Vet. Med. Assoc.* **2015**, *246*, 868–876. [CrossRef]
36. Madularu, D.; Mathieu, A.P.; Kumaragamage, C.; Reynolds, L.M.; Near, J.; Flores, C.; Rajah, M.N. A non-invasive restraining system for awake mouse imaging. *J. Neurosci. Methods* **2017**, *287*, 53–57. [CrossRef]
37. Cojean, O.; Vergneau-Grosset, C.; Masseau, I. Ultrasonographic anatomy of reproductive female leopard geckos (*Eublepharis macularius*). *Vet. Radiol. Ultrasound* **2018**, *59*, 333–344. [CrossRef]
38. Hildebrandt, T.B.; Saragusty, J. Use of Ultrasonography in Wildlife Species. In *Fowler's Zoo and Wild Animal Medicine*; Elsevier BV: Amsterdam, The Netherlands, 2015; Volume 8, pp. 714–723.

MDPI
St. Alban-Anlage 66
4052 Basel
Switzerland
Tel. +41 61 683 77 34
Fax +41 61 302 89 18
www.mdpi.com

Animals Editorial Office
E-mail: animals@mdpi.com
www.mdpi.com/journal/animals

www.ingramcontent.com/pod-product-compliance
Lightning Source LLC
LaVergne TN
LVHW070201170726
843515LV00007B/1972

* 9 7 8 3 0 3 6 5 7 3 2 8 1 *